W0112639

Wireless Game Development in C/C++ with BREW™

Ralph Barbagallo

Wordware Publishing, Inc.

Library of Congress Cataloging-in-Publication Data

Barbagallo, Ralph.
   Wireless game development in C/C++ with BREW / by Ralph Barbagallo.
     p. cm.
   ISBN 1-55622-905-4
   1. Computer games--Programming. 2. C (Computer program language).
   3. C++ (Computer program language) 4. Application software--Development.
   5. Wireless communication systems. I. Title.
   QA76.76.C672 B36 2002
   794.8'15265--dc21
                              2002154121
                                  CIP

ISBN 1-55622-905-4

10 9 8 7 6 5 4 3 2 1
0302

All inquiries for volume purchases of this book should be addressed to Wordware Publishing, Inc., at the above address. Telephone inquiries may be made by calling:

(972) 423-0090

Dedication

To the memory of Vito and Mary

Contents

Foreword

Millions of people around the world consider their cell phone a vital communication tool. For them, it is not just "nice to have" but a necessity in their personal and professional lives. Now millions of PC and console gamers around the world are quickly discovering that they can also take their adventures on the road—and play anytime, anywhere. QUALCOMM's BREW™ platform is a boon to game lovers everywhere.

The BREW platform's growing popularity among the world's leading wireless carriers and handset manufacturers, along with its active international community of top game publishers and developers, has stimulated the rise of a promising market opportunity. People are paying for compelling content, and games are some of the most popular.

This book serves as a valuable resource for developers planning to create games for the BREW platform. Inside you'll find information regarding the BREW platform and the application development process—from the BREW SDK™ to the TRUE BREW™ testing process and much more.

We know you will enjoy learning about the BREW platform and wish you success in your quest to create the ultimate BREW game!

Mike Yuen
Director
Developer Relations
QUALCOMM® Internet Services

Acknowledgments

This book is the end result of a year's journey into the heart of darkness of wireless gaming development. During the wild and unpredictable duration of this book's development, I learned a lot—not just about wireless technology but also about the sometimes utterly nonsensical practices and people of the business world.

Jim Hill and Wes Beckwith at Wordware Publishing deserve a lot of credit for spending time and resources on a book about a subject that nobody really knew anything about at the time. Their vision and flexibility has made this book possible.

Also, credit is due the people who read over the manuscript and checked it for technical and factual accuracy. This includes another pioneering BREW developer, John Szeder of Seismic Studios, and my old college buddy from the U-Mass Lowell Computer Science days, Dave Sowsy. Also, my editor, Beth Kohler at Wordware, has proven invaluable in pointing out all of my lingering sloppiness. Not to mention the good people at Qualcomm and Porter Novelli, who offered invaluable support in gathering information. This includes the inimitable Adriana Saldaña, Nick Mendoza, Steve Albers, Mike Yuen, Alison Graves, and Michele Bakic. Also, thanks goes out to the artists who provided artwork, photos, and other support including Roger Seto, Chris George, and Brian Schmitt.

Many people helped in providing content including pictures of applications and information about various products and services. Thanks goes out to Ilkyu Song and Jay Jung at GameVIL, Barry Sohl from DemiVision, Jerry Graham and Peter Bernard at Insignia Solutions, F. Conrad Hametner III at DWANGO, Ann Garner at nGame, Don DeLucia at Mattel, Jonathan Correa formerly of Mattel, Nicholas Dickey at Dynamo Development, and Matt Sponer for some last-minute byte-packing information. Special mention goes to the crew at Monkeystone Games, including but not limited to Tom Hall, John Romero, Stevie Case, and Lucas Davis. Good luck, guys!

Of course I cannot forget my parents, Ralph and Marie Barbagallo, without whom I probably would have ended up as a random assortment of molecules in a tree or a bowl of soup or something. Random thanks goes to others who helped contribute to the book or otherwise deserve mention such as James A. Frost, Karl Hornell, Doug McCartney, Jennifer Olsen and Daniel R. Huebner at *Game Developer* magazine, everyone at Wireless Gaming Review, the team at Nuvo Studios, and anyone else I forgot.

Finally, I cannot forget to thank the artists whose CDs playing in the background have helped the creation of the book go much smoother than it would have otherwise. This includes Kool Keith, M.F. Doom/Monsta Island Czars, 7L and Esoteric, People Under the Stairs, Non Phixion, Necro, Pete Rock and CL Smooth, Ultramagnetic MCs, Digable Planets, Curtis Mayfield, Black Box Recorder, Diamond D, De La Soul, A Tribe Called Quest, Nas, Public Enemy, Moka Only, Del/Hieroglyphics, Chill Rob G, Scienz of Life, The Jungle Brothers, Kraftwerk, Masta Ace, Sam Cooke, Godfather Don, Outkast, Percee-P, KRS ONE, and the list goes on and on.

Introduction

We are now in the 21st century—the astounding world of the future! An era of flying cars. Tele-transportation. Jet packs. The utopian existence dismissed as a mere science fantasy conjured up in the minds of novelists and dreamers has come about!

Well, perhaps not.

One thing we do have that a few short decades ago was cutting-edge sci-fi is the mobile phone. Just like the trusty communicator on *Star Trek*, a device that fits in the palm of our hand can interact with people thousands of miles away as we traverse the alien landscape of the planet we call Earth. Now, I never saw Captain Kirk lose his connection with the *Enterprise* when passing under a bridge, but hey, that's the 23rd century. We still have a little way to go.

Originally, mobile phones were just that—phones. But the relentless pace of computer technology has transformed the mobile phone into a complicated computing device. Sure, it is not going to rival the power of your desktop machine anytime soon, but the latest models coming out from major manufacturers contain increasing amounts of computing power to drive their personal digital assistant-like features. In fact, some phones have merged the functionality of the Palm and PocketPC platforms to become full-fledged phone-PDA hybrids.

These newer devices include one major feature that we coders can truly appreciate: They are programmable. Previously, when you purchased a mobile phone, you got the features it came with and nothing more. This usually included a simple phonebook utility, perhaps a clock with an alarm function, and maybe even a primitive game or two. Newer phones can actually download small programs that add to the phone's built-in suite of applications. Among the various types of programs available, games are proving to be very popular. There are not many available just yet, but the mobile phone gaming industry is projected to grow rapidly as we progress into the decade. The key to this potentially massive industry is the wide audience.

It seems odd that only a few short years ago a mobile phone was considered a luxury. Now the technology is cheap enough that it is almost ubiquitous. Everyone from grade-schoolers to my grandmother has a mobile phone. Millions of people have adopted the technology and, in many cases, are replacing their traditional land-line phones with a mobile.

This makes the audience for games potentially huge. We have already seen the global public's thirst for mobile data products with such services as Japan's iMode or SMS messaging in Europe. The ability to download games and entertainment applets to your phone could become just as common in the years to come as downloading new ringtones.

Some major companies are becoming interested in this growing market. There are many new upstarts in the arena of mobile gaming, but even the old guard of traditional game software publishers is starting to take notice. Right now, the money generated is insignificant to a corporate giant such as Electronic Arts. The average revenue generated from a mobile phone game is tantamount to a rounding error on the bar tab at their last E3 party. However, once it is proven that a large paying audience is ready for this content, mobile gaming development will be in demand.

The great thing about these games is that by the very nature of the device, the projects are limited in scope. You are dealing with a miniscule amount of computing power and very primitive display technology. It is not necessary to harness the resources of massive teams, hundreds of thousands of dollars of equipment, and licensed technology to create a mobile phone title. A resourceful programmer in her basement could create the next big mobile phone hit. With the limited resolution and color depth of most of these devices, she could probably do the artwork herself too! Compare this to the multi-million dollar half decade-long death marches at your average major game company, and you may see this as a healthy alternative to mainstream corporate game development.

What is BREW and How Does It Fit into All of This?

Qualcomm's BREW, or Binary Runtime Environment for Wireless, is a new hardware platform, SDK, and distribution model designed to address the new era of programmable mobile phones. Although not specifically created for game developers, the BREW SDK has enough features to write games for any BREW-equipped phone.

Who Does This Guy Think He Is Anyway?

Who am I, and why did I decide to write this book? I am a professional game programmer who has worked on a variety of platforms ranging from the PC to the PlayStation 2. I saw mobile phone games as a largely untapped market that I could tackle with a minimum amount of resources. Shortly after my foray into unknown territory, I developed my first wireless phone game using Qualcomm's BREW technology. During the process of creating my game, I found that there was an interest in mobile phone game development but not really any central place to get information. Instead of writing yet another book on 3D graphics or DirectX, I figured focusing on an entirely new gaming market would be an interesting challenge.

Who Is This Book For?

This book is for game programmers interested in mobile phone application development. In fact, it can also be a handy tool for those developing non-game applications. Writing games is often a fun and productive way to get familiar with any new SDK, and BREW is no exception. Knowledge of C is assumed, but you do not necessarily have to be a total guru.

One thing to know is that this book is not supposed to teach you how to program. It is also not a general book about programming games. Instead, I focus on how to use the BREW SDK to apply existing game development techniques in the realm of mobile gaming. This book highlights features of the SDK of interest to game developers and then puts them in simple examples to demonstrate their use. These example programs are not meant to be frameworks for commercial games or a sterling example of how to create your own engine. There are plenty of books already on the shelves that cover the software development process as a whole. Instead, use this book to apply your existing skills as a game developer to mobile devices.

How Do I Use This Book?

This book is a combination of reference-style instructional chapters and working examples. Each major functional block of the BREW SDK that is relevant to game development will be explained in detail along with some code examples on the usage of the major features. The later chapters provide complete working examples of games that feature many of the techniques described in previous chapters. Therefore, it is possible to sit down and read some chapters, while others will require you to work through examples at your computer.

Although parts of this book may seem overly reference-oriented, this is not a comprehensive reference volume on BREW. There are way too many features in the BREW SDK to fit in this book alone. Instead, I cover the major features that will be useful in the creation of a BREW game. For more minute details, the BREW SDK documentation should be sufficient.

Why C?

The vast majority of the code in this book is written in C, but you can develop BREW applets in C++. In fact, I provide instructions on how to use BREW with C++ later in the book. Do not get me wrong—I am a C++ fan. However, most of Qualcomm's examples in the SDK are written in C. Therefore, I feel it is better to use the same language for this book. That way, you can easily use code from the examples in your own applications.

Coding Conventions

I am a big fan of Hungarian notation, so you will see a lot of it in this book. Basically, I prefix my floats with f, my ints with n, my pointers with p, and so on. I also suffix my structs with _s, typedefs with _t, etc. Also note that even though the BREW SDK uses the m_ notation to identify members of structs, I do not. Instead, I reserve this notation for members of C++ classes. That way, I can tell the difference between using a struct and a class. Sure, they are almost the same thing, but I feel it is important to know the difference between the two. You will get used to it.

Companion Files

The companion files, available at www.wordware.com/brew, include all of the source code discussed in the book, as well as Robin Burrow's useful Mappy tile map editor and a demo of Cosmigo Pro Motion. There are also a few other useful files, including bitmaps that have various palettes that may be used by some color BREW handsets.

How Do I Use the Companion Files?

The code in the companion files is organized in folders according to chapters in the Source directory. Simply copy the folders to your hard drive, and you should be able to compile them with Visual C++ or, in some cases, the ARM BREW Builder. This book assumes that you have installed the BREW SDK to the default location on your C drive. BREW Developer Studio projects require the inclusion of a few source files from the folder in which the BREW SDK is installed. If you have put the BREW SDK anywhere other than the default installation folder, you will have to edit the workspaces to reflect this.

Conclusion

You have now been introduced to the book, the author, and the justification for my existence. Well, sort of. Take a deep breath. Prepare for a journey into an entirely new area of game development.

Chapter 1

A Crash Course in Wireless Gaming

Introduction

As both a professional game developer and rabid fan, I have seen the game industry go from its humble, spare-bedroom beginnings to its current state of multi-million dollar budgets, huge teams, and massive profits.

To the casual observer, it may seem that games are getting bigger and more complicated every year. You may even hear industry veterans complaining about the death of the early days, when a single programmer could code a game all by herself and the process of bringing a game from a concept on the back of a napkin to the store shelf was measured in months instead of years. If you catch these dinosaurs waxing nostalgic about the "good old days," you might be able to shake them out of their misery with a simple observation.

Those days are still here!

The games of the late '70s and '80s may seem simple compared to today's skeletal animated polygonal extravaganzas, but they still can be highly entertaining. The fact is, there was a large audience for these simple and entertaining games back then. It is easy to see that the audience is still there.

What was fun then is, in many cases, still fun now. It is not as if these classic games suddenly stopped being entertaining. However, the mainstream gaming audience is no longer the primary target for these sorts of games. There is an entirely new generation of gamers that are receptive to the simple and fun concepts of yesteryear.

You can see this by walking the aisles of your favorite software store. Just check out the massive amounts of so-called "budget" games lining the shelves. There are plenty of puzzle titles and simple arcade challenges offered at low "impulse" prices that appeal to this crowd. It is not totally unheard of to see an occasional compilation disc of these sorts of games climb its way up the charts alongside major, so-called "A-title," releases.

With the rise of portable computing in the form of personal digital assistants, such as the Palm and PocketPC, a new market for low-end gaming has appeared. Because of the limited nature of the hardware, PDAs and other small devices are often only capable of running the same sorts of games that were considered state of the art ten or so years ago. This is a grand opportunity for the independent developer to create simple games with a small amount of resources. With inherently low overhead brought on by the restrictive device capabilities, these games have the potential to be profitable with a relatively small amount of sales.

The new generation of programmable mobile phones provides a brand new avenue for this kind of independent development. With capabilities often far below even the simplest PDAs available today, even fairly advanced mobile phone games require development resources attainable by the average lone-wolf developer.

Not only are the development costs associated with wireless development much less than PC, console, or PDA titles, but the distribution is cheap as well. Unlike retail games that require CDs to be duplicated, boxes to be constructed, and products to be shipped and distributed to stores, there is no physical object that changes hands on a mobile phone. It's all bits—no atoms. The product exists as a chunk of binary data on a carrier's server. The customer chooses which product she wants, and that chunk of data is sent over the air to her handset. The cost associated with producing the physical media is completely absent from the wireless gaming business model, keeping overhead low.

To most of the world, the wireless gaming market is brand new. Some major traditional game publishers have begun developing products for this arena. Yet, many of the major players in this new industry are relative unknowns and start-ups. They are getting in at the ground floor of what is poised to become a major source of revenue in the coming years.

Because of the nascent state of the market, there are a lot of questions to be answered. These largely concern the technology used and business models for creating a viable revenue stream. Once it is clear how to actually make money in this area, the major publishers are sure to show more interest. Right now, the market is wide open for small and independent developers to make their mark.

Mobile Phone Gaming: Past, Present, and Future

Wireless gaming is not a new concept. Games have been available on mobile phones for quite some time. Here we will illustrate the various phases of mobile phone game development, as well as the major technologies involved in the past, present, and future of mobile phone gaming.

Embedded Games

I am sure that a lot of you are familiar with Snake, the primitive action game embedded into many Nokia phone handsets. Games such as these can be classified as the first generation of mobile phone entertainment. While the typical embedded can be a good way to waste time, it is programmed into the phone's firmware. This means that you are stuck with the same version of Snake until you get a new phone. Some newer models come with multiple games encoded in the chipset. However, this does not change the fact that you cannot swap or add new games later.

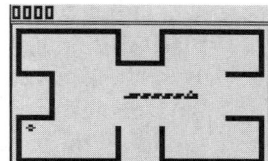
Figure 1-1: Nokia's embedded Snake game

Wireless Application Protocol

Today you see a lot of wireless carriers promoting data services. What they usually mean by this is services using the *Wireless Application Protocol*, or *WAP* for short. WAP is a protocol standard created in the late '90s by a consortium of wireless phone companies. This consolidated their individual efforts to create a mobile Internet experience into one comprehensive open standard. Phones that support WAP feature a special browser, called a microbrowser, that can access web pages crafted specifically for the limited capabilities of a mobile phone.

Instead of using HTML as the World Wide Web does, these pages are created in the *Wireless Markup Language*, or *WML* for short. Creating a page in WML is similar to HTML, but the functionality is very basic and suited for the small screens and simple interfaces of most mobile phones. WML supports very limited graphics, basic text formatting, and even some client-side scripting using a language called WML Script.

Note: In the United States, most of the original so-called WAP services were actually something called the Wireless Web. Wireless Web deviates from the WAP standard in minor ways. Usually this means that Wireless Web sites use a different markup language, HDML, but the same type of binary protocol scheme that WAP employs. Most carriers provide gateways that automatically translate WML WAP pages into HDML for older Wireless Web microbrowsers.

Not long after the debut of WAP, sites from E*Trade, CNN, Yahoo, and other major Internet content providers were made available to WAP users worldwide. Of course, games also became an early favorite among WAP users.

Creating a WAP site is very similar to making a traditional web site. The only major difference is that instead of generating HTML, your server application creates WML pages. This can be done through the use of common web technologies, such as CGI scripts, JSP, PHP, ASP, or any other tool that you may use to create dynamic database-driven web sites. This limits most WAP games to simple turn-based forms interaction, similar to those seen in the early days of web gaming.

Figure 1-2: This is an example of a basic WAP page using gelon.net's emulator.

There are a wide variety of games available using WAP technology. If you open up the games listing on your carrier's WAP portal, you may find anything from classic card and gambling games to multiplayer space combat and strategy contests. Some wireless game publishers have seen massive amounts of visitors to their WAP game sites, despite the rather unimpressive capabilities and unwieldy interface of most WAP browsers. Even though the visuals of most games are relegated to postage stamp-sized monochrome graphics accompanied by plain text, a small audience of game players has accepted the limitations and spent millions of minutes playing these simple games.

Although it is technologically possible to create games using WAP, most companies have found it to be not very economical. In most cases, wireless carriers do not host your server application and do not provide any air-time royalties or other means of billing for your service. This situation is not likely to change, with many carriers putting less of a focus on WAP-style applications in favor of new technologies.

Figure 1-3: nGame's Carrier Force WAP game has some basic bitmap graphics.

iMode

The customer base of WAP pales in comparison to its Japanese equivalent, NTT DoCoMo's iMode. iMode is a similar technology to WAP, aside from a few major differences. The biggest difference is that iMode resides on a packet-switched network, as opposed to most WAP carriers' circuit-switched technology. What this means is that WAP users pay by the minute, while iMode users pay by the kilobit. Therefore, Japanese iMode users can read the same page of information all they want without incurring additional costs. It is not until they fetch a new page that they are charged for the information downloaded. With WAP, you are charged by time. So regardless if there is any network activity or not, you are blowing precious minutes as you browse for information.

Figure 1-4: An example of an iMode page, using Wapprofit's emulator available at www.wapprofit.com. In this case, it is displaying Hudson's MiracleGP game.

Not only does this make iMode more economical for the end user, but it makes the browsing and playing experience much more enjoyable. You are no longer bound by the pressures of time—you are free to read and browse for as long as you like.

Also, iMode uses a markup language familiar to every web developer: HTML. iMode actually uses a derivative called Compact HTML, which is a tailored subset of HTML for iMode handsets. The side effect of this is that iMode pages can be viewed in a normal web browser. You can view WAP pages on your PC, but to do so requires a special plug-in or emulator.

Another major difference between iMode and WAP is that iMode is not an open standard. It is owned and controlled by NTT DoCoMo, a large Japanese telecommunications company. With WAP, you have to negotiate with each carrier individually to get your game linked on their customer portals. Since iMode is operated by only one carrier, there is only one organization to appeal to. Of course, with the massive popularity of iMode in Japan, actually getting a prominent listing on their customer portal is proving to be a difficult task for many. If you can get decent portal placement, it is possible to make a significant amount of money off of iMode sites using NTT DoCoMo's centralized billing system. Unlike WAP, there is an existing business model for content developers that actually works.

The handsets that NTT DoCoMo uses for iMode has far more advanced than the handsets commonly available in the U.S. or Europe. NTT DoCoMo subsidizes new phone technology, which means they take a loss on the price of the phone in order to stimulate customer growth. The phones available in Japan are much sleeker with advanced features such as music, color, and even full motion video. As a result of NTT DoCoMo's subsidization, they are cheaper to acquire than they would be otherwise.

iMode has enjoyed massive success in Japan, partly because of the above-mentioned attributes and also because it has become a replacement for the desktop. Traditional desktop PCs are nowhere near as popular in Japan as they are in the U.S. or Europe. Therefore, iMode's net access has acted as sort of a substitute for the traditional desktop Internet experience that we are used to in the U.S. and Europe.

Figure 1-5: An example of a basic iMode news portal site

Just how popular is iMode? Recently, NTT DoCoMo said it had 26 million subscribers for iMode. Each of iMode's minor competitors that provide similar services had around 6 million users each when this figure was revealed. Around this time, SprintPCS, the top WAP provider in the United States, announced the accumulation of 1.5 million Wireless Web subscribers.

Programmable Phones

If that's the current situation, then what's the next wave? WAP and iMode are both network resident technologies. This means that the application resides on a server somewhere on the Internet that sends data to your handset. The new trend in wireless data services is programmable phones that are able to download and run applets right on the handset. These applets range from expense accounting programs and instant messengers to, of course, games. Right now, there are two major technologies in this area: Sun's slimmed-down version of Java, J2ME, and Qualcomm's BREW.

J2ME

J2ME, or Java 2 Micro Edition, is a subset of Sun's powerful Java language tailored for small devices. With the addition of various sets of packages (known as profiles and classes), it is possible to write Java applets, known as MIDlets, for a growing number of phones supporting Java technology.

Major handset manufacturers such as Motorola and Nokia have been releasing phones with this technology in both Europe and the U.S. Japan has its own flavor of Java phones, using slightly different standards that are both technologically impressive and highly popular in that country.

Note: In Japan, NTT DoCoMo's Java service is known as iAppli. Handsets that use iAppli have their own custom version of Java developed before Sun completed the MIDP standard. NTT DoCoMo's competitors, such as J-Phone, also have their own custom Java virtual machines embedded on handsets with all sorts of unique features not found in the MIDP specification, including 3D graphics and MIDI music.

For years in Japan, and now in Europe and the U.S., you have been able to download applets right over the wireless network to your handset. This means that if you are waiting in line for a movie or bored out of your skull on a long bus ride, you can simply download a game to your phone and start playing.

Figure 1-6: DWANGO's Samurai Romanesque is a popular iAppli-based Java game in Japan.
Copyright DWANGO Company, Ltd. All rights reserved.

Java presents a unique solution to wireless development because of its "Write Once, Run Anywhere" philosophy. Java is an interpreted language, not compiled directly for the hardware. The source code is compiled into byte-codes that are executed by a *virtual machine*, or *VM* for short. This means that a program written in Java can run on a variety of different hardware platforms. It is necessary for the device manufacturer to provide a VM capable of running Java code for their specific hardware. Of course, as we have seen from the various bugs and functional differences in Java VMs in web browsers, there are bound to be differences between Java implementations on different handsets. In fact, there are already several bugs in Motorola's virtual machine that differ from the standard implementation of J2ME's VM. Also, various handset manufacturers have provided their own custom extensions to Java. This often means that you must tailor your code for each handset's unique features.

Figure 1-7: An example of an MIDP-based J2ME game, nGame's BeatReactor

Overall, J2ME is a solid platform for mobile application development. Its platform-agnostic philosophy makes supporting a wide variety of different handsets somewhat painless. Also, you can develop a complete applet using only free tools provided by Sun and the handset manufacturers. It seems that most major carriers worldwide are rolling out plans to support Java. Therefore, a bright future for the technology is assured.

BREW

BREW, or *Binary Runtime Environment for Wireless*, is a mobile applications standard from Qualcomm. Originally, BREW was Qualcomm's internal technology for developing embedded software for their own handsets. Now, it has been made available to anyone interested in wireless development. BREW is actually three things: a hardware platform, a C/C++ SDK, and a distribution model.

For the hardware platform, BREW requires an ARM CPU-based chipset. ARM processors are very popular among portable device makers because of their hefty computing capabilities and low power consumption. ARM chips have been used in many popular devices, including Microsoft's PocketPC and Nintendo's GameBoy Advance. Qualcomm has designed several tiers of chipsets offering increasing levels of speed and features for BREW handsets.

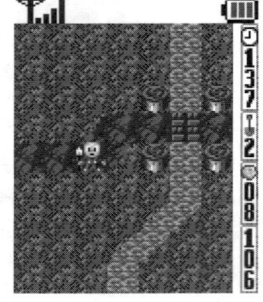

Figure 1-8: Monkeystone's Gold Digger is a fine example of what kind of gaming is possible with current BREW technology.

BREW's SDK is a set of C/C++ libraries for creating programs to run on BREW hardware. The usage of this SDK is this book's primary focus. The SDK provides major blocks of functionality in such areas as GUI, graphics, networking, sound, and much more. Major functionality is continually being added to the SDK in revisions. Also, handset manufacturers can add their own extensions to deal with custom technology not in the general BREW specification. With Qualcomm's continual support, BREW remains a flexible standard for accommodating upcoming mobile technologies.

Finally, BREW specifies a distribution model for getting your applets out to customers. Qualcomm has taken a page from iMode's book by providing a centralized billing model that can make it easier to reap the financial rewards of mobile development. Each BREW applet must be tested and certified as bug-free. Then it is offered for distribution by BREW carriers around the world. In exchange for Qualcomm's billing and distribution

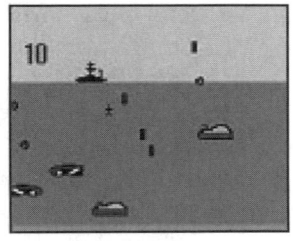

Figure 1-9: nGame's UBoat is an example of a very simple BREW game.

services, they take a 20 percent cut of the applet price. The process of selling your applets via BREW's distribution network is detailed in later chapters.

One of the major differences between BREW and J2ME is that BREW is compiled native code. Because you are operating closer to the hardware, it is possible to squeeze more performance out of BREW applets on chipsets similar in power to those on J2ME devices. Also, because of BREW's low-level functionality, it is possible to write a Java VM on top of BREW. In fact, this is already being done. Therefore, it is possible to run both native BREW applets and Java MIDlets on BREW phones in the future. Java on BREW will be discussed later in the book.

BREW is also a proprietary standard. It relies on licensed compilers, registered developer programs, and other mechanisms familiar to those accustomed to developing on consoles from Sony and Nintendo. Although the costs associated with being a licensed developer are far less than developing for the aforementioned consoles, BREW remains one of the more expensive options when developing wireless entertainment. However, the SDK and emulator are available for free, so you can get started developing today at no cost. It is only when you want to launch your applet on real hardware that money comes into play.

The Future

There are many other mobile technologies on the horizon. One increasingly popular option is the merging of PDAs and mobile phones—the so-called "smartphone." Models such as Kyocera's 6035 or Samsung's SPH-I300 provide the full functionality of a Palm PDA with a mobile phone. Microsoft has similar plans for its PocketPC with the Stinger project. Could these hybrid devices make BREW and J2ME applets obsolete in favor of native PalmOS or PocketPC applications? This is a possible scenario. However, these devices are currently bulky and expensive alternatives to an inexpensive BREW or Java phone.

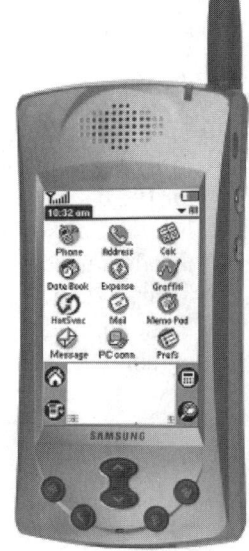

Figure 1-10: Samsung's SPH-I300 is a Palm-based smartphone.

The Business Model

How does the business side of things work? At the moment, this is all rather cloudy. Japan and South Korea have shown that it is possible to be profitable creating and selling mobile games. The business models in Europe and the U.S. have yet to emulate the success of Asian companies and are instead still struggling to find a way to generate revenue. With the release of new BREW and Java handsets as well as the launching of new digital networks, this is slowly changing.

The business of wireless games usually involves three parties: developer, aggregator, and carrier. The developer is you: the person or company that performs the technical task of programming and designing a mobile game. The aggregator is what is known as a publisher in traditional game development. The aggregator usually acquires content from a number of different developers and releases them on wireless phone networks operated by a carrier. The carrier is the wireless phone service provider, such as Verizon Wireless or SprintPCS.

In the case of downloadable applet-based games, a fee is usually charged for the user to download the applet. A portion of this fee goes to the carrier and the rest to the aggregator. The aggregator then gives a portion of their slice to the developer. As you can see, there are many hands in the pie. Therefore, a lot of developers have chosen to also be aggregators and publish their own content.

Conclusion

Mobile gaming is set for an explosive growth period. It may take awhile for the carriers to actually realize this, but consumers have shown an interest, as we have seen in the runaway success of NTT DoCoMo's iMode service. This chapter has given you a rundown of the current and future technologies facing game developers targeting mobile phones. Although there are a number of competing and complementary technologies, this book will introduce you to Qualcomm's BREW platform as a powerful tool in the creation of mobile games.

Chapter 2

Introducing BREW

Introduction

This chapter will introduce the reader to Qualcomm's BREW platform. Sure, you may have read the brief description in the introduction of this book, but there is a bit more to it. This chapter is not required reading if you just want to get down and code. However, it is a good idea to know what you are getting into before devoting some significant effort to the commercial development of BREW applets.

The Big Picture

Binary Runtime Environment for Wireless, or BREW for short, is a new technology developed by the wireless telecommunications giant, Qualcomm. Originally intended as an internal library for developing software on their own custom handset hardware, Qualcomm has now released BREW to the world.

This means that you can now use their comprehensive code library to develop applets that run directly on BREW-enabled phones. Several major handset manufacturers, such as Samsung, Kyocera, and Motorola, are eagerly creating dozens of new handsets with all sorts of interesting features and increasingly powerful capabilities.

BREW is not just a software technology. Along with the SDK and hardware specification, Qualcomm has created a content distribution and billing system that corrects many of the issues that developers have had with previous wireless content delivery schemes—most notably, how to actually make a profit.

A Note About Versions

Currently, there are three major versions of BREW: 1.0, 1.1, and the new 2.0 release. BREW 1.0 is what the bulk of this book is about. Why focus on the oldest version? Because most of the handsets available from carriers supporting BREW are running BREW 1.0. It will take some time for BREW 1.1 to become widely available on commercial hardware. BREW 1.1 is a relatively minor upgrade of the original release anyway. The BREW 2.0 SDK has recently been released with advanced multimedia functionality and many major updates. However, 2.0-supporting handsets are few and far between at the moment. This book includes sections on 1.1 and 2.0 that outline the new features if you wish to develop for the advanced versions of the BREW platform. However, it may be worthwhile to play it safe and stick with BREW 1.0 if you are planning on creating a commercially available application. BREW 1.0 code will run fine on 1.1 handsets and beyond. However, code written for later versions of BREW will not work with older handsets.

The Hardware

You may wonder which phones support BREW. After all, you would like to know what your game will finally run on when all is said and done. There are quite a few different BREW handsets available in the various markets in which BREW has already been launched. Once you become a member of Qualcomm's BREW Developer Extranet, you can download extensive information and emulator files for current and upcoming handsets. Here we will detail a few of the commonly available and eagerly anticipated devices in the U.S. market. See Qualcomm's web site for more comprehensive and up-to-date information.

Current Handsets

In the United States, Verizon Wireless launched with two basic handsets. From here, they have scheduled a gradual launch of increasingly more powerful hardware. The two basic handsets are as follows:

Kyocera QCP3035

This is the first BREW phone available in the U.S. and also the cheapest. Dubbed a "smartphone" by Kyocera, this BREW-equipped handset features many options found on more sophisticated PDAs. These include improved contact list functionality and advanced e-mail and data communications options, as well as the ability to install and run BREW applets. It is nothing that will make you toss your Palm or PocketPC out the window, but there is a significant advance from the previous generation of mobile phone technology.

This being the lowest-end model, it sports a rather unimpressive array of technical specifications. The first thing you will notice is the display; it is 1-bit. The only colors available are black and white. Couple this with the rather miniscule resolution of 89x99 pixels, and you have quite a challenge to create visually appealing games. Compounding this problem is the exceedingly slow refresh rate of the display hardware. A full-screen redraw will take about half a second. Using tricks such as dirty rectangles and other ways to minimize redraws can help things. But it definitely is a challenge to squeeze a decent frame rate out of this device.

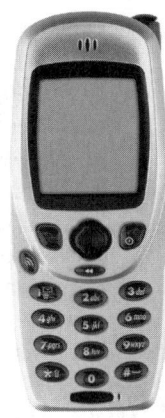

Figure 2-1: The Kyocera QCP3035

The memory footprint is also quite small. Although there is a grand total of 90 kilobytes available, BREW occupies 30 kilobytes of this space. All of your data, code, and heap allocations have to fit in 60 kilobytes. Tactics for dealing with low memory situations will be presented in later chapters.

As with most BREW phones, the standard keypad is augmented by a tiny joypad-like controller below the screen. By pressing the pad in the center, you can select menu items and perform other tasks. This joypad is quite sensitive and can be something of a pain to deal with in any application that requires precise movement. Chances are that this will be the case with any game. I know it was with mine!

Overall, the QCP3035 is a rather unimpressive device. However, this is the cheapest of all the BREW handsets; therefore it is likely to be very popular. The 3035 and other entry-level phones are a major consideration in developing BREW applications.

Sharp Z-800

The Sharp Z-800 is one of the higher-end BREW handsets. Perhaps the most noticeable enhancement of this phone is its vibrant color screen. With a higher resolution of 128x144 and an 8-bit display, it is quite possible to develop games with rather impressive visuals (well, impressive for a phone, that is). Of course, the LCD screens used in most phones are not of the highest quality, and this Sharp model is no exception. Moving objects tend to streak, as they

Figure 2-2: The Sharp Z-800

leave a ghosting after-image from their previous position. This phone still has some refresh rate speed problems as well.

In addition to the enhanced display, the Z-800 also has much more RAM and storage space. You have about 200K of RAM to hold your in-use resources, code, and heap allocations. In addition to the increased RAM capacity, the Z-800 has 32 megabits of storage for BREW applets and other data.

These are just a few of the readily available handsets. More phones are promised in the future, some of which are said to include 16-bit displays, MP3 playback, and other more advanced features. Qualcomm's BREW web site has more current information on handset hardware.

Future Hardware

Verizon promises some impressive devices for release by the end of 2002 and beyond. They range from mid-range grayscale handsets to 16-bit color monsters. Some of these may already be available by the time you read this.

LGE VX-10

The LGE VX-10 is one of the first 3G BREW phones available in the States. If this phone looks familiar, it is because it is basically the same device as the VX-1 with BREW installed on it. It also uses Verizon Wireless' upgraded 1XRTT service known as the Express Network. This means you will have faster data transfers, a far cry from the full bandwidth that 3G will offer in the near future.

Figure 2-3: The LGE VX-10

The LGE VX-10 has a fairly large screen at 120x100 but can only display graphics in four shades of gray. As for memory, it has 200K of RAM, 650K of static memory for applet storage, and 2K of stack. It also has support for MIDI sound and ringers. I have seen this handset in action, and its performance is rather impressive for such a low-end device.

Motorola T720

Motorola is a surprising new entry in the realm of BREW handset manufacturers. Their first device, the T720, is quite an impressive debut. Most notably, it sports a 4096 color screen with a resolution of 120x130 pixels. It also has MIDI support for audio. The device also sports 400K of RAM, 1.5 MB of flash memory storage for applets, and

Figure 2-4: The T720

around 4K of stack space. As with all modern BREW devices, this uses 1XRTT for high-speed network access.

Audiovox/Toshiba CDM-9500

The Audiovox/Toshiba CDM-9500 is another higher-end device with a color screen and lots of memory. As seems to be the standard with high-end devices, the CDM-9500 sports a large active-matrix TFT color screen. It has a resolution of 144x158 pixels and a color depth of 16 bits. 65,536 colors is a far cry from the two colors on the 3035, I must say. The CDM-9500 has 250K of RAM, around 2.5 MB of file system memory for applet storage, and a 4K stack as well as support for high-speed 1XRTT communication.

Figure 2-5: The Toshiba CDM-9500

Phones in Other Markets

BREW has been quite a success in South Korea, and as a result, there are dozens of handsets available with varying capabilities. Check the BREW Developer Extranet for the details and emulator files for all of these models. Some are quite advanced with heaps of memory and multimedia capabilities.

The SDK

The second piece of the BREW platform is the SDK. I will not go into excruciating detail here, as the rest of the book focuses on how to use the SDK to develop applets. However, I will discuss how it fits into the BREW platform.

The BREW SDK is a library that programmers can use to write an application for the BREW platform using C or C++. The vast majority of the sample code is written in C, which is the reason this book is focused on using C. However, instructions are provided on how to use C++ in later chapters.

The SDK covers major blocks of functionality, including graphics, networking, sound, text display, GUIs, and most any other area that you could imagine. It is a constantly evolving beast, and Qualcomm is providing updates to the SDK, based on input from developers.

The BREW SDK comes with a number of tools to aid in the development of applets. Perhaps the most prominent of them is the emulator. This emulator is a program that allows you to run your code on your Windows desktop instead of having to compile for the real hardware. This makes it a bit easier to quickly debug and test code. It is also possible to configure the emulator to simulate many different types of memory configurations, screen resolutions, and bit depths. However, there is no getting around having to test your application on real hardware, as the emulator is not terribly accurate.

The Distribution Model

The thing that differentiates BREW from other similar technologies, such as J2ME, is the distribution model. Not only is BREW a technology for developing mobile phone applets, but it is also a distribution system for getting your content out to paying customers. BREW removes a lot of the confusion surrounding how to get your content to customers by simplifying the process with a central billing and distribution model.

The short of it is this: Qualcomm certifies your applet as "True BREW" in an independent testing lab. By being certified as "True BREW," it signifies to carriers that your applet is bug-free and ready to go. (A carrier is a company such as Verizon Wireless that provides wireless phone service to its customers.)

This testing process costs money. Once certified, you can decide how much you want to charge for your applet. The applet is then made available to carriers that support the BREW platform. For every sale at a carrier, Qualcomm takes 20 percent of the cost that you decided upon and gives you the remaining 80 percent. The carrier can mark up the price of the applet for its own profit. We will go further into the publishing details in later chapters.

Conclusion

You are now primed on Qualcomm's BREW platform. This chapter has glossed over the basic elements of BREW, as well as the devices that support it. Each major topic discussed in this section will be elaborated upon in greater detail. But it should be a bit easier to wrap your head around the subject matter after taking in this brief overview.

Chapter 3

BREW Gaming Examples

Introduction

Now that you are familiar with the concept of mobile games and BREW, what kinds of games are out there? Despite the short time BREW has been on the market, there are a surprising number of games already available. By looking at the following commercial games, you may get a good idea as to what kind of game you would like to develop. (All copyrights and trademarks associated with these games are property of their respective owners.)

Action Games

Although current BREW handsets have a rather limited capacity for high-speed arcade action games, there have been several attempts at the genre. With creative programming and optimization techniques, it is possible to create fun action challenges for BREW devices.

Jewels & Jim

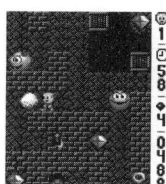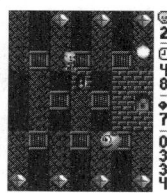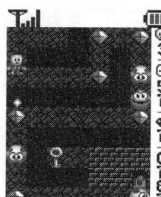

Figures 3-1 to 3-3: Jewels & Jim

Jewels & Jim is an action game from Monkeystone Games where you maneuver Jim through 99 different levels collecting gems and battling monsters called Gloobs. There are also doors that need to be unlocked and switches that can trigger events affecting the layout of the map.

Puzzle/Board Games

Considering the limited graphics capability and constrained interface of the typical BREW mobile phone, puzzle and board games are a perfect match for the platform. Puzzle games usually rely on exceedingly simple game mechanics and limited controls, making them a prime candidate for mobile systems. Board games also have these same properties, often with multiple players involved. There are a number of popular puzzle and board games for BREW on the market right now.

Quick Chess

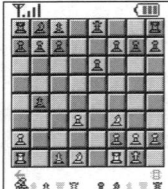

Figure 3-4: Quick Chess
Copyright © 1991-2002
Dynamo Development,
Inc.

It is not hard to imagine chess being one of the first board games for BREW. Dynamo Development's Quick Chess is a single-player chess game with all of the basic features. Quick Chess manages to fit a full game of chess including a robust AI inside a small package designed for BREW handsets.

Dinc

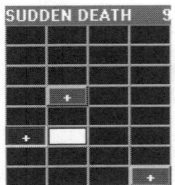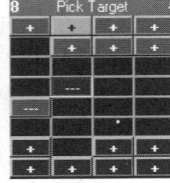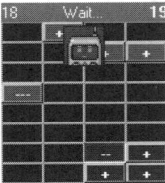

Figures 3-5 to 3-7: Dinc

Dinc is a puzzle/board game from Monkeystone similar to Checkers. In Dinc, you must remove your opponent's pieces from the board until you capture them all and knock her out of the game. As you can see, Dinc has a very simple graphical approach, which is characteristic of many puzzle games.

Strategy Games

Strategy games can be strictly turn-based statistical analysis type affairs typically associated with war-gaming. Or, they can almost be a hybrid of action and strategy by having an interface and feel similar to an action game but with the complex planning and thinking gameplay required by a pure strategy game.

Gold Digger

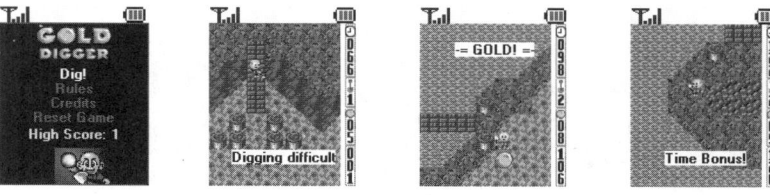

Figures 3-8 to 3-11: Gold Digger

Gold Digger from Monkeystone is one of those hybrid games where you have both action and strategy elements manifested in one game. Once again, you play the part of Jim, fearless adventurer and digging enthusiast. Jim must dig for gold in ten different maps under the pressure of time. The various terrain types affect how long it takes to dig, which can impact your score.

Sports Games

Sports have always been a popular game genre on both consoles and the PC. Mobile gaming is no exception. There are several sports games on the market simulating a variety of popular sports, ranging from football to basketball and everything in between.

Handy Baseball

Figures 3-12 to 3-16: Handy Baseball

South Korean developer GameVIL's Handy Baseball is a vivid example of sports gaming with BREW. Baseball 2002 is a cartoon-styled baseball simulator, including realistic pitching, hitting, and fielding action. There is also some management and statistics gameplay for the armchair coach.

Adventure/RPG Games

Another popular category is adventure and role-playing games. Creating games with long-term goals and quests requires quite a bit more development time than your average arcade or puzzle game; however, a few entries into this exciting genre for BREW devices have popped up recently.

Last Warrior

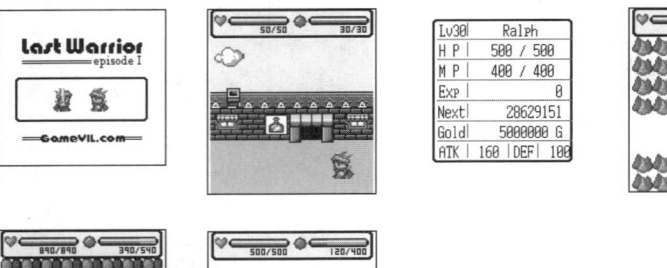

Figures 3-17 to 3-23: Last Warrior

South Korean developer GameVIL's innovative Last Warrior is a classic console-style role-playing game. The player assumes the role of Giyom, a warrior battling across the fantasy world of Loren. The story has three episodes filled with various different vicious monsters, countryside villages, and dangerous dungeons. There is also a random map feature for massive replay value.

Casino Games

Casino games have the same appeal as puzzles. They often have very simple rules and require only very basic controls. Therefore, a number of casino titles have shown up on BREW devices as of late. Since gambling games are to be consumed in short bursts, they are an ideal genre for the typical mobile gaming consumer.

MonkeyMaster Slots

Figures 3-24 to 3-25: MonkeyMaster

Monkeystone's MonkeyMaster Slots is a prime example of casino gaming at its most basic. You simply press 1, 2, or 3 to bet $1, $2, or $3 in virtual money and give the machine a whirl.

Software Toys

Since the early days of Activision's Little Computer People and Maxis' SimCity, the so-called "software toy" has been a popular genre of game. Although not a game in the most strict sense, a software toy is usually an entertaining application with no set goal or even stringent rules. These sorts of applications have proven to be very popular on Japanese networks, so it is no surprise that several have begun to appear on the BREW platform.

Magic 8 Ball®

Figure 3-26: Magic 8 Ball®

As described by Mattel*:

Now you can carry along your very own personal and professional consultant wherever you go. The Magic 8 Ball is available at your fingertips and will never run out of answers. Simply ask a yes or no question and the Magic 8 Ball will turn and give you an answer. Never again will you be left with a decision to make on your own!

Date Ball™

Figure 3-27: Date Ball™

As described by Mattel*:

From the makers of the Magic 8 Ball is the Date Ball. Your personal consultant for your dating questions. Simply ask a yes or no dating related question and the Date Ball will turn and give you an answer. Never again will you be left with making a dating decision on your own! Wherever you go you can always consult the Date Ball.

Card Games

Card games come in various types. From the traditional casino-style gambling games to the collectible card type, the simple graphical nature and slow-paced gameplay style of most card games make this type a good match for the average BREW handset's interface and capabilities. There are a number of different card games currently available for BREW.

UNO®

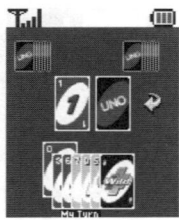

Figure 3-28: UNO®

As described by Mattel*:

Now you can experience the fast unpredictable fun of UNO—everyone's favorite family card game! UNO is easy to learn and quick to play. Simply match card color, number or word. The first player to rid them-selves of all their cards is the winner. When you have one card left, yell "UNO"! Keep track of your score while competing against up to 3 virtual opponents. Watch out for those special command cards like

Reverse, Skip, Draw 2 and the Wild Draw 4 card! You never know when they'll pop up.

Multiplayer Games

Because all BREW handsets can connect to the Internet via the same mobile data network used for voice calls, playing games against remote opponents is a natural application for the technology. The engineering, testing, and design time required for multiplayer games is exceedingly more complicated than the average single-player action or puzzle title. However, you will learn in this book how to use BREW's socket functions to create your own multiplayer experiences.

Conclusion

In this chapter you have seen several examples of BREW games. This should give you a good idea of what is popular in mobile gaming, as well as the possibilities of developing games with BREW. If you want to see what the absolute latest is in BREW gaming, you can check out sites like Wireless Gaming Review (www.wirelessgamingreview.com) for up-to-date reviews, news, and previews.

* Magic 8 Ball®, Date Ball™, and Uno® are trademarks owned by and used with permission from Mattel, Inc. © 2002 Mattel, Inc. All Rights Reserved.

Chapter 4

Getting Started

Introduction

There are a few things that you must do to start developing for the
BREW platform. This section details the tools you need, where to get
them, and how to configure Microsoft's Visual Studio to compile your
BREW projects.

The Tools of the Trade

Before we get into the details of acquiring and setting up the BREW
SDK, perhaps it is time for a brief overview of the tools both necessary
and helpful for the development of BREW games.

Microsoft Windows 2000 or XP

Right now, Qualcomm's BREW development tools only work on
Microsoft's Windows 2000 or XP operating systems. Although no support for Linux or MacOS have been announced, it has been hinted that
alternative operating systems may be a possibility for future tool
releases. However, it is possible to run some third-party tools, such as
ARM's BREW Builder, on various flavors of UNIX, but for our purposes
we will use Windows exclusively for development.

Microsoft Visual C++

Right now, the only compilers supported for use with BREW and its
emulator is Microsoft's Visual C++ 6 or Visual C++ .NET. Throughout this book, I use VC++ 6.0. Although VC6 has not seen a major
update since 1998, it is still a very flexible and robust development tool.
It is also fairly expensive. For students, it is possible to get deep

discounts on educational versions of Visual C++. For professionals, it may be worth looking into a subscription to the Microsoft Developer Network for discounts on not only this tool but the entire suite of Microsoft compilers, tools, and operating systems. When it comes to compiling your code for a real handset, you actually need to use a completely different compiler, the ARM BREW Builder. We will get into that later in the book.

IBM WebSphere

There is actually an alternative IDE for use with BREW. IBM's WebSphere is an IDE that has been used in the past for Java development. At the BREW 2002 Developers Conference in San Diego, IBM was showing off a new BREW version of this tool. Unfortunately, you still need to have Visual Studio installed, as it uses Microsoft's compiler to create DLLs for use with BREW's emulator. Therefore, the advantage to WebSphere is sort of nonexistent. Still, if you prefer WebSphere's interface and somewhat more advanced support for native BREW compiles, this product may suit some of your needs.

Jasc Paint Shop Pro

There are a number of different programs useful for drawing bitmap graphics for use with mobile games and other applications. I find that Jasc's Paint Shop Pro is not only well suited for this task but extremely cheap (around $99). A full-featured, time-limited demo can be downloaded from www.jascsoftware.com.

Cosmigo Pro Motion

Back in the early days of game development, a popular paint program among pixel artists was Electronic Arts' Deluxe Paint. When Electronic Arts stopped supporting this immensely popular tool, old copies of the package became quite a commodity. Cosmigo's Pro Motion is a paint program that includes the familiar interface of Deluxe Paint along with a host of new features for animation, tile creation, and image editing that make it another popular package for mobile game bitmap graphics. It is also very inexpensive with a "Lite" version for only $19.99 and a full edition at the seemingly arbitrary $58. A demo is included with the companion files (www.wordware.com/brew) and at www.cosmigo.com/promotion.

There are a wide variety of other tools in image editing, source control, music composition, sound editing, and other areas useful for BREW game development. For the purposes of this book, we will mainly use Visual Studio 6 and keep the reliance on third-party tools to a minimum.

Becoming a Registered Developer

Before you can download the BREW SDK, you need to become a registered developer. At the most basic level, all this requires is filling out a form on Qualcomm's web site (www.qualcomm.com/brew). There are no fees or any other hefty requirements. You can, however, pay for more advanced levels of developer support. All you need for now is the most basic level of developer status, which is entirely free.

After you are registered, you can then download the latest revision of the BREW SDK. For the purposes of this book, download BREW Version 1.0. The code in this book should work with the newer versions, but I make no guarantees. This archive comes with all the library code, tools, and documentation necessary to get started. Once downloaded, simply double-click the icon and the installation procedure will begin.

What Is in the SDK?

The SDK contains everything that you need to develop a BREW applet on the Microsoft Windows platform. A different set of tools and utilities is needed to actually compile your applet for actual handset hardware. We will detail that in later chapters. For now, this section describes what you get with the BREW SDK.

BREW Emulator

The BREW emulator is a tool that mimics the behavior of an actual handset. Once your code is compiled, you can run it in the emulator with a little preparation. The emulator attempts to behave like a real handset, including using replaceable graphic "skins" to look like a number of popular phones. It can also be configured for a variety of screen sizes, bit depths, and memory footprints to adhere to the standards of any number of handsets.

BREW Device Configurator

The Device Configurator is the tool that allows you to create device profiles for the emulator. Device profiles specify the screen size, bit depth, memory footprint, and other attributes. These properties are then saved in a QSC file that is used in the emulator to simulate the operation of various handsets.

BREW MIF Editor

A MIF file is a file used by BREW to identify an applet. This also includes the icon image that is used on the "desktop" of the phone. The MIF file specifies various attributes of the applet that identifies what kind of application it is, what security permissions it uses, and a host of other criteria. Every BREW applet needs a MIF file to run in the emulator or on a real handset.

BREW Resource Editor

The BREW Resource Editor is a tool used in the creation of resource files. A resource file is a collection of all the images, strings, GUI components, and other "resources" that your application needs to run. While many of these can be loaded up as separate files or manually added in code, the resource file gives an easy and convenient way to manage your applet's data.

BREW Documentation

The BREW SDK comes with a series of comprehensive documentation that details the operation of each of the aforementioned tools, as well as the individual functions in the SDK. They are provided as Adobe Acrobat PDF files that must be viewed with Adobe's Reader program.

BREW Examples

Perhaps even more useful than the documentation are the BREW Examples. The BREW Examples are a series of projects set up to show the complete functionality of the BREW SDK. The code is well documented and a valuable resource for seeing how the SDK works. There is an example for pretty much every segment of the SDK—from graphics to networking. The vast majority of the code is written in C. Each of the examples is accompanied by an appropriate binary that runs in the emulator.

BREW SDK Libraries

Most importantly, the BREW SDK comes with the libraries needed to test and run your applet. These libraries are designed to integrate into Microsoft's Visual Studio and are strictly for development in C or C++. This book assumes the usage of Visual Studio.

Using the Emulator

Perhaps the best way to see what BREW is all about is to play around with the emulator. But first, let us define the term "emulator." One of my favorite reference web sites, dictionary.com, defines the term as it relates to computer science like so:

> *"To imitate the function of (another system), as by modifications to hardware or software that allow the imitating system to accept the same data, execute the same programs, and achieve the same results as the imitated system."*

Emulation in regard to games is most commonly found in programs that allow modern machines to run the very same programs that older computers and game consoles ran, which means you can take the same binary information off of, say, a ColecoVision cartridge and use it with an emulator that is running on top of your Windows-based PC. The emulator accurately mimics the intricate timings, CPU instructions, and various other hardware behaviors so that the native program code runs as if it were on the real hardware itself.

BREW comes with a program that they call an emulator but I call a simulator. As explained in the definition of the term, emulation usually means that you can run exactly the same compiled binary code on the emulator as on the real hardware. In the case of BREW, you actually compile your code specifically for either the emulator program or the actual hardware. Therefore, it is not an emulator. It simulates the operation of a handset by running specially compiled Windows DLLs that link into the emulator's own routines for drawing, sound, and other such operations. If you want to run the code on real phone hardware, you have to use a different compiler to create a binary suitable for the actual handset.

That is not to say that the emulator is useless. Compiling your code for the handset is a potentially expensive and laborious process that we will get into later in this book. It is far more convenient to make quick Windows builds than it is to compile and upload your code to a handset. However, because the emulator does not accurately represent the real hardware, you have to keep this in mind when programming your game.

Running Applets

With that lengthy explanation out of the way, we can get into the actual usage of this tool. Simply start the emulator program by clicking on its icon in either the Explorer window or Start menu. You will see a screen similar to Figure 4-1 but perhaps with a different phone display.

This is the "desktop" of the phone—if you want to call it that. From here, you can select various applets or, if it were an actual handset, make a call. Although this does not accurately represent the operation of a real BREW phone, the desktop metaphor is similar across all models.

To operate the phone, you have two options. You can either use the keyboard or actually click on the image of the handset to press buttons on the phone. The black, disc-looking thing is the control pad. It works very much the same as a digital pad on any of the major consoles. As an alternative, you can use the arrow keys. By pressing the pad in the center, you issue a "select" command. The keyboard alternative is the Enter key.

To browse the applets that are available, simply push the directional keys to scroll through the list of applets and hit Select to start any one of them. Each applet in the list is accompanied by an icon as well as a name. The icon sometimes identifies the applet's function by the image used. For instance, all game applets have an icon of falling colored bricks as a default image. This image is, of course, modifiable by the applet's author.

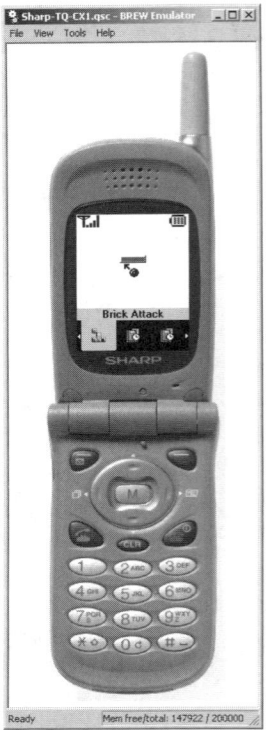

Figure 4-1: The emulator in action

Most of the applets that come with BREW are rather boring. You will find all sorts of "enterprise" and business applets, such as appointment schedulers, tip calculators, and other such utilities. However, the SDK comes with two games: Brick Attack and Space Dudes.

The creatively titled Brick Attack is, as you can probably tell, a clone of the late '70s Atari classic Breakout. If you have been living in a cave for the past 20 years or were perhaps born long after the heady days of fat laces, Run DMC, and the McDLT, you may need an introduction to this arcade classic. The idea is simple: Move the paddle at the bottom of the screen to bounce the moving ball into the bricks above. Yes, you may doubt it, but this was major entertainment in the '70s and '80s.

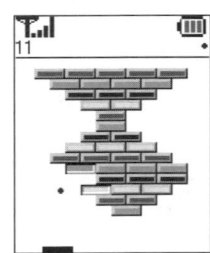

Figure 4-2: Brick Attack in action

The second game is Space Dudes, a tribute to Taito's dominating arcade hit of the '70s, Space Invaders. Simply move your spaceship at the bottom of the screen and shoot missiles up at the descending aliens. Yes, we were starved for entertainment in those days.

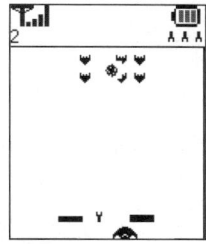

Obviously, the key here is simplicity. These games are not exactly groundbreaking examples of graphics or gameplay. Yet, when you are dealing with the miniscule amounts of memory and processing

Figure 4-3: Space Dudes in action

power in these phones, keeping it simple is almost a necessity to make any kind of functional applet.

Feel free to play around with the other applets. They are good examples of BREW's abilities in various areas, such as GUIs, sound, graphics, and networking. All of the source code to these applets is available, as well as a host of other example programs that can be run in the emulator, which explains how various SDK features are used.

The Device Configurator

One of the most useful features of the emulator is its ability to change the type of phone it mimics. This way, you will see how your applet will look and operate on a variety of different handsets with varying attributes in number of colors, amount of memory, screen size, and so on.

BREW comes with a set of default phone types. These attributes are stored in QSC files. To load these QSC files, do the following:

1. Click on the **File** menu.

2. Select **Load Device** from the File menu.

3. Pick any one of the QSC files to see how the phone looks.

While we are down there, I might as well show you how to make your own QSC file. By use of the Device Configurator, you are able to make any number of hardware variations. That way, when a new phone is released, it is possible to create a mock-up for the emulator that gives a vague idea of how your applet looks.

1. Click on the **Device Configurator** icon in the Explorer window or through the Start menu.

2. We will start by opening an existing QSC file. Click on the **File** menu and select **Open**.

3. Select a file from the **Devices** folder, which is directly underneath the root of the BREW directory. Select the **7GP_256Color_ Screen.qsc** file.

Modifying Attributes

Here you see all the different attributes that are modifiable. In the case of a Screen profile (denoted by the Screen suffix of the filename), you can modify the resolution and size of the display. These values are designated in the Pixels Per Row and Pixels Per Column fields.

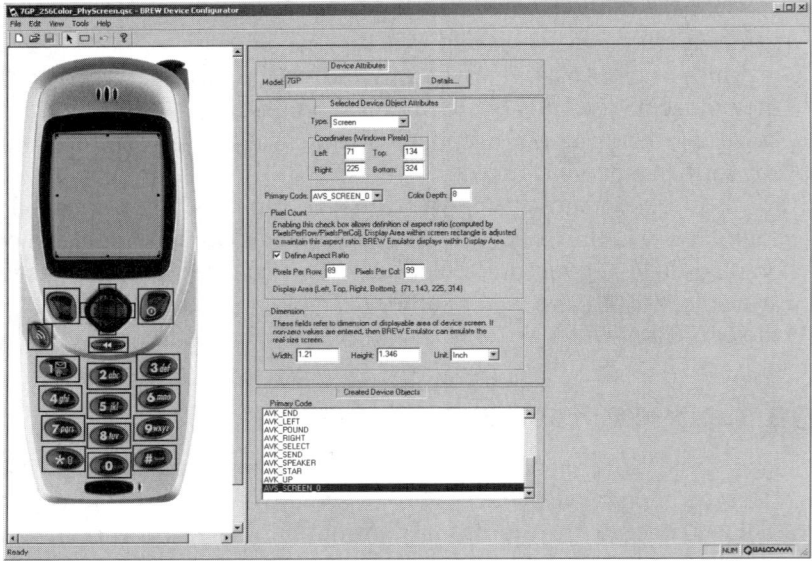

Figure 4-4: The Device Configurator screen with a profile loaded

The Color Depth field governs how many colors are available on the given device. Currently, BREW supports 1-, 2-, 4-, and 8-bit color depths. This represents 2, 4, 16, and 256 colors, respectively. Most real devices that I've seen are either 1- or 8-bit. Higher bit depths, such as 16-bit and beyond, are on the horizon, however.

Although there are many other attributes that you have control over, perhaps the last important one is the heap size. This is how much RAM is available on the handset. This is different from the static RAM which is used to store the actual applets. By default, all profiles have a heap size of 98,304 bytes. In my experience, the entry-level hardware has perhaps a third of that memory. To edit this and alter other values, click on the Details button.

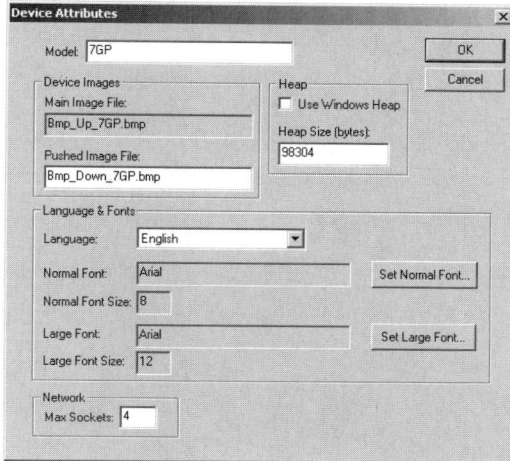

Figure 4-5: The Details window

Here you can also change the image used for the phone's display as well as the native language of the phone if you are making an applet for other countries. But right now, we are mainly concerned with the heap size. As you can see, the Heap Size text box is filled with the default value of 98,304 bytes. This can be changed to suit any hardware that you may need.

Finally, to use this custom profile, simply save out the QSC under another filename and load it up in the emulator as previously explained.

Conclusion

In this chapter I've showed you what comes with the BREW SDK and how to get a taste of things to come with the emulator. You also learned how to create your own custom device profiles to mimic various different handsets.

I cannot stress enough that this application is not a real emulator. As you begin testing your applet on real handsets, you will encounter many differences. The emulator is far more tolerant of illegal pointer access and memory leaks. In some cases there are outright bugs in the BREW implementation on either the hardware or the emulator. Also, the performance of the actual hardware varies greatly from phone to phone. You may find that your applet runs excruciatingly slow on one handset but faster than expected on another.

Luckily, I have already encountered many of these land mines in the development of my own commercial BREW projects. Therefore, I can hopefully help you avoid making these frustrating mistakes in future chapters.

Chapter 5

Your First BREW Application

Introduction

Enough with the preamble! It is time to actually start programming. In this chapter you will compile your first BREW applet and learn the process of getting it to run in the emulator. In this chapter we will create a very simple applet that prints the text "Hello World" on the screen. The code for this applet can be found in the Source\Chapter 05 folder in the companion files.

The Ingredients of a BREW Applet

BREW applets have a set structure. This means that certain elements must be present in order to have a functional BREW applet. First, there are several different files associated with a BREW applet.

The BID File

The first ingredient in a BREW applet is the *class ID*. This is a unique number used to identify an applet. It is possible to write applets that call functions inside of other applets—they do this by referencing the desired applet's class ID. This unique ID is used by the applet, either as a definition or through the inclusion of a BID file. The BID file is really just a regular header file with a single define for the class ID number. For a commercial applet, the class ID must be acquired through Qualcomm's web site. For a yearly fee, you can generate a number of class IDs for use in your applets. The online service ensures that no two applets have the same ID number. For now, we can make up our own. However, when you want to release your applet commercially, you must have a valid ID.

The Applet File

The applet file is the actual executable that contains your compiled applet code. It is compiled either for the emulator or the actual handset hardware. Emulator binaries are stored in the form of a Windows DLL, while the actual compiled native handset code is stored in files with the .mod extension.

The Resource File

Although completely optional, resource files contain data, such as images, text strings, and GUI element layouts, to be loaded from within your applet. It is possible to use standard BREW file I/O to load bitmaps and other files; however, resource files are much more manageable. Resource files have the extension .bar and are created using BREW's Resource Editor. For our purposes, we do not need a resource file for this simple applet. We will discuss the usage and creation of them in a later chapter.

The MIF File

Each applet must also have a MIF file. The MIF file identifies the applet and also stores its icon and things such as the title, copyright information, and other data associated with the applet. The class ID is also set in the MIF file so that it is referencing your applet binary.

Creating the Ingredients

So how do we create all of these files? Fortunately, BREW has a series of tools for dealing with most of them. The others, such as the BID, are easy to build from modified example code. The source code for these examples is found in the Source\Chapter 05 folder in the companion files.

Creating the BID File

The BID file contains our class ID and must be included in the source code of the applet. This ID must be acquired from a VeriSign account when delivering an applet for commercial distribution. For our purposes, we just have to make up a number that is not used by any of the example applets. If you look in the Chapter 05 folder, you will find a BID file already made for you.

Using the MIF Editor

As described previously, the MIF file identifies the applet and associates it with various information, such as its icon image, copyright info, and other data. To create the MIF file, you must use Qualcomm's included MIF Editor. To start the editor, simply click on the BREW MIF Editor icon from the Start menu or in the BREW folder's Explorer window.

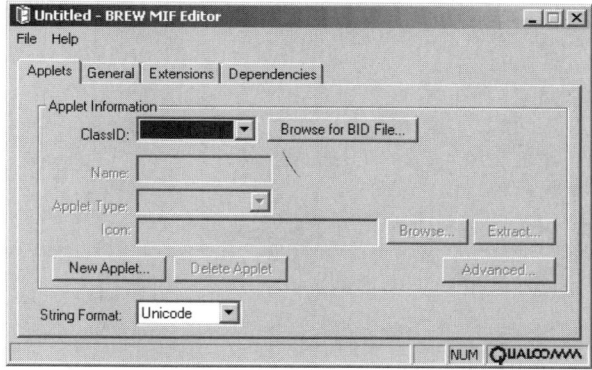

Figure 5-1: The MIF Editor

As you can see, the MIF Editor has several tabs. On each tab are different applet properties that you can create or modify. We will set up our applet here.

The first thing we need to do is define the class ID. You will notice a drop-down list box where you can add your own class IDs. Why does it allow multiple IDs? This is because it is possible to call functions inside of other applets. Therefore, you can create an *applet suite*, which contains many different applets that call code inside each other. For now, we will only be using one.

Setting the Class ID

To set the class ID, simply click the Browse for BID File button. This will bring up a file dialog where you can select the BID file that contains the class ID for your applet. For now, use the BID file found in the Chapter 05 directory.

Once the class ID has been selected, the rest of the fields become active. Now it is possible to edit the applet name, as well as the icon and other data. The Name field is the applet name that appears on the phone's applet menu. The Type field will identify the kind of applet it is.

An applet can be identified as a tool, a game, or a variety of other gen-res. To set the icon, simply click on the Browse button and find a graphic image that you want as the icon. This image is stored inside the MIF file, so it does not matter if you delete or move this file later. For the sake of our example, set up the values under this tab as seen in the following screen shot:

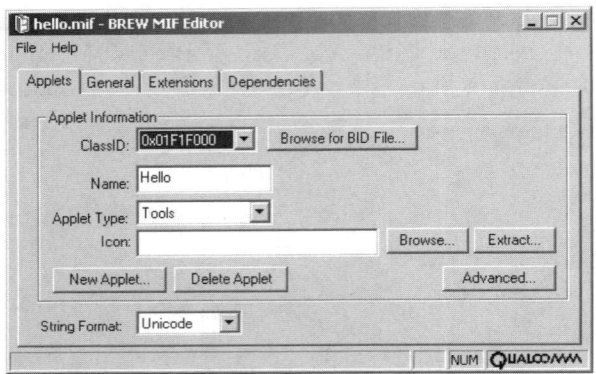

Figure 5-2: The MIF file for our applet

Setting the General Attributes

The General tab is where you can set the author and copyright informa-tion. It also has an important series of check boxes in the section titled Privilege Level. The privilege level tells BREW which operations your applet will perform. For instance, if your applet will read and write files, the File box must be checked. If you are going to use any TCP/IP net-working, the Network box must be checked. In the emulator, your code will sometimes run regardless of your permissions. However, on real hardware if you do not have the File box checked and you try to do a file operation, the function will not do anything. It is important to make sure your MIF file has the proper permissions in order for your applet to behave properly on real hardware. Set up your General tab screen like so:

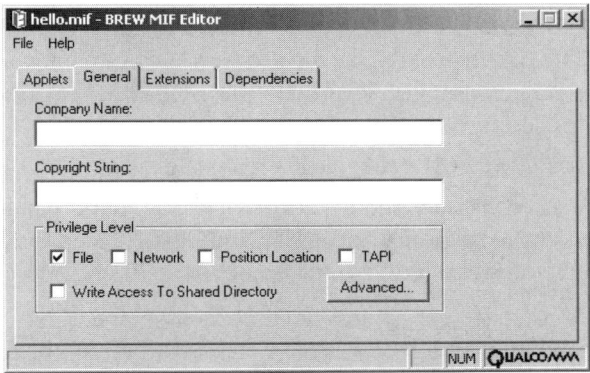

Figure 5-3: The General tab for our MIF file

We do not have to worry about the other two tabs. They are used for advanced functions that are not necessary at this moment. Simply save the MIF file by using the File menu item. Name the file Hello.MIF.

Creating the Project

Now we are on to the actual source code. Of course, there are still some preparatory steps to perform before we can actually start writing real code. A proper Developer Studio project must be created. Unfortunately, the BREW 1.0 SDK does not have a Visual Studio Wizard to set up a project. Therefore, you must follow these simple steps to create a proper workspace:

1. Select **File|New** from the menu in Visual Studio.
2. If not already active, click on the **Projects** tab.
3. Select **Win32 Dynamic-Link Library** from the projects list.
4. Type in the directory and project name in the appropriate text boxes to the right.
5. Click **OK** and then **Finish** on the next screen.

    You now have an empty DLL project. The next step is to set up the project's properties to compile correctly for the BREW emulator.

6. Click on **Projects|Settings** from the menu.
7. Select the **C/C++** tab.
8. Add the string **AEE_SIMULATOR** to the Preprocessor Definitions text box. This tells the BREW SDK that we are compiling for the emulator and not real hardware.

Now that the project settings are complete, we have to add a few files from the BREW SDK to the project. These are the AEEAppGen.c and AEEModGen.c files. They are both located in the BREW\src folder. The process of adding these files is simple:

1. Click on **Project|Add to Project|Files** from the menu.

2. Browse for the src directory under your BREW installation and select the files **AEEAppGen.c** and **AEEModGen.c**. Click the **OK** button, and they will be added to your project's file listing.

Now we have the workspace set up and can start coding. The first thing you should do, if you have not already, is copy the BID file to your project's directory. This is stored in the Chapter 05 folder of the companion files. This BID file will be included in the primary source file of your project. To create the source file, do the following:

1. Click on **File|New** from the menu.

2. Select **C/C++ Source File**, and then type **hello.c** into the Name text box.

3. Click **OK**.

You now have the first source file of your application. This will remain the only source file in this example. The final hello.c file is included in the companion files. Use the code from that file to create this applet.

If you look at the code, you will see that right after we include the header, we include the BID file. Sure, it's not a .h, but C's preprocessor lets us include any kind of file we want. It is up to the compiler to determine if it is legitimate code. If you look in the BID file, you will see a single define, which is the number we are using as our class ID.

Programming the Applet

BREW applets have a specific form and design to them. Certain functions are required, and certain rules must be adhered to so that your applet works properly. Before we can actually write a line of code, we need to be aware of these elements of BREW programming.

Restrictions

Because of the limited nature of the handset hardware, BREW imposes some restrictions on what can and cannot be used in a BREW program. Also, you have to be aware of some hard limits when dealing with memory and stack space. Here is the laundry list of BREW restrictions:

■ **No globals or static data**—There are no global variables or static data arrays allowed in BREW. Variables and data that need to be available to the program as a global can be stored in the applet's internal structure. For applets that require large tables, this data will have to be dynamically allocated and read in from a file instead of compiled in with the source. It may be annoying, but it actually forces you to think in a data-driven manner.

- **No floating-point operations**—Because we are dealing with very low-power and inexpensive CPUs, current BREW hardware does not support a floating-point unit, or FPU for short. This means that you can only use integer numbers. This may seem like quite a restriction. However, it is possible to support fractional values with integers through the use of fixed-point math. Fixed-point math is explained in Appendix F. If you use floating-point values, your code will actually be compiled and executed in the emulator. However, it will not work on an actual handset.

- **No STDLIB functions**—Another broad restriction is the lack of standard library calls. BREW substitutes the functionality for the Standard C Library functions with BREW-specific functions and macros. There are functions for string manipulation, file I/O, memory allocation, and most other common operations normally found in the Standard C Library.

- **Only 500 bytes of stack**—Yeah, you read that correctly. You only have 500 bytes of stack space on a low-end BREW handset. This changes depending on the amount of memory on the phone, but current standards outline a minimum of 500 bytes. What this means is that local variables and function calls have to be watched. Highly recursive functions will not fit in such a small stack space. Allocating a lot of variables on the stack will cause major problems. Therefore, a good rule is to use the smallest data type available. If you are using an integer, but it only has to go to a number smaller than 255, use a byte instead of an int. Depending on how the packing goes, that could be a 3-byte savings. With only 500 bytes available, that is nothing to sneeze at. The same rule applies for function arguments. Again, the emulator will not complain about stack limitations because your host PC's stack is much larger. However, if you overrun your stack on real hardware, you will be knee-deep in obscure error messages and mysterious crashes.

- **All strings are wide character**—This is not really a restriction but rather a notice of sorts. Because of BREW's international audience, the SDK supports many different languages. Certain languages have more than 256 symbols and therefore cannot be properly represented with normal single-byte ASCII characters. It is for this reason that BREW uses wide characters to represent text. These characters are defined as 16 bits instead of 8, allowing for 65,536 different symbols, although in some cases two 8-bit characters are merged into one 16-bit wide character. Regardless, there is more than enough room for any language's character set. There are functions to manipulate standard byte-sized strings, but all of the controls and other text-display functions take wide character, or AECHAR, types as arguments.

The Applet Structure

Each applet is defined by a structure defined in the applet's header file. The structure must begin with the **AEEApplet** structure defined in the AEEAppGen.h header. The **AEEApplet** structure contains the basic variables needed by BREW to run any applet. You put any globals that you need after this structure. If you look in the hello.h file in the Chapter 05 folder, you can see how we lay out our applet structure, as shown here in Listing 5-1.

Listing 5-1: The applet structure

```
typedef struct myapp_s
{
    AEEApplet    a;    //applet header
} myapp_t;
```

In this case, we only have the **AEEApplet** structure in our custom structure, **myapp_t**. This is because we are creating the most basic applet possible. Therefore, we need no additional data structures or variables. We simply need to provide the structure that BREW requires. In a more complicated applet, we would have different variables and structures below the **AEEApplet** declaration.

BREW does not care about any information after the **AEEApplet** structure. It only needs to be told about the total size of your applet structure on creation so it can allocate enough memory. In lieu of global variables, the applet accesses variables inside of your applet structure. The applet structure is passed by the OS to the event handler. This structure can then be passed to any functions that you call from the event handler. Therefore, it can be accessed in the same way that globals would be. They just happen to be neatly organized in your applet structure.

Applet Construction and Destruction

All BREW applets must have a function called AEEClsCreateInstance.
This function is defined like so:

```
AEEClsCreateInstance(AEECLSID ClsId, IShell * pIShell, IModule * po, void ** ppObj)
```

The ClsId variable is the number defined in the BID file (or at least it
should be). The code inside AEEClsCreateInstance checks if this is the
same before proceeding. The rest of the variables are merely passed on
to a BREW function call, but they deserve mention.

The pIShell pointer is a pointer to BREW's shell interface. BREW is
made up of several interfaces, each one containing functions of a certain
common use. The shell is composed of several functions for general
utility. The pointer, po, points to the IModule interface. This interface
lets you create and destroy applets, otherwise known as modules.
Finally, the ppObj pointer simply points to your allocated applet structure (or rather is a pointer to that pointer). The function call actually
allocates the memory for the applet internally, and the pointer is then
assigned to the pointer beginning the allocated memory chunk.

These pointers are passed in as NULL, and it is up to the
AEEClsCreateInstance function to initialize them. This is done through
a simple function call:

```
boolean AEEApplet_New(int16 nSize, AEECLSID clsID, IShell * pIShell, IModule * pIModule,
    IApplet **ppobj, AEEHANDLER pAppHandleEvent, PFNFREEAPPDATA pFreeAppData)
```

As you can see, the first argument is the size. If you look at the source,
we pass the size of our applet structure. This is the size of our applet
structure that includes the AEEApplet structure inside. This tells
BREW to allocate enough memory in the entire applet structure upon
creation.

Next, you will see that we simply pass the pointers to be initialized
out to this function. However, the two last arguments, pAppHandle-
Event and pFreeAppData, are unique.

These two pointers are function pointers. The pointer pAppHandle-
Event points to our event handler function. Our event handler is defined
in the function Hello_HandleEvent. This function is called every time an
event occurs that belongs to this applet. It is responsible for the main
execution path. The final argument, pFreeAppData, is a pointer to the
function responsible for cleaning up after our applet right before it exits.
This argument is actually optional, and in this case, we will just pass
NULL, as our simple applet will not have any memory to clean up upon
exit.

Listing 5-2 shows the complete source to our AEEClsCreate-
Instance function.

Listing 5-2: Creating the applet instance

```
int AEEClsCreateInstance(AEECLSID ClsId, IShell * pIShell, IModule * po, void ** ppObj)
{
    *ppObj = NULL;

    if( ClsId == AEECLSID_HELLO_BID )
    {
        if(AEEApplet_New(sizeof(myapp_t), ClsId, pIShell, po, (IApplet**)ppObj,
           (AEEHANDLER)Hello_HandleEvent, NULL) == TRUE)
        {
            return(AEE_SUCCESS);
        }
    }
        return(EFAILED);
}
```

As you can see, we first detect if we are creating the proper class ID.
This is always good form, as in some weird cases it may be possible that
your code receives a creation call destined for a different applet. Then,
we call the now familiar **AEEApplet_New** function without the aforementioned arguments. If this call fails for some reason, we return the
appropriate error message.

A Word About Errors

Many BREW functions return error codes, as seen in Listing 5-2. In that
case, we return **AEE_SUCCESS** if the function has executed properly
and **EFAILED** if not. You can find the error codes defined in the header
AEEError.h. There are many different error codes that describe why a
particular function call has failed. These reasons can range from bad
function parameters to a lack of memory on the device. It is useful to
get acquainted with these error codes; they can really help when tracking down crashes and other bugs.

Note: When it comes to the final product, ignoring errors is not
a good idea. You must expect the unexpected. Have a way to ele-
gantly quit out of the application and/or warn the user upon
receiving an error from a BREW function call.

Event-based Programming

Before we go into actual code, the basic flow of a BREW applet must be
discussed. More specifically, we need to detail the event-based nature of
a BREW applet. Event-based means your BREW program responds to
messages, or events, sent from the operating system. This may be an
unfamiliar concept to you if you are used to DOS or basic UNIX

environment programming. There is no main loop in a BREW program. Instead, the phone's operating system notifies your code of certain events.

These events can be triggered by a number of actions. For instance, when the user pushes a button, an event telling your applet that a button has been pressed is sent. The event is sent as an argument to a function in your applet called an event handler. This function checks which event is being sent and acts accordingly. There are many different events for all sorts of occasions. You can pick and choose which ones your applet responds to; however, certain events must be handled in order for your applet to perform properly.

The Event Handler

As discussed previously, the event handler is the function that all events are sent to by the BREW OS. This function is defined in our hello example as:

```
boolean Hello_HandleEvent(IApplet * pi, AEEEvent eCode, uint16 wParam, uint32 dwParam)
```

The **pi** pointer points to a structure of type **IApplet**. This is a pointer to the applet interface. This pointer is actually a pointer to our own applet structure. However, it is of type **AEEApplet**. It merely needs to be cast to our own internal applet structure, as we do with the following line of code:

```
myapp_t * pApp = (myapp_t*)pi;
```

Now, **pApp** allows us access to the custom fields that we defined in our own structure, **myapp_t**. You can then pass this pointer around to all of your functions called from the event handler. In essence, it acts as your applet's global variables.

The next two arguments identify the type of event that we are receiving. The integer **eCode** is the event code. Each type of event has a specific code that is a number defined in header AEE.h. If you look through this file, you can see all sorts of events for such things as keypresses, switching the power off, and others. The second variable, **dwParam**, is additional data required for certain events. For instance, with a keypress, **eCode** will identify the event as a keypress and **dwParam** will identify which key is being pressed.

The basic events we need to worry about at this point are the following:

- **EVT_START**—This is the start event. It is sent after the applet is created in our previously defined **AEEClsCreateInstance** function. This is useful for one-time initialization of variable and other data.

- **EVT_STOP**—This event is sent right before the applet is destroyed. This is supposedly why the last argument to AEEApplet_New is optional. Your cleanup code can be executed upon this message instead of the callback. However, in some instances, an EVT_STOP message is not sent before the applet is shut down. I find that EVT_STOP is almost completely useless because of this. I prefer to put all my cleanup code in the callback.

- **EVT_SUSPEND**—A suspend message is issued when the applet is put in the background. This can happen for a number of reasons. Perhaps the user has received a phone call. Or the keyguard has kicked in. Whatever the case may be, this event must suspend the operation of the applet. This usually includes things like canceling any timers that may be currently running or stopping any ongoing network communications.

- **EVT_RESUME**—This is the event sent in response to an EVT_SUSPEND. The resume is sent whenever the activity that initially paused the applet has ceased. The resume function must do things like redraw the screen, start up any timers that were previously cancelled, and resume network communications.

For the purposes of our applet, we only need to handle the start message. The start message will display the text. Because we are using the cleanup callback, we do not need a stop event handler. Because we have no timers or other background operations, we do not need to suspend anything either. Although it is absolutely necessary to handle the EVT_RESUME message in a real applet, this code is strictly for example. So, we will ignore it for now.

Because the sole purpose of our applet is to display the words "Hello World," the start handler will do just that: display the text. Let's look at the event handler code:

Listing 5-3: The event handler

```
static boolean Hello_HandleEvent(IApplet * pi, AEEEvent eCode, uint16 wParam, uint32
    dwParam)
{
    myapp_t* pApp = (myapp_t*)pi;
    AEEApplet * pMe = &pApp->a;
    AECHAR szBuf[] = {'H','e','l','l','o',' ','W','o', 'r', 'l', 'd', '\0'};
    //wide character string

    switch (eCode)
    {
        case EVT_APP_START:
        //clear screen (default color is white)
        IDISPLAY_ClearScreen(pMe->m_pIDisplay);

        //draw the text
```

```
IDISPLAY_DrawText(pMe->m_pIDisplay, AEE_FONT_BOLD, szBuf, -1, 0, 0, 0,
        IDF_ALIGN_CENTER | IDF_ALIGN_MIDDLE);

//update screen
IDISPLAY_Update(pMe->m_pIDisplay);

//we've successfully handled this message
return(TRUE);
break;
}
    return(FALSE);
}
```

The first thing you will notice is the rather strange-looking creation of the actual "Hello World" text. We are creating and initializing an array of **AECHARs**. The **AECHAR** is a wide-character structure. That is, BREW uses 16-bit characters instead of the traditional 8-bit char type for most functions involving string data. This is to accommodate foreign languages that have more characters than 8 bits can represent. BREW's wide-character system will be explained in detail later on. All you need to know for now is that this is a simple way to create a wide-character string.

Moving on to the message handler, you can see that we only act upon the **EVT_APP_START** message. First, we clear the screen with the function **IDISPLAY_ClearScreen**. This function is in the **IDisplay** interface. All it does is fill the screen with the color white. As with all interface function calls, you have to pass a pointer to the interface that you are using as the first argument. In this case we pass a pointer to our **IDisplay** interface that is stored in the default applet structure.

Next, we get to the actual drawing of the text. **IDISPLAY_DrawText** is another function in the **IDisplay** interface. In general, we pass the string we want to draw, the position, and some other attributes. Here are the arguments:

```
IDISPLAY_DrawText(IDisplay * po, AEEFont Font, const AECHAR * pcText, int nChars, int x,
    int y, const AEERect * prcBackground, uint32 dwFlags);
```

The first argument is a pointer to our interface. The second is the type of font that we want to use. There are a number of them defined in the header AEEDisp.h. Next, pass the wide-character (**AECHAR** type) array as the text we want to draw. The nChars argument is the length of this string. If you pass –1 to this argument as we do, BREW will determine the length for you. The arguments x and y are the pixel coordinates relative to the upper left-hand corner of the text. Finally, we have the flags. There are many different flags that you can combine to change the style of the text. In this case, we use the flags **IDF_ALIGN_CENTER** and **IDF_ALIGN_MIDDLE** to position the text in the center of the screen.

This negates the use of the x and y coordinates, which is why we pass 0 to both of those arguments.

Next is perhaps the most important function when dealing with BREW's graphics: IDISPLAY_Update. This function essentially executes all the drawing commands that you have called. You will not see the result of your text drawing or any other graphical operation until you call IDISPLAY_Update. IDISPLAY_Update is an expensive function, which means it takes up a lot of processor time to draw the entire screen. This is why all your drawing operations should be batched together instead of calling IDISPLAY_Update after everything you draw.

That's it! The applet is finished. This may seem like quite a number of steps for a simple Hello World program. But once you get used to developing with BREW, these steps will become second nature. Here is a complete code listing to better illustrate the scope of a simple BREW applet.

Listing 5-4: The entire applet source code

```
int AEEClsCreateInstance(AEECLSID ClsId, IShell * pIShell, IModule * po, void ** ppObj)
{
      *ppObj = NULL;

      if( ClsId == AEECLSID_HELLO_BID )
      {
          if(AEEApplet_New(sizeof(myapp_t), ClsId, pIShell, po, (IApplet**)ppObj,
              (AEEHANDLER)Hello_HandleEvent, NULL) == TRUE)
          {
              return(AEE_SUCCESS);
          }
      }

      return(EFAILED);
}

static boolean Hello_HandleEvent(IApplet * pi, AEEEvent eCode, uint16 wParam, uint32
    dwParam)
{
      myapp_t* pApp = (myapp_t*)pi;
      AEEApplet * pMe = &pApp->a;
      AECHAR szBuf[] = {'H','e','l','l','o',' ','W','o','r','l','d', '\0'};
      //wide character string

      switch (eCode)
      {
          case EVT_APP_START:
          //clear screen (default color is white)
          IDISPLAY_ClearScreen(pMe->m_pIDisplay);

          //draw the text
          IDISPLAY_DrawText(pMe->m_pIDisplay, AEE_FONT_BOLD, szBuf, -1, 0, 0, 0,
              IDF_ALIGN_CENTER | IDF_ALIGN_MIDDLE);
```

```
            //update screen
            IDISPLAY_Update(pMe->m_pIDisplay);

            //we've successfully handled this message
            return(TRUE);
            break;
    }
            return(FALSE);
}
```

Running Your Applet

Now the process of running your applet begins. Getting your programs
up and running in the emulator is not very difficult. However, there are
some caveats along the way. Certain files have to be placed in certain
ways. Usually, I create a customized batch file that copies my DLL,
resource, and MIF files to their appropriate location after a successful
build. We will analyze the steps involved so that you know how to do
this manually. You can add your own post-build step in Visual Studio to
automatically do this later.

The emulator needs several files to run your applet properly. At the
most basic level, this is your applet DLL and its associated MIF file. If
you click on Tools|Settings, you will see where it pulls these files.

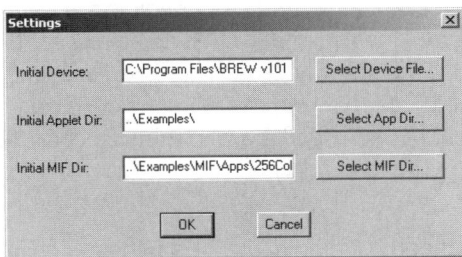

Figure 5-4: The Settings dialog

If you have not played around with your settings, you should have these
values in your dialog. This dialog has three buttons: Select Device File,
Select App Dir, and Select MIF Dir.

The device file is the QSC file that is used to define the handset
properties. You can have the emulator use any QSC as a default so that
you no longer have to manually load the device every time you wish to
test your program. Simply click the Select Device File button and
browse for the QSC file that you wish to use.

The initial applet directory is where the emulator will look for all its
applets. By default, this is set to the Examples directory. This is why
the emulator lists a number of different applets in the menu upon
startup. It picks the applet files out of the sub-folders in the Examples
folder.

Of course, each applet has to have a corresponding MIF file in order to be properly recognized by the emulator. By default, the emulator is set to the Examples\MIF\Apps\256Color directory. It is necessary to have a different MIF file for each color depth because each version is considered a completely different applet. Therefore, each version must have a different class ID. Also, the icon stored in the MIF file can be of any supported bit depth. So it does not make sense to have a MIF file with an 8-bit icon graphic also be the same for the 1-bit version.

Note: It is possible to display an 8-bit icon on a 1-bit device when browsing via the "desktop." However, having an 8-bit color icon on a device that cannot display it is a massive waste of space. Also, you may fail the True BREW certification process if your MIF file is not properly set up for the device on which your application is testing.

In order to get your applet to run, you need to set up a directory for your applet DLL and MIF file. The simplest way is to make a new folder in BREW's Examples directory and add your MIF to the default MIF folder as well. However, this can soon become an organizational night-mare. So we will create our own folder structure.

Create a folder anywhere and call it Applets. Then create a sub-folder with the same name as your applet DLL. In this case, the folder should be called hello. This sub-folder is where you will put your DLL and any other associated files that you may have. These may include resource files, external files loaded from within the code, and other data.

Warning: Having any extraneous symbols, such as periods, in your folder names can cause the emulator to not be able to find your applet.

Create a folder in that directory called MIF. It is a good idea to make the three sub-folders that the BREW examples use. So, under MIF, make three sub-folders: 256Color, 4Grey, and Mono. You will most likely be using only 256Color and Mono, however.

Then click on the Select App Dir button and browse for the direc-tory that you created called Applets. It is important not to pick the directory that actually has your applet DLL in it. Then, finally click on the Select MIF Dir button and pick any of the folders that you created underneath your Applet\MIF folder. For simplicity's sake, select the 256Color sub-folder. Close the emulator to save all of your changed settings.

Now you have to copy the files over. If you have successfully compiled your applet, you should have a file called hello.dll in your project's folder. Take this DLL and copy it to your Applets/hello sub-folder. Now take the MIF file that you created earlier in this chapter and copy it to the Applet\MIF\256 folder.

When you start up the emulator, the icon list at the bottom of the display should display only the icon for your Hello applet. If you select this applet, you should see it run with a display similar to Figure 5-5.

Figure 5-5: Hello World in action

If your applet does not run, it can be for a number of reasons. If your BID file and the class ID in your MIF file do not match up, the emulator will not be able to locate and run your applet. If you copied your applet to the default examples folder but are using a BID file and class ID from one of the examples, there will be a conflict between the two and your applet may not run.

Anyway, it is not the most impressive thing in the world. But remember my motto: Baby steps. Soon after this chapter, it will be time to put on your running shoes!

Conclusion

Although this was a huge chunk to digest, this chapter has introduced you to programming in BREW. You have learned about the basics of dealing with BREW's SDK. By keeping the rules and restrictions outlined in this chapter in mind when developing BREW code, deploying your applet on actual hardware should be a lot easier. Now you are ready to tackle bigger issues.

Chapter 6

The Anatomy of BREW

Introduction

Now that you know the basics of BREW programming, it is time for a useful rundown of all the major components of BREW 1.0. Not all of these components will be covered in this book; however, it is a good idea to know the different pieces of BREW so that you know where to look if you need some added functionality.

The BREW Interfaces

As we saw in the previous chapter, BREW is organized into a collection of various interfaces. Each interface governs a specific area of functionality, usually described by the name. Later on in this book, you will learn how to create your own custom interfaces to encapsulate a reusable game engine or any other shared code. Here we will list each standard BREW interface and give a brief description of its contents.

- **IApplet**—This is the base interface for all BREW applets. It encapsulates the basic functionality of an applet—mostly including message handling. You will rarely, if ever, delve into this interface.

- **IAStream**—The IAStream interface is used for asynchronous stream data. A stream is basically a mechanism to access data as a continuous flow of bytes. The OS automatically buffers reads and writes, leaving the stream access transparent to the programmer.

  For instance, if you want to serve up a new level for your game from a remote web server, you can retrieve it as a stream. The file will be received by the handset in packets, but you can begin processing the received bytes immediately. As you work on the bytes taken from the stream, more data is being received continuously until all of it has been received or the stream is closed. This interface is used in other modules for this type of data access.

■ **IBase**—IBase is the base interface from which most objects in the BREW SDK are derived, including IApplet. Therefore, this interface implements a lot of the functions that are common to all BREW objects. This includes the Release() mechanism for object references and memory management.

■ **IControl**—IControl is the base interface from which all GUI controls are derived. Among the functionality provided by the interface are the GUI's event handling and redraw capabilities.

■ **IDatabase**—BREW contains a handy database system for storing organized records of information. The IDatabase interface controls the creation, destruction, and access of records in a database.

■ **IDateCtl**—One of the many GUI controls in the BREW SDK, the IDateCtl allows for easy display and entry of date information.

■ **IDBMgr**—The IDBMgr interface manages the creation and destruction of BREW databases.

■ **IDBRecord**—Each BREW database consists of individual records. Each record is composed of different fields of varying data types and sizes. The IDBRecord interface provides access to the record and its associated fields.

■ **IDialog**—Although individual GUI controls can be displayed manually, it is also possible to collect a series of different controls into dialogs. The IDialog interface provides a few basic functions, including access to the individual controls associated with a given dialog.

■ **IDisplay**—The IDisplay interface provides basic graphics and drawing functionality. This includes simple shapes, text, and bitmap images. You will become very familiar with this interface as you develop BREW gaming applications.

■ **IFile**—Because BREW does not allow for any Standard C Library calls, file functionality is provided through the IFile interface. Its functions essentially mimic those of the Standard C Library's file access operations.

■ **IFileMgr**—While IFile gives access to individual files, the IFileMgr interface gives access to BREW's entire file system. This interface allows the creation and destruction of files and directories, and other powerful file manipulation operations.

- **IGraphics**—IGraphics provides more complicated drawing operations than IDisplay. This includes a wide variety of shapes and geometry, including triangles, polygons, and arcs. It also provides viewport and coordinate-system functionality for advanced geometric graphic techniques.

- **IHeap**—Another Standard C Library replacement, the IHeap interface gives access to BREW's memory heap. Here you can perform BREW's equivalent to malloc(), free(), realloc(), and other memory management functions typically found in the Standard C Library.

- **IImage**—The IImage interface provides high-level access to imaging functions, such as the display and animation of BMP files. Personally, I avoid the use of this interface in favor of straight bitblts. This will be discussed at length in Chapter 8.

- **IMemAStream**—Similar to the IAStream interface, this allows stream operations to be performed on chunks of allocated memory.

- **IMenuCtl**—Another item in BREW's repertoire of GUI controls, the IMenuCtl interface provides the ability to create lists of selectable text and graphical items. This interface is probably the most useful of all BREW's GUI controls.

- **IModule**—It is possible to access other applets from inside yours. These external applets are typically referred to as modules. Think of them as libraries or DLLs. BREW development discourages statically linked libraries, and instead you can group common functionality into various other modules that can be shared by a suite of applets. Most modules and so-called "classes" in BREW are built on top of the IModule interface.

- **INetMgr**—The INetMgr interface gives access to BREW's robust networking capabilities. Here it is possible to query BREW's network subsystem for information, such as the handset's IP address, as well as open sockets.

- **INotifier**—When using multiple modules in an applet, it is often necessary for them to communicate with each other via messages. The INotifier interface allows one module to send event messages to another's handler.

- **IShell**—The IShell interface is different from most in that it is not confined to one strict area of functionality. Instead, it is sort of a grab bag of different operations. These range from loading resources to getting information on the handset's capabilities.

■ **ISocket**—Used in conjunction with the INetMgr interface, the ISocket interface allows for the transmission and reception of data from a TCP or UDP socket.

■ **ISound**—ISound is sort of the audio equivalent to IDisplay. It provides functionality for basic sound operations, including simple beeps, tone lists, and vibration functions. It also allows control over the handset's volume levels.

■ **ISoundPlayer**—ISoundPlayer is the more advanced sound interface. It allows the loading and playing of MIDI and MP3 files. It is also possible to pause, seek, and even alter the playback speed of an input stream.

■ **IStatic**—Another item in BREW's collection of GUI control interfaces, the IStatic allows for the display and scrolling of large amounts of non-modifiable text.

■ **ITAPI**—The ITAPI interface gives access to actual phone operations. TAPI is, in fact, a common acronym, which stands for Telephone Application Programming Interface. There are TAPI modules for a variety of different platforms aside from BREW. In our case, ITAPI includes functionality for placing phone calls, as well as using SMS messaging services. This interface will not be used in the course of the book, but it is very useful for other non-game applications.

■ **ITextCtl**—The ITextCtl interface is an editable text box control for BREW's GUI. It is fairly similar to text boxes that you may be familiar with from web page forms or just about any other GUI API.

■ **ITimeCtl**—ITimeCtl acts in a similar manner to IDateCtl, except it is used for the display and entering of time data instead of date.

■ **IViewer**—The IViewer interface is somewhat of a mystery. It has exactly the same functions as IImage. It may be used in later revisions for a more advanced and format-neutral image display system.

Conclusion

Aside from these interfaces, there are a host of helpful functions, macros, and data structures. You will be introduced to these as we show how to use some of the interfaces, as well as walk through practical examples. Now that you are familiar with the basic anatomy of BREW, we will forge ahead into more programming.

Chapter 7

Using Resources

Introduction

One of the most convenient things about the BREW SDK is the resource file. If you are familiar with Windows programming, you may already be familiar with the concept. A resource file is a suitcase stuffed with all the external data that your applet needs. Using resource files makes it easy to create custom versions for different screen resolutions or languages. Simply create a new resource file with different graphics or text strings and you can use largely the same code for multiple versions of your applet. It is also possible to use multiple resource files per applet for the same purpose. This chapter will discuss what is in a typical BREW resource file and how it is used in your code.

Resource Types

There are currently three resource types in the BREW SDK: text, image, and dialog. Other resource types are on their way, but for now the resource file system is limited to these three types.

Text

Text resources are strings of wide characters to be used as text in your applet. Of course, it is possible to compile text right in with your code, but this makes it inconvenient to change later. In some cases, you may have non-programmers in charge of managing the text for your applet. For instance, a translator may need to rewrite text when bringing your product to a non-English-speaking region. Therefore, having the text separated in a resource file makes it easier for the person doing the translation to edit the text without touching and recompiling the code.

All text in BREW is in 16-bit wide-character format to support non-English languages. Currently, BREW supports English, Japanese, Korean, simplified Chinese, and Portuguese. More languages are to follow. This means that string data takes up double the space when using BREW. Keep this in mind when looking for ways to make your applet smaller.

Images

Image resources are BMP graphics. More image types are to follow, but for now, only Microsoft BMPs are supported in BREW. It is possible to load BMPs as a normal file using BREW's file system. However, packaging BMPs in a resource file means that you only have to keep track of one collection instead of a lot of individual files.

Dialogs

Finally, dialog resources are collections of GUI controls defining user interfaces. These dialogs can later be instantiated inside your applet code for user input and control. Dialogs and GUI controls can be created and assembled in code as well. However, the usage of resource files lets more than just programmers edit your applet's interface if necessary. This could come in handy for localizing foreign languages or refitting interfaces for smaller screen resolutions.

Creating a Resource File

The BREW SDK ships with a convenient resource editor that is used for the creation and management of resource files. We will create a simple resource file with one of each type of resource.

Click on the BREW Resource Editor icon in the Start menu or the BREW folder in Explorer.

You will see the main screen of the editor. Each heading in the panel to the left can be clicked to access the resource file's list of resources under each particular category. The panel to the right displays each resource and some basic attributes. The individual resource can be clicked on for modification or more details. Let's start off by making a new string resource.

Creating a Text Resource

Click on New Resource | New String from the menu. This can also be accessed via the shortcut Alt+S.

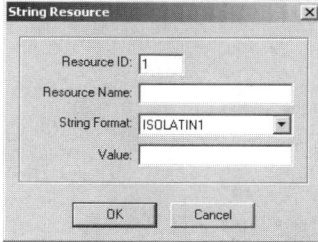

Figure 7-1: The String Resource dialog

You will now see the String dialog. This includes two fields that are common to all resources: Resource ID and Resource Name. The resource ID is automatically generated by the editor and is the number by which it will be referenced in the code. As you can see here, the editor has already numbered this as 1. You can manually enter your own resource ID number. However, it is best to let the editor handle it, as it ensures that there are no two resources with the same ID. The resource name is the label under which the resource ID is defined in the header. We will explain the header and other files generated by the Resource Editor shortly.

The String Format drop-down list box allows you to specify the encoding type for your string. This is useful when creating text for foreign language versions of your applet. For now, the default value of ISO/Latin is just fine. Finally, you can enter the desired text in the Value field. For those at a loss for words, feel free to fill out your fields like so:

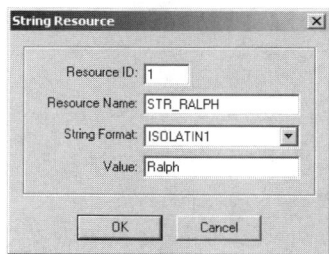

Figure 7-2: The String Resource window filled in

Once you click OK, you will see your string resource in the right panel. If you double-click on the resource, you will once again see the String Resource dialog box. However, it is filled out with the values that you specified this time.

Creating an Image Resource

Let's add an image to our resource file. The process is similar to adding text as shown in the previous section. Click on New Resource|New

Image from the menu. This can also be accessed by the shortcut Alt+I. You will see the following dialog:

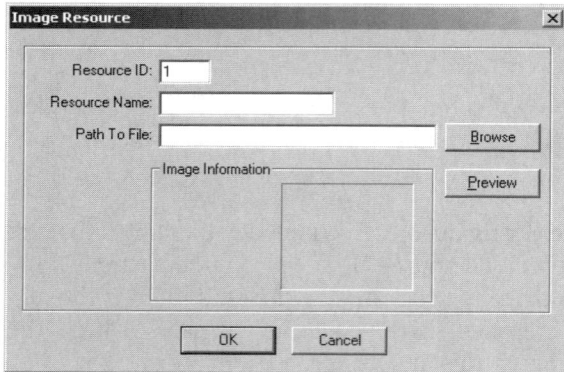

Figure 7-3: The Image Resource dialog

Here you can set the resource ID and name as usual. However, now you can select the desired image file by clicking on the Browse button. Simply use the file dialog to find the desired BMP and click OK. The editor may complain about your image being too big or the bit depth being wrong. Ignore these warnings, as they really are only suggestions and very poor ones at that.

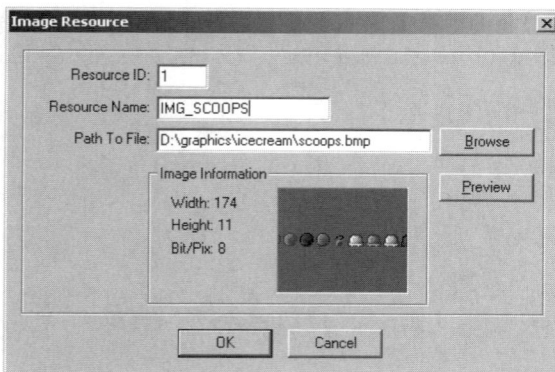

Figure 7-4: The Image Resource dialog filled in

As you can see, we have selected a picture and it is now viewable in the preview window. The preview window does not actually scale your image. So, if your image is larger than the preview window, it will be cut off. But rest assured that the file is intact inside of the resource archive.

Creating a Dialog Resource

Finally, we will create a dialog resource. These are predefined dialogs that you can set up for customized user interfaces. We will go more in depth about GUIs once we get to Chapter 10. For now, I will give a brief overview of how this editor works. Bring up the dialog menu by clicking on New Resource | New Dialog from the menu. This can also be accessed by the shortcut Alt + D.

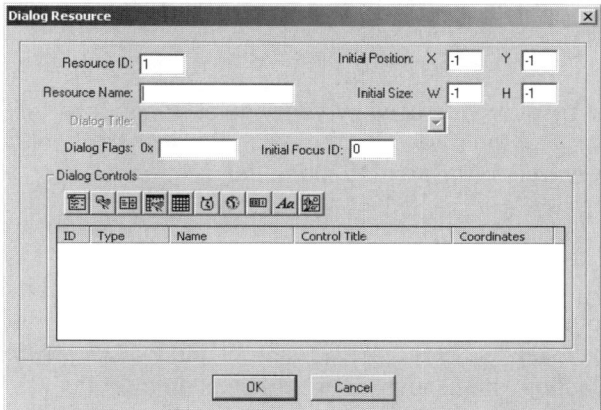

Figure 7-5: The Dialog Resource dialog

Here you see our old friends Resource ID and Resource Name. Also, there are text boxes to control the positioning and other attributes of the actual dialog. In the Dialog Controls area are a bunch of buttons, each of which represents a GUI element. By clicking on any of these buttons, you add the respective GUI element to the dialog. Each button brings up another dialog that allows the user to modify the attributes for that given GUI element. The actual creation in the Resource Editor and usage of dialog elements in code will be further detailed in Chapter 10.

Generating the Resource File

Now that the resource file has been created in the editor, it is time to generate the actual resource file itself. Actually, not just one file is generated. Instead, a series of files are created that each have a different purpose. They are as follows:

BRI

The BRI file is an intermediate resource file. It is not used by your applet but rather by the Resource Editor when modifying an existing resource archive. This is the file that you open with the Resource Editor when you want to modify the contents of your applet's resources. It is not required for the applet's execution. Therefore, this file does not need to be loaded into the phone or placed in your emulator's applet directory when running your code.

BAR

This is the actual resource file used by your applet. It contains the actual binary data of all the resources that you included with the tool. This is the file required for use with your applet.

Header File

A C header file will also be generated by the Resource Editor. It has the same name as your resource file appended with the text "_res" and the .h extension. For instance, if we made a resource file for our previous example called "hello," the header file would be named "hello_res.h." This header contains all the defines that you set up in the Resource ID and Resource Name fields for each resource. This needs to be included in your applet's code in order to access resources properly.

To create these files, you must first save your resource archive by clicking on the File|Save menu item. This saves out the BRI intermediate file. You must then click on the Build|Build QUALCOMM .BAR/.H Files menu item to generate the header and actual BAR archives.

Conclusion

You now know how to create and maintain your own resource files. The usage of resource files will become necessary in later examples in this book. Resource files are not only an easy way to store data for your applet, but they also make it much easier to create customized versions. For instance, if you need to localize your game for various countries, you can merely include a new resource file with new text in the native language of the country for which you are delivering. This also goes for supporting multiple color depths and resolutions. It is easy to provide a different resource file for each phone type. At the very least, I have provided both monochrome and color resource archives in my projects.

Chapter 8

Bitmap Graphics

Introduction

There was a time when games had no graphics (ahh yes, the glory days of text adventures). The graphics were the best of all—images conjured up in your imagination. Well, at least that was the theory.

Today, graphics are seemingly the overriding factor in a game's quality. With BREW, this is no exception. From drawing bitmap images to simple geometric shapes, BREW has quite the array of functionality for providing your game with impressive imagery. However, the meager computing power and memory restrictions on most BREW handsets do present a formidable obstacle in the quest for good visuals. Not all is lost. This just means that you will have to use some inventive optimization tricks explained later in the book.

Regardless, BREW has adequate facilities for the display of both bitmap and geometric graphics. This chapter will focus on bitmaps, with geometric graphic primitives the focus of the next chapter.

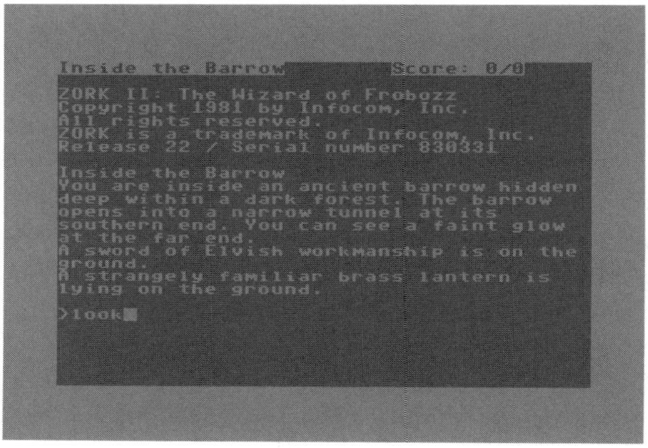

Figure 8-1: In a distant time known as the early 1980s, many games had no graphics at all. This is a screen shot from one of my old favorites, Infocom's Zork II.

Bitmaps

Perhaps the most common graphical object is the bitmap. A *bitmap* is a chunk of data that represents an image as a series of bits for each pixel (hence the name, bitmap). Bitmaps are stored in a variety of common file formats with which you are most likely familiar. These include PNG, GIF, PCX, and the native BREW format, BMP.

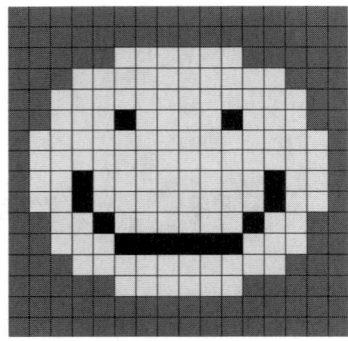

Figure 8-2: A bitmap is a 2D grid of pixels.

BMP Files

Qualcomm did not invent the BMP. In fact, it is a Microsoft file format. But for some reason, Qualcomm has decided to make BMP the file of choice for bitmap images in BREW 1.0. This makes it easy for us because the BMP format is so widely supported. Just about any art or drawing package can read and write BMP files.

Palettes

In version 1.0 of the SDK, all of BREW's graphics use palettes. This means that every pixel in a bitmap image is an index into a list of colors, otherwise known as a *palette*. Usually, the actual colors in the palette are stored in the BMP file. With BREW, these palettes are built into the hardware itself. If the BMP file does not have the exact palette as defined in the hardware, BREW will attempt to use the colors in the hardware palette that are the closest match to the ones in the file. Naturally, with such a limited amount of colors, the matches are often not much of a match at all. This does make it rather difficult to bring some images into BREW, as the default palettes simply do not have the color range needed. Your best bet is to design your bitmaps using the built-in palettes. Unfortunately, most handsets seem to have different hard-coded palettes, and few if any of the manufacturers have publicly released the palette information to developers. Early BREW documentation details a few different default palettes that may or may not be supported by any given device. These palettes are included in the companion files.

So, what is the BREW palette? Well, there are three basic palettes in BREW. These are 1-bit, 4-bit, and 8-bit palettes. By "bit," I mean how many bits are used as the lookup value into the palette. In the case of a 1-bit image, you can have either a 0 or 1—therefore two colors. In 4-bit mode, you can have 16 colors, and in 8-bit mode it is possible to have

256 colors. However, 50 of these 256 colors are reserved, and therefore only a palette of 216 colors is available.

Right now, the most popular handsets use monochrome 1-bit or color 8-bit displays. So why use 4-bit graphics? Since you are dealing with extreme memory limitations, it may be wise to use 4- or even 1-bit graphics with an 8-bit or greater display. These formats are considerably smaller than 8-bit. So it is possible to work around the extreme memory limitations of a handset with smaller image files. In many cases, 16 colors is pretty much all you need for such small graphics on a tiny display.

Displaying Bitmaps the Easy Way

BREW contains a number of functions for dealing with BMP images directly. These include simple functions to load BMPs out of a resource file as well as for the display and animation of these images. After all of this discussion, you probably want to see something on the screen, right?

Loading

The easiest way to load a bitmap is to store it in a resource file and use BREW's resource functions to extract it. It is possible to load a BMP straight from a file, but it is much easier to deal with resources. For the sake of completeness, we will show both examples here.

The function we use to load an image from the file system is in the IShell interface. The function **ISHELL_LoadImage** has the following syntax:

```
IImage * ISHELL_LoadImage(IShell * pIShell, const char* pszResFile)
```

As usual, the Shell interface has to be passed as the first argument. Just about every function that belongs to a given interface takes a pointer to the interface as the first argument. The second argument is the file-name of the BMP file. BREW assumes that this file is in the same directory as the executable, so all directory paths start from the application's root directory. This function then returns a pointer to an **IImage** object, which is allocated from that file. This image must be deallocated at some point or you will have a memory leak.

Loading an image from a resource file is not much more difficult. The function call looks like this:

```
IImage* ISHELL_LoadResImage(IShell* pIShell, const char* pszResFile, int16 nResID)
```

The new arguments are **pszResFile**, which is the filename of the resource file, and **nResID**, which is the number of the resource that you wish to get. These numbers are defined in the header that is generated in the Resource Editor. This header should have the same name as the resource file to which it belongs but with an .h extension. You will see an example of resource ID usage in Chapter 14.

Both of these functions return a pointer to an IImage object. This is an interface just like IShell, but it also contains the BMP image data allocated from the file or resource loaded. All IImage interface calls take the interface of the bitmap that you want to display as the first argument.

Drawing

This brings us to the next task at hand—drawing the image on the screen. The **Draw** function is simple. It looks like this:

```
IIMAGE_Draw(IImage * pIImage, int x, int y)
```

The **IImage** interface pointer is the pointer to the image that we want to draw. In our case, this is the pointer returned from the loading functions. The **x** and **y** variables are the coordinates of the image. This is the location relative to the upper-left corner of the image. If you provide coordinates that are off-screen or so close to the edge that part of the image is off-screen, BREW will clip the image for you.

However, simply calling the **Draw** function will not put your image on the screen. As explained in the Hello example, the **Update** function must be called in order for BREW to execute your drawing commands. So after the **Draw** call, you must use the following function:

```
void IDISPLAY_Update(IDisplay * pIDisplay)
```

As you can see, it is a simple function. It merely takes a pointer to its own interface as an argument. **Update** is an expensive operation. Therefore, it is a good idea to call this function only once after all of your drawing operations have been executed. It is not necessary to call **Update** after every single drawing operation.

Cleanup

Once we are done with a BMP image, how do we remove it from memory? We use the **IIMAGE_Release** function:

```
uint32 IIMAGE_Release(IImage * pIImage)
```

This "releases" our reference to that particular image's interface. It is possible to have an interface with more than one "reference" to it. A reference just means that some part of the code is using that interface. When you release an interface, you tell BREW that you are not using it anymore. If there are no more operations referencing that interface, it is destroyed and the memory it occupied is returned to the operating system. All BREW interfaces have a **Release** function used for deallocating memory and resources.

In the case of the images, and many other interfaces, we only have one reference to them. As soon as we call **Release** on that interface, it is destroyed and all the memory that image occupied is returned to the system. This is how we clean up the memory that is allocated by the loading of the BMP. If this is not done for every BMP loaded into memory, we will have a memory leak. If enough memory leaks are left hanging around, the applet will refuse to run because too much memory is taken up by references that were not properly released.

Animation

One great thing about the Image interface is the built-in ability to animate. It is possible to create a large BMP consisting of several uniform-sized frames of animation and have BREW automatically cycle through them. Here is how it works.

First, you have to create a suitable bitmap. If you look in the companion file directory Source\Chapter 08\animation, you will find the file anim.bmp. The image looks like this:

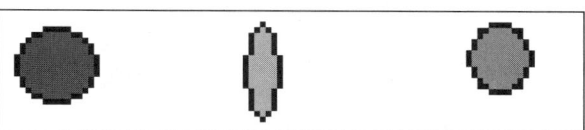

Figure 8-3: The animation frames in one wide BMP

As you can see, we have several frames of a colored ball changing shape. If you calculate the dimensions of the bitmap, you will find that it is 96x32 pixels. By eyeballing the bitmap, you can tell that there are three frames. So by dividing 96 by 3, you get a width of 32 pixels per frame. That means each frame's dimensions are 32x32. If we tell BREW the dimensions of the individual frames, it is intelligent enough to slice up the image into frames and flip through each one at a default time interval. We will do this next.

Now that we have the bitmap in the format that we want, we simply need to tell BREW the dimensions of each frame. This is done through the **IImage_SetParm** function. This function also allows us to set various other attributes that can alter the display and behavior of the image. For now, we will focus on animation. Here is the full syntax of the function:

```
void IIMAGE_SetParm(IImage * pIImage, int nParm, int p1, int p2)
```

The first argument in this function is the interface to the image for which we are setting the parameters. The second argument, nParm, is the type of parameter that we are setting. If you look in the header AEEImage.h, you will find the definitions of all the parameter types. The next two arguments, **p1** and **p2**, are the parameter values that we are setting.

So, which parameters do we use and how do we set them to create an animation? This is simple. First, we need to tell BREW what the width is of each frame. Previously, we determined that each frame is 32 pixels in width. So we set this parameter as follows:

```
IIMAGE_SetParm(pImage, IPARAM_CXFRAME, 32, 0);
```

We are assuming that we have loaded the image into the pointer, **pImage**. As you can see, the parameter definition for the frame width is **IPARAM_ CXFRAME**. This is just one of the different parameter types that are defined in AEEImage.h.

If we set no more parameters, we have a legitimate animation. BREW automatically calculates the number of frames by taking the width of the frame and dividing it by the total width of the bitmap. This is the same process that we went through when examining the bitmap ourselves.

The last thing to do is start the animation. This is done by the simple function call **IIMAGE_Start**. The syntax is as follows:

```
void IIMAGE_Start(IImage * pIImage, int x, int y)
```

The first argument is, of course, the pointer to the image interface we created upon loading the bitmap. The second pair of arguments are the x and y coordinates of the top left-hand corner of the image. This operates in the same fashion as **IImage_Draw**. There is no need to call **IDISPLAY_Update** because BREW's animation system updates the screen at the interval of 150 milliseconds. This means that the animation switches to the next frame after a second and a half of displaying the current one.

In many cases, a different animation speed is required. There is a way to do this via **IIMAGE_SetParm** with the **IPARAM_RATE** parameter. For **p1**, pass how many milliseconds that you want between frame changes. The second parameter, **p2**, is unused. For example, if we wanted to change to a quicker 50 millisecond rate, we would use the function like this:

```
IIMAGE_SetParm(pImage, IPARAM_RATE, 50, 0);
```

The animation system remains a handy feature for very simple applets. In the example for this section, we will create a very simple slide show using multiple frames of a single BMP. The code is found in the Source\Chapter 08\animation directory in the companion files.

A Simple Animation Example

We are going to create a simple applet that loads and displays an animating BMP using the functions described previously. Later in this chapter, we will contrast and compare alternate methods to display bitmap graphics. However, this example is a good introduction to the basic functionality of the **IImage** interface.

Our applet contains a standard applet structure definition in animation.h, with the addition of a pointer to an **IImage** interface:

Listing 8-1: The applet structure

```
typedef struct myapp_s
{
    AEEApplet    a; //applet header
    AEEDeviceInfo di;

    IImage*      pImage; //our animated BMP image
} myapp_t;
```

Here we have the standard applet structure included, as well as the **AEEDeviceInfo** structure. This is used to record the capabilities of the device. We will cover the usage of this structure when we talk about the applet initialization. Finally, we have a pointer to an **IImage** interface. We will use this to point to our loaded BMP.

Getting to the actual applet code, we have a standard **AEECls-CreateInstance** function that we have seen in all applets so far:

```
int AEEClsCreateInstance(AEECLSID ClsId, IShell * pIShell, IModule * po, void ** ppObj)
{
    *ppObj = NULL;

    if(ClsId == AEECLSID_ANIMATION_BID)
    {
        if(AEEApplet_New(sizeof(myapp_t), ClsId, pIShell, po, (IApplet**)ppObj,
            (AEEHANDLER)Animation_HandleEvent, (PFNFREEAPPDATA)Animation_CleanUp) == TRUE)
        {
            return(AEE_SUCCESS);
        }
    }

    return(EFAILED);
}
```

As you can see, we pass a message handler and cleanup function to **AEEApplet_New**. We will start by looking at the message handler:

```
static boolean Animation_HandleEvent(IApplet * pi, AEEEvent eCode, uint16 wParam, uint32
    dwParam)
{
    myapp_t* pApp = (myapp_t*)pi;
    AEEApplet * pMe = &pApp->a;
```

```
switch (eCode)
{

case EVT_APP_START:

        //get info about the handset (resolution, etc.)
        ISHELL_GetDeviceInfo(pMe->m_pIShell, &pApp->di);

        //clear screen (default color is white)
        IDISPLAY_ClearScreen(pMe->m_pIDisplay);

        pApp->pImage = ISHELL_LoadImage(pMe->m_pIShell, "anim.bmp");

        //set up image animation properties
        IIMAGE_SetParm(pApp->pImage, IPARM_CXFRAME, 32, 0);
        IIMAGE_SetParm(pApp->pImage, IPARM_NFRAMES, 3, 0);
        IIMAGE_SetParm(pApp->pImage, IPARM_RATE, 100, 0);

        //start the animation
        IIMAGE_Start(pApp->pImage, pApp->di.cxScreen / 2, pApp->di.cyScreen / 2);

        //we've successfully handled this message
        return(TRUE);
        break;
}

return(FALSE);
}
```

The only message we handle is **EVT_APP_START**. This is because
we already handle the cleanup in a callback, so we do not need
EVT_APP_STOP. Because of the automatic nature of BREW's BMP
animation facilities, we do not need to do anything other than load the
image, set its properties, and let it do its thing.

Looking at the **EVT_APP_START** handler, we can see that upon
applet start, we will perform a few housekeeping tasks as well as load
our image into memory. The first new function call here is
ISHELL_GetDeviceInfo:

```
void ISHELL_GetDeviceInfo(IShell * pIShell, AEEDeviceInfo * pi)
```

This is a simple function used to get a description of the capabilities of
the device. We pass it a pointer to the **AEEDeviceInfo** structure that we
placed in our applet structure. **AEEDeviceInfo**, as defined in
AEEShell.h, looks like this:

```
typedef struct
{
    uint16      cxScreen;        // Physical screen size (pixels)
    uint16      cyScreen;        // Physical screen size (pixels)
    uint16      cxAltScreen;     // Physical screen size of 2nd display (Pager)
    uint16      cyAltScreen;     // Physical screen size of 2nd display (Pager)
    uint16      cxScrollBar;     // Width of standard scroll bars
    uint16      wEncoding;       // Character set encoding (UNICODE, S_JIS,
                                 // KSC5601, etc.)
    uint16      wMenuTextScroll; // Milliseconds that should be used for auto-scroll
```

```
    uint16        nColorDepth;       // Color depth (1 = mono, 2 = grey, etc.)
    EmptyEnum     unused2;
    uint32        wMenuImageDelay;   // Milliseconds that should be used for
    uint32        dwRAM;             // Total RAM installed  (RAM)
    flg           bAltDisplay:1;     // Device has an alternate display (Pager)
    flg           bFlip:1;           // Device is a flip-phone
    flg           bVibrator:1;       // Vibrator installed
    flg           bExtSpeaker:1;     // External speaker installed
    flg           bVR:1;             // Voice recognition supported
    flg           bPosLoc:1;         // Position location supported
    flg           bMIDI:1;           // MIDI file formats supported
    flg           bCMX:1;            // CMX audio supported
    uint32        dwPromptProps;     // Default prompt properties
    uint16        wKeyCloseApp;      // Key to close current app
    uint16        wKeyCloseAllApps;  // Key to close all apps (AVK_END is default)
    uint32        dwLang;            // Used by Resource Loader - See AEE_LNG_XXX below

    //
    // NOTE: In order to use the following fields, you MUST fill in the wStructSize
    // element of the structure before passing this to the GetDeviceInfo call.
    //

    uint16        wStructSize;       // Size of this structure.
    uint32        dwNetLinger;       // Network PPP linger timer
    uint32        dwSleepDefer;      // Active non-sleep
    uint16        wMaxPath;          // Max length of file path
} AEEDeviceInfo;
```

BREW's comments pretty much sum up what each member of this structure describes. Because each handset has different capabilities, it is necessary to detect such attributes as screen resolution, color depth, and amount of RAM if you intend to have your applet run on many different devices without compiling the code specifically for each handset. In our case, we are really just interested in the resolution variables cxScreen and cyScreen.

Further along in the source listing, we clear the screen. You cannot depend on the screen to be clear when your applet begins. In many cases, whatever was on the screen before you ran your applet will remain until you clear or draw over it. This may be the phone's "desktop," a wallpaper image, the current time display, or whatever runs on your phone's display when not in use.

In this case, we are loading them from the local file system instead of using a resource file. Later in this book, you will see examples of using resource files for graphics. For now we will keep it simple. Load the images like this:

```
pApp->pImage = ISHELL_LoadImage(pMe->m_pIShell, "anim.bmp");
```

The anim.bmp file contains all the frames of our animation in a 96x32 bitmap. This means there are three 32x32 frames total. We need to set the parameters to notify BREW of the dimensions of each frame and the animation rate that we want, which is what we do in the next several lines of code:

```
IIMAGE_SetParm(pApp->pImage, IPARM_CXFRAME, 32, 0);
IIMAGE_SetParm(pApp->pImage, IPARM_NFRAMES, 3, 0);
IIMAGE_SetParm(pApp->pImage, IPARM_RATE, 100, 0);
```

First, we set the **IPARM_CXFRAME** parameter that tells BREW that each frame is 32 pixels wide. Then, we tell BREW that there are three frames total with the **IPARM_NFRAMES** parameter setting. Finally, we set our animation rate to 100 milliseconds between frames with the **IPARM_RATE** setting.

Now that we have set up the animation, we need to start it with the following code:

```
IIMAGE_Start(pApp->pImage, pApp->di.cxScreen / 2, pApp->di.cyScreen / 2);
```

This starts the animation process and places the image at our desired location. In this case we use the screen resolution that we got from the **AEEDeviceInfo** structure to place it roughly around the center of the screen. From this point on, BREW will automatically redraw the screen with the next frame every 100 milliseconds. When the final frame is reached, it will loop back to the first and start all over again.

Finally, we need to discuss the **CleanUp** function:

```
void Animation_CleanUp(myapp_t* pApp)
{
    IIMAGE_Stop(pApp->pImage);
    IIMAGE_Release(pApp->pImage);
}
```

After stopping the animation, call the function **IIMAGE_Release**. That is all we need to do in order to clean up all of our allocated resources (namely, the one animating image). As you can see, BREW's built-in animation system is quite easy to use.

Displaying Bitmaps the Hard Way

Now that you know how to use the Image interface, I will show you how to use images the "hard" way through device-native bitmaps. It is not really that difficult, but native bitmaps do not have many of the conveniences of the **IImage** interface. The advantage is that they use a bit less memory and are more flexible in their use, as a lot of the display implementation is up to the programmer.

Although the BMP is the file format of choice for BREW, it is not the format that the actual handset uses internally to display graphics. The actual native bitmap format is up to the individual manufacturer to decide. This is why bitmaps converted to this format are called "device-dependent" bitmaps. They are dependent on the actual hardware implementation. Because BMPs are not the device-dependent format on the handset, even when using the **IImage** interface, the BMP is converted to a device-dependent bitmap behind the scenes when displayed on the screen. By using device-dependent bitmaps directly, you avoid this extra step and unnecessary duplication of memory.

Creating a Device-dependent Bitmap

To create a device-dependent bitmap, you still need to load an image from either a file or a resource. Once you have the **IImage** interface, you can then convert it to a device-dependent bitmap using the handy little macro **CONVERTBMP**, which returns a pointer to your device-dependent image. The syntax is like this:

```
void * CONVERTBMP(void * pSrcBuffer, AEEImageInfo * pii, boolean * pRealloc)
```

The first argument to **CONVERTBMP** is the source buffer (that is, the BMP data we are converting into a device image). Luckily the BMP image is stored inside the **IImage** interface. We just have to find it and send the macro a pointer to it. This is a bit of a kludge, but it works.

The first byte of the **IImage** interface structure contains the size of the header. The header contains all sorts of information—most of it not available to the user because BREW's **IImage** functions handle it internally. Beyond this header is the raw BMP data that is in the actual file. **CONVERTBMP** needs this information to create a device-dependent bitmap. Therefore, we must assign a pointer to the BMP data by adding the value of the first byte in the **IImage** structure to the address of the **IImage** pointer itself. This will advance the pointer past the header to the beginning of the BMP data. Assuming the pointer **pImage** points to our loaded image resource, this is done like so:

```
byte * pDataBytes;
pDataBytes = (byte *)pImage + *((byte *)pImage);
```

As you can see, we take the contents of the first byte of our image and add it to the image pointer to make our new raw BMP pointer. Sure, this is ugly. But at least it works. A somewhat easier way is to use the file I/O functions to load a raw BMP file straight into a memory buffer. However, using the **IImage** interface allows you the convenience of pulling BMPs out of resource files.

The next argument is a pointer to an **AEEImageInfo** structure. This is filled with various characteristics about the image, including the size, color depth, and other attributes.

Finally, a pointer to a Boolean argument, **pRealloc**, is passed. The value of this Boolean will be True if this function allocated memory or False if no memory allocations occurred in the conversion process. This means that simply releasing the interface to the loaded image is not enough when cleaning up after your applet. The pointer to your device-dependent bitmap must also be freed via the macro **SYSFREE**. This is a very important variable because if it is not properly handled, you will have memory leaks on real hardware. This and more hardware-specific issues are discussed in Chapter 18.

Note: Using CONVERTBMP on an actual device almost always allocates memory. When this happens, the memory occupied by the original IImage you converted is now redundant as we are going to always blt the resultant device-dependent bitmap. As a result, you can release the original IImage to conserve memory. Keep in mind that this can lead to severe memory fragmentation issues as you are releasing resources in first in, first out (FIFO) order, which can lead to many small holes in memory that will not be able to be filled on subsequent allocations.

The return value is a pointer to raw data that is the device-dependent image. It is unwise to write to or otherwise modify this data because its format is potentially different on every handset. If you wish to directly modify the pixels of an image, you must modify the source BMP and reconvert it to a device-dependent bitmap.

Drawing a Device-dependent Bitmap

Now that you have this raw chunk of image data, how do you display it? This is done using BREW's display interface function, **IDISPLAY_BitBlt**. If you are wondering what "BitBlt" actually means, the term "Blt" is short for Block Image Transfer. A *blt* is basically a quick method of copying a rectangular image (block) from one chunk of memory to another. However, you will find that with many handsets, this is anything but fast. The syntax of the function is as follows:

```
IDISPLAY_BitBlt(IDisplay * pIDisplay, int xDest, int yDest, int cxDest, int cyDest, const
    void * pbmSource, int xSrc, int ySrc, AEE_RasterOp dwRopCode)
```

Let's break it down. The first argument is, of course, a pointer to the display interface. Secondly, we have the **xDest** and **yDest** arguments; this is the location of the upper left-hand corner of the location to which you want to blt this image. The arguments **cxDest** and **cyDest** are the width and height, respectively, of the image. If set to –1, both of these arguments assume the total width and height of the image. The **pbmSource** pointer is the pointer to our raw image data that we created with **CONVERTBMP**. The xSrc and ySrc variables determine the upper left-hand corner of the image that we want to copy over. By using these arguments and the cxDest and cyDest arguments, we can create what is called a *source rectangle*. That means we can blt over only a smaller portion of the total image to the screen. This is exactly how BREW's built-in animation facilities work. For each screen update, it blits from a small rectangle inside the larger picture to grab an individual frame. The process is detailed in Figure 8-4.

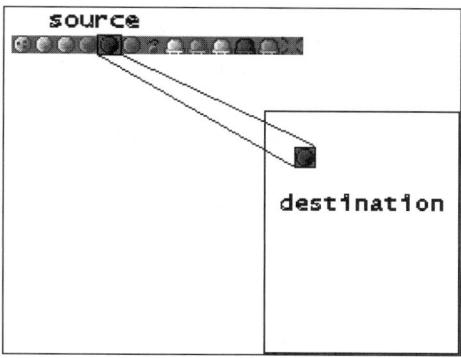

Figure 8-4: This is the relationship of the source rectangle to the destination. We blit a source rectangle from a single strip of sprite images to a destination rectangle on the screen.

Finally, we have the dwRopCode argument. This is a value that is defined in the AEE_RasterOp enumeration. These Raster Operation Codes can alter the drawing behavior of the blt call. It is possible to perform various logical operations on each bit with the data already drawn on the screen for interesting effects. Perhaps the most useful of all raster operations is the AEE_RO_TRANSPARENT value.

Pixel Transparency

There are many cases in common graphics operations where you want only some pixels of an image to show. This is where pixel transparency comes in. The pixel transparency mode, denoted by the raster operation define AEE_RO_TRANSPARENT, skips over drawing pixels of a special color. In the case of an 8-bit image, if a pixel is pure purple with a Red, Green, and Blue color value of 255, 0, 255, it will not be drawn. This color is typically referred to as the "magic color." This ugly shade of purple may seem odd. But it is ugly for just that purpose—it is assumed that it will not be used in most normal images. Therefore, it is safe to reserve this color for this special purpose.

By not drawing purple pixels, it is possible to have irregularly shaped images, despite the inherently rectangular size of a bitmap. The pixels that are not drawn show through whatever is underneath the image's rectangle, allowing a bitmap to be stamped on top of any background without ugly borders. Figure 8-5 explains the process in visual terms.

Using **BitBlt**'s raster operations argument is not the only way to achieve this. You can also do pixel transparency via the regular **IImage** interface. It is one of the various parameters that are accessible through the **SetParm** command.

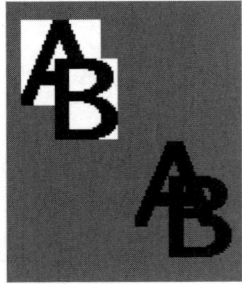

Figure 8-5: This is an example of pixel transparency. The two letters in the upper left-hand corner have a non-transparent white background. As you can see, this does not allow the background to show through the holes in the letters, and the B overlaps the A. In the lower right-hand corner of the image, the two letters are drawn with pixel transparency. The background color and underlying letter show through.

The Blt Advantage

So why use blts when you have to go through all of this hassle to convert the image and write special-case code to clean up yourself if the conversion process allocates memory? First, device-dependent bitmaps are smaller than images. The **IImage** interface contains a lot of unnecessary variables relating to parameters, animation settings, and other features that are not necessarily in use. Also, the variable source rectangle allows you to efficiently pack arbitrary graphic images in one large BMP. This saves memory as well, since each individual BMP has redundant palette and header information. It may not seem like much, but when developing applications for such a small device, every byte counts.

Two Image Examples

To illustrate the usage of both bitmap drawing methods, I have created a simple slide show applet that toggles between two different images. One version uses the **IImage** interface, and the other uses low-level bitblts. In both cases we are taking it to a slightly more complicated level by properly handling the resume event and using BREW timers to update the screen at regular intervals. These projects are included with the companion files in the folders Source\Chapter 08\slideshow and Source\Chapter 08\slideshow2, respectively.

An Image Slide Show

The first step in both of these examples is to create the applet class. Let's take a look at the applet structure that we are going to create:

Listing 8-2: The applet structure

```
typedef struct myapp_s
{
    AEEApplet    a; //applet header
    AEEDeviceInfo di;

    int      nImage;
    IImage*  pImages[MAX_IMAGES];
} myapp_t;
```

After the mandatory **AEEApplet** structure, we have a few variables that we are going to use in our applet. The first is **di**, which is an **AEEDeviceInfo** structure. We have already seen this in the animation example.

The final pair of variables are used for containing our loaded images. The first, **nImage**, is an integer that counts how many images we have loaded. Finally, we have an array of **IImage** pointers where our pointers to the loaded BMP images will go. The array is defined as **MAX_IMAGES** in size, which in this case is defined as 2. Feel free to modify this value when experimenting with your own images.

Now that we have defined our applet structure, we need to create our applet class. We should all be familiar with this code by now.

Listing 8-3: Creating the applet instance

```
int AEEClsCreateInstance(AEECLSID ClsId, IShell * pIShell, IModule * po, void ** ppObj)
{
    *ppObj = NULL;
    if(ClsId == AEECLSID_SLIDESHOW_BID )
    {
        if(AEEApplet_New(sizeof(myapp_t), ClsId, pIShell,po,(IApplet**)ppObj,
        (AEEHANDLER)Slideshow_HandleEvent, (PFNFREEAPPDATA)Slideshow_CleanUp)  ==
        TRUE)
            {
                return (AEE_SUCCESS);
            }
    }

    return (EFAILED);
}
```

This is a standard applet construction function. We pass pointers to our message handler and custom **CleanUp** function to **AEEApplet_New**. The **CleanUp** function will be detailed later in this example.

Now that we have created the applet class, we will look at the message handler and each message individually. First, look at the message handler in its entirety.

Listing 8-4: The event handler

```
static boolean Slideshow_HandleEvent(IApplet * pi, AEEEvent eCode, uint16 wParam, uint32
    dwParam)
{
```

```
myapp_t * pApp = (myapp_t*)pi;
AEEApplet * pMe = &pApp->a;

switch (eCode)
{
    case EVT_APP_START:

        //get info about the handset (resolution, etc.)
        ISHELL_GetDeviceInfo(pMe->m_pIShell, &pApp->di);

        //clear whole screen
        IDISPLAY_ClearScreen (pMe->m_pIDisplay);

        //load our images
        pApp->pImages[0] = ISHELL_LoadImage(pMe->m_pIShell, "circle1.bmp");
        pApp->pImages[1] = ISHELL_LoadImage(pMe->m_pIShell, "square1.bmp");

        pApp->nImage = 0;

        //draw the first image
        IIMAGE_Draw(pApp->pImages[pApp->nImage], 0, 0);
        IDISPLAY_Update(pMe->m_pIDisplay);

        //Set the timer...let's go!
        ISHELL_SetTimer(pApp->a.m_pIShell, DISPLAY_DELAY, (PFNNOTIFY)Slideshow_Timer,
            pApp);

        return(TRUE);
        break;

    case EVT_APP_SUSPEND:

        //If we've paused the applet for some reason, kill the timer
        ISHELL_CancelTimer(pApp->a.m_pIShell, NULL, NULL);
        return(TRUE);
        break;

    case EVT_APP_RESUME:

        //When the applet is re-activated, redraw and start the timer
        IIMAGE_Draw(pApp->pImages[pApp->nImage], 0, 0);
        IDISPLAY_Update(pMe->m_pIDisplay);
        ISHELL_SetTimer(pApp->a.m_pIShell, DISPLAY_DELAY, (PFNNOTIFY)Slideshow_Timer,
            pApp);

        return(TRUE);
        break;
}

return FALSE;
}
```

Now we need to look at each message individually. Naturally, the first stop is the **EVT_APP_START** message in Listing 8-4. After the familiar **IDISPLAY_ ClearScreen** call, we will load our BMP images as in the previous example. What we do for each image is put the resultant pointer into our array of **IImage** pointers. A simple filename and interface pointer is all we need to load the BMP from an external file.

After clearing our **nImage** variable, which tracks the bitmap we are currently displaying, we will draw the first bitmap of the pair and update the screen like so:

```
IIMAGE_Draw(pApp->pImages[pApp->nImage], 0, 0);
IDISPLAY_Update(pMe->m_pIDisplay);
```

Remember, if you want to see the results of your graphics operations, whether it be drawing bitmaps, text, or geometric primitives, you need to call **IDISPLAY_Update**. It is only necessary to call **IDISPLAY_Update** once per screen refresh, not after each drawing operation.

Finally, set up the timer. Although timers will be explained in depth later in this book, it is time for a brief introduction. A timer is a mechanism where you tell BREW to call a function at a specified point in the future. In this case, we want to call the function **Slideshow_Timer** 1000 milliseconds from now. We do this like so:

```
ISHELL_SetTimer(pApp->a.m_pIShell, DISPLAY_DELAY, (PFNNOTIFY)Slideshow_Timer, pApp);
```

The **DISPLAY_DELAY** definition is defined as 1000. It is a good idea to use definitions instead of raw literals because it is much easier to change a single definition than every single reference to that value in your code. Once this function is called, BREW knows that it has to call **Slideshow_Timer** 1000 milliseconds from now. Of course, those math wizards in the audience will know that 1000 milliseconds is one second.

Since we have discussed timers, we might as well detail the **Timer** function:

Listing 8-5: The timer callback

```
void Slideshow_Timer(myapp_t * pApp)
{

    //Toggle between the two images, draw image, and start
    //timer again for next image.

    AEEApplet * pMe = &pApp->a;

    ++pApp->nImage;

    if (pApp->nImage >= MAX_IMAGES)
        pApp->nImage = 0;

    IIMAGE_Draw(pApp->pImages[pApp->nImage], 0, 0);
    IDISPLAY_Update(pMe->m_pIDisplay);

    ISHELL_SetTimer(pApp->a.m_pIShell, DISPLAY_DELAY,(PFNNOTIFY)Slideshow_Timer, pApp);
}
```

This is the function that gets called by BREW's timer system. It advances the number of the image that we are displaying and then draws the current image. This code is fairly similar to the drawing code in our **EVT_APP_START** handler. The only trick here is that we have to call **ISHELL_SetTimer** again to make sure that this function gets called again in 1000 milliseconds.

Although it is overkill for a simple demonstration applet, we properly handle the suspend and resume events in our message handler. In a real-world situation, your applet can be suspended at any time. For example, if someone is playing your game and gets a phone call, your applet will be suspended when the user accepts the call. In this case, an **EVT_APP_SUSPEND** message will be sent to your message handler. It is up to you to suspend your applet's activity, which in this case means canceling the timer. Our **EVT_APP_SUSPEND** message looks like this:

Listing 8-6: The EVT_APP_SUSPEND handler

```
case EVT_APP_SUSPEND:

    //If we've paused the applet for some reason, kill the timer
    ISHELL_CancelTimer(pApp->a.m_pIShell, NULL, NULL);
    return(TRUE);
    break;
```

As you can see, all we do is cancel the timer. In most suspension cases, you have to cease all asynchronous activity. This includes stopping any timers and perhaps killing network connections, pausing music playback, and halting other operations that run asynchronous of your program's operation.

Note: Be careful with your return values. You must return True from the EVT_APP_SUSPEND and EVT_APP_RESUME handlers. If you do not return True from EVT_APP_SUSPEND, your applet will not receive a corresponding EVT_APP_RESUME. If you do not return True from EVT_APP_RESUME, your applet will receive an EVT_APP_START message directly after.

The opposite of the **EVT_APP_SUSPEND** message is **EVT_APP_ RESUME**. This message is sent when your applet is reactivated after a suspension (for instance, if the user finishes her call that previously interrupted your applet's operation and returns to the game). In this case, you have to resume the asynchronous tasks that were halted in **EVT_APP_SUSPEND**'s handler, as well as redraw the screen. When an applet resumes, the screen is still filled with the drawing operations of the previous application. In this case, it may have the phone display up.

Therefore, you must redraw your entire screen to properly resume the applet's operation. Our EVT_APP_RESUME handler looks like this:

Listing 8-7: The EVT_APP_RESUME handler

```
case EVT_APP_RESUME:

    //When the applet is re-activated, redraw and start the timer
    IDISPLAY_ClearScreen (pMe->m_pIDisplay);
    IIMAGE_Draw(pApp->pImages[pApp->nImage], 0, 0);
    IDISPLAY_Update(pMe->m_pIDisplay);
    ISHELL_SetTimer(pApp->a.m_pIShell, DISPLAY_DELAY, (PFNNOTIFY)Slideshow_Timer, pApp);
    return(TRUE);
    break;
```

As you can see, we redraw the screen by first clearing it and then drawing the current image. We can then update the screen and set our timer again. If we wanted to be totally accurate, we would save how much time we had left until the next timer event and then set the timer's period to be this modified value, but this applet is not timing critical.

Finally, we have to deal with the cleanup of the applet once it is exited:

Listing 8-8: The CleanUp function

```
void Slideshow_CleanUp(myapp_t* pApp)
{
    //delete our loaded BMPs

    int i;

    for (i = 0; i < MAX_IMAGES; i++)
    {
        IIMAGE_Release(pApp->pImages[i]);
    }
}
```

Here we simply release each loaded **IImage** interface stored in our array of pointers. That's it. We pass this function as a function pointer to our **AEEApplet_New** call, as detailed at the start of this example. Therefore, it will be called when the applet is destroyed, and it is not necessary to handle the EVT_STOP message.

A BitBlt Slide Show

I have also developed a version of this simple applet that uses device-dependent bitmaps instead of high-level **IImage** objects. The code is largely the same, except for a few differences. I will illustrate these differences in detail. The first difference is in the definition of our applet structure:

Listing 8-9: The applet structure

```
typedef struct myapp_s
{
    AEEApplet       a;                          //applet header
    AEEDeviceInfo   di;                         //device info (resolution, etc)
    int       nImage;                           //current image displayed
    void *    pRawImagePtrs[MAX_IMAGES];        //Array of device-dependent bitmap pointers
    byte      pMallocFlags[MAX_IMAGES];         //Flags for handset deallocation
} myapp_t;
```

This is pretty much the same, except we have an array of void pointers to store our images. Also, we have an array of bytes labeled **pMallocFlags**. This array is used to store the memory allocation flag when using the CONVERTBMP macro. This will be explained later.

Next, we should look at the **EVT_APP_START** message handler in our **Slideshow2_HandleEvent** function:

Listing 8-10: The EVT_APP_START handler

```
case EVT_APP_START:

    ISHELL_GetDeviceInfo (pMe->m_pIShell, &pApp->di);
    IDISPLAY_ClearScreen (pMe->m_pIDisplay);  // Clear whole screen

    //load our images
    SlideShow2_LoadImage(pApp, 0, "circle1.bmp");
    SlideShow2_LoadImage(pApp, 1, "square1.bmp");

    //start on the first image
    pApp->nImage = 0;

    //do the first screen refresh
    IDISPLAY_BitBlt(pMe->m_pIDisplay, 0, 0, -1, -1, pApp->pRawImagePtrs[pApp->nImage],
        0, 0, AEE_RO_COPY);
    IDISPLAY_Update(pMe->m_pIDisplay);

    //start the timer ticking....
    ISHELL_SetTimer(pApp->a.m_pIShell, DISPLAY_DELAY, (PFNNOTIFY)Slideshow2_Timer, pApp);

    return(TRUE);
    break;
```

Besides the usage of **IDISPLAY_BitBlt** to draw our images, the only other difference is in the usage of **SlideShow2_LoadImage** to load our bitmaps. Because processing a device-dependent bitmap is slightly more complicated than simply loading a BMP from a resource or file, I decided to encapsulate this process in the function **SlideShow2_LoadImage**:

Listing 8-11: The BMP loading routine

```
void SlideShow2_LoadImage( myapp_t * pApp, int nIndex, char * pszFileName)
{
    byte * pDataBytes;
    int nSize            = 0;
    IFile * pIFile       = NULL;
```

```
IFileMgr *  pIFileMgr        = NULL;
boolean bVal                 = FALSE;
FileInfo * pFileInfo         = NULL;
AEEImageInfo * pImageInfo     = NULL;

//We have to malloc these instead of using stack variables because
//on a real handset we only have 500 bytes of stack!
pFileInfo = MALLOC(sizeof(FileInfo));
pImageInfo = MALLOC(sizeof(AEEImageInfo));

//create the file manager object and load desired bitmap
ISHELL_CreateInstance(pApp->a.m_pIShell, AEECLSID_FILEMGR,(void **)&pIFileMgr);
pIFile = IFILEMGR_OpenFile(pIFileMgr, pszFileName, _OFM_READ);

//bad craziness
if (!pIFile)
{
    IFILE_Release(pIFile);
    IFILEMGR_Release(pIFileMgr);
    FREE(pFileInfo);
    FREE(pImageInfo);
    return;
}

//find out how big the file is, and load it in
IFILE_GetInfo(pIFile, pFileInfo);
nSize = pFileInfo->dwSize;
FREE(pFileInfo);

pDataBytes = MALLOC(nSize);

IFILE_Seek(pIFile, _SEEK_START, 0);
IFILE_Read(pIFile, pDataBytes, nSize);
IFILE_Release(pIFile);

//run CONVERTBMP macro to convert our BMP to a device-dependent bitmap
pApp->pRawImagePtrs[nIndex] = CONVERTBMP(pDataBytes, pImageInfo, &bVal);

//See if we have allocated memory in the process
if (bVal)
{
    pApp->pMallocFlags[nIndex] = 1;
}
else
    pApp->pMallocFlags[nIndex] = 0;

    //clean up
    IFILEMGR_Release(pIFileMgr);
    FREE(pImageInfo);
}
}
```

This is quite a difference from the simple process of creating an IImage object. This function loads a BMP file into an allocated chunk of memory. We then run the CONVERTBMP macro on this chunk of memory and save the resultant pointer in our previously defined array. Let's go in depth and explain the function line by line.

After declaring our variables, we have to allocate space for our two structures: **AEEImageInfo** and **FileInfo**. Why not declare them as stack variables? This is because the stack on BREW handsets is traditionally very small. In some cases your stack can be as small as 500 bytes. Therefore, allocating large structures on the stack can cause your applet to crash when running on real hardware. When using a number of large structures in a function, it is a good idea to allocate them on the heap instead. All we need to do is make a few calls to BREW's substitution for the Standard C Library's **malloc** call:

```
pFileInfo = MALLOC(sizeof(FileInfo));
pImageInfo = MALLOC(sizeof(AEEImageInfo));
```

Next, we have to create our **FileManager** object and open our image files and read them into memory:

```
ISHELL_CreateInstance(pApp->a.m_pIShell, AEECLSID_FILEMGR,(void **)&pIFileMgr);
pIFile = IFILEMGR_OpenFile(pIFileMgr, pszFileName, _OFM_READ);
```

Before we proceed, we need to determine if this file operation was successful. By handling the worst-case scenario, we make this function robust and able to properly exit without crashing, if for some reason we cannot read the desired file:

```
if (!pIFile)
{
    IFILE_Release(pIFile);
    IFILEMGR_Release(pIFileMgr);
    FREE(pFileInfo);
    FREE(pImageInfo);
    return;
}
```

Here we release our file and manager object and then deallocate all of our allocated memory using BREW's equivalent for the Standard C Library's **free** call.

Now, we have to read the entire file into memory. This process includes finding out how large the file is, allocating enough storage on the heap to hold the data, and reading the file in its entirety. BREW provides a handy function used to get all sorts of information about a file: **IFILE_GetInfo**.

```
int IFILE_GetInfo(IFile * pIFile, FileInfo * pInfo)
```

This function first takes a pointer to the interface of the file that we currently have open. It also takes a pointer to a **FileInfo** structure. This is where the function will record the various attributes of this file. The structure looks like this:

```
typedef struct _FileInfo
{
    FileAttrib    attrib;
    uint32        dwCreationDate;
```

```
    uint32          dwSize;
    char            szName[MAX_FILE_NAME];
} FileInfo;
```

Here we can retrieve the filename, size, date, and individual attributes. The **attrib** variable is an enumerated type defined in AEEFile.h that has values signifying whether the file is read-only, hidden, or whatever the case may be. What we are really interested in is **dwSize**, which is the size of the file in bytes.

In the next chunk of code, we will find out the size of the file using the now familiar **IFILE_GetInfo** function, allocate enough memory, and read it in:

```
IFILE_GetInfo(pIFile, pFileInfo);
nSize = pFileInfo->dwSize;
FREE(pFileInfo);

pDataBytes = MALLOC(nSize);

IFILE_Seek(pIFile, _SEEK_START, 0);
IFILE_Read(pIFile, pDataBytes, nSize);
IFILE_Release(pIFile);
```

As you can see, we get the size from the **dwSize** member of our **FileInfo** structure. We then release the memory allocated for the **FileInfo** structure once we are done with it. Next, we seek to the beginning of the file and read it into our newly allocated chunk of memory pointed to by **pDataBytes**. Finally, we release the file, as we no longer need to refer to it.

Now that we have the raw BMP read into memory, we need to convert it into a device-dependent bitmap using the macro **CONVERTBMP**, as detailed earlier. This is done with a single line of code:

```
pApp->pRawImagePtrs[nIndex] = CONVERTBMP(pDataBytes, pImageInfo, &bVal);
```

Here we are setting the desired entry in our array of device-dependent bitmap pointers to the results of the **CONVERTBMP** macro. The first argument is the pointer to our buffer containing the raw BMP data. Next is a pointer to our **AEEImageInfo** structure, which will be filled with the statistics of this image. The structure looks like this:

```
typedef struct _AEEImageInfo
{
    uint16  cx;
    uint16  cy;
    uint16  nColors;
    boolean bAnimated;
    uint16  cxFrame;
} AEEImageInfo;
```

The first two members are the most important. The **cx** and **cy** variables determine the width and height of the bitmap. The **nColors** member tells how many colors are used in this image. The **bAnimated** and cxFrame members relate to animated bitmaps signifying if it is animated and how many frames it has.

Getting back to CONVERTBMP, the final image is a pointer to a Boolean, bVal. This Boolean is set to True if the CONVERTBMP process has allocated memory. Otherwise, bVal will be set to False. What you will find is that this Boolean will always be set to False when running on the emulator. But in the case of running a native binary on real hardware, CONVERTBMP will frequently allocate memory in the conversion process. Therefore, we need to record the state of this Boolean in our array, pMallocFlags, as defined in our applet structure. We do this in the following code chunk:

```
if (bVal)
{
    pApp->pMallocFlags[nIndex] = 1;
}
else
    pApp->pMallocFlags[nIndex] = 0;
```

In this code fragment, we set the value to 1 or 0, depending on the state of this Boolean value. We use this array in the cleanup process to determine if we need to deallocate additional memory for native bitmaps. We will take a look at the CleanUp function next:

```
void Slideshow2_CleanUp(myapp_t * pApp)
{
    //delete our loaded BMPs

        int i;

        for (i = 0; i < MAX_IMAGES; i++)
        {
        //If the flag has been set (allocation in CONVERTBMP), we need to
        //use SYSFREE instead.  This only happens on a real handset.
        if (pApp->pMallocFlags[i] == 0)
            FREE(pApp->pRawImagePtrs[i]);
        else
            SYSFREE(pApp->pRawImagePtrs[i]);
        }
}
```

When we allocate memory, we use the FREE macro. However, in the special case of CONVERTBMP, we need to use the SYSFREE call instead. Therefore, based on the status of our pMallocFlag entry, we will call either FREE or SYSFREE.

The rest of this code is nearly identical to that of the previous example. As you can see, using device-dependent bitmaps requires a lot more work. However, the speed and efficiency benefits are good enough that you should consider using them instead of high-level IImage objects.

Conclusion

In this chapter you have learned how to load, display, and animate bitmaps. BREW provides basic bitmap facilities that are suitable for most mobile games. Future editions of BREW promise full-blown sprite support, double buffering, and other goodies. For now, there is an adequate range of options for most bitmap graphics.

Chapter 9

Text and Geometric Graphics

Introduction

Now that we have seen how to display bitmaps, it is time to move on to BREW's geometry and text display features. Geometric graphics are simple shapes such as rectangles, circles, and triangles that can be drawn in various shapes, sizes, and colors to create compelling visuals. These geometric scenes often take up less space than a comparable bitmap. This chapter will focus on the usage of these geometric drawing operations, as well as BREW's ample text drawing facilities that were briefly illustrated in our simple Hello World program from earlier in this book.

Drawing Text

Before we get into drawing shapes and patterns, we will focus on the simple task of displaying text. BREW provides a few simple but flexible functions for the display of simple lines of text. The task of drawing strings of text is far simpler than handling bitmaps. There is one function that you must know in order to display simple chunks of text: IDISPLAY_DrawText.

```
void IDISPLAY_DrawText(IDisplay * pIDisplay, AEEFont Font, const AECHAR * pcText,
    int nChars, int x, int y, const AEERect * prcBackground, uint32 dwFlags)
```

We already have detailed this function in our first Hello World applet. After the interface pointer to the display, we will pass in an enumeration of one of the available fonts with the Font argument. Next we will pass a pointer to a wide-character string of text via pcText, followed by the length of the string with nChars. Obviously, we will pass the x and y coordinates of the string with the x and y arguments. We will then pass a clipping rectangle with the prcBackground pointer, which determines the area in which the text will draw. Any portions of text that extend out of this box will not be drawn. Finally, we will pass in some bit flags via dwFlags, which govern different properties of the font including the positioning.

Fonts

There are a few basic fonts available to all BREW programs. This is determined by the Font argument, which is one of a series of values defined in BREW's headers. Most devices are severely limited in fonts, so there are minimal choices. These font numbers are defined in AEEDisp.h. Check Table 9-1 for the details.

Table 9-1: Standard font types

Font Type	Description
AEE_FONT_NORMAL	Normal font used for text drawing
AEE_FONT_BOLD	Darker bold font
AEE_FONT_LARGE	Font with letters larger than normal

It is up to the device manufacturer to decide how these fonts will look and if the device will support them. Therefore, the dimensions of your string may be different, depending on the device that the applet is running on. For now, we will use the emulator as a reference, but keep this in mind when laying out text for an applet that will eventually be implemented on a real handset.

Position and Clipping

As detailed earlier, the x and y parameters determine the pixel coordinates of the string. The prcBackground variable is an AEERect structure that allows you to specify a clipping rectangle for the text. Any portions of text that fall outside of the rectangle will not be drawn. If the argument is NULL, it uses the entire screen as the clipping area. The coordinates of the text are relative to the upper left-hand corner of the string (that is, if we are not using any positioning bit flags to center or otherwise manipulate the position of the text).

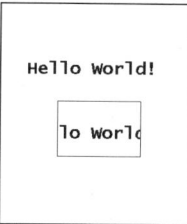

Figure 9-1: This image shows normal text above and the same string drawn inside of a clipping rectangle below. Clipping rectangles can be used to clip geometric graphics as well.

Style

The final argument is a concatenation of bit flags that affect the drawing style. These flags can be used to center, underline, and invert, among other attributes. The flags can be any combination of single values from the four categories detailed in Tables 9-2 through 9-5.

Table 9-2: Text horizontal alignment styles

Alignment Style	Description
IDF_ALIGN_LEFT	The text will have its left edge up against the left edge of the clipping rectangle.
IDF_ALIGN_CENTER	The text will be centered horizontally in the middle of the clipping rectangle.
IDF_ALIGN_RIGHT	The text will have its right edge up against the right edge of the clipping rectangle.

Table 9-3: Text vertical alignment styles

Alignment Style	Description
IDF_ALIGN_TOP	The text's top edge will be against the top edge of the clipping rectangle.
IDF_ALIGN_MIDDLE	The text will be vertically centered in the middle of the clipping rectangle.
IDF_ALIGN_BOTTOM	The text's bottom will be against the bottom of the clipping rectangle.

Table 9-4: Text format styles

Format Style	Description
IDF_TEXT_UNDERLINE	The text will be drawn with a line underneath.
IDF_TEXT_INVERTED	The text will be drawn backwards, as if shown in a mirror.

Table 9-5: Rect format flags

Flags	Description
IDF_RECT_FRAME	The text will be surrounded by a line drawn around the border of the clip rectangle in the current frame color.
IDF_RECT_FILL	The clip rectangle will be filled with the background color.
IDF_RECT_INVERT	The colors inside the rectangle will be inverted (black changes to white and vice versa).

Only one flag can be used from each category. For instance, having a style flag set to (IDF_ALIGN_CENTER | IDF_ALIGN_MIDDLE | IDF_TEXT_UNDERLINE | IDF_RECT_FRAME) will display the text centered vertically and horizontally in the center of the screen and underlined with a box around it. Although there are few options, by using these bit flags in various combinations you can come up with a few handy basic text styles.

Limitations

There are a few limitations with text drawing. Although it is possible to define a clip region, it is impossible to wrap the text inside of this rectangle. Therefore, IDISPLAY_DrawText is only good for displaying very short strings. I find it handy for things like score displays and brief in-game messages but nothing more. For displaying long blocks of text, it is better to use the static GUI control. This and other GUI components are described in Chapter 10.

What Are Geometric Primitives?

What is a geometric primitive? In traditional computer graphics, a *primitive* is a basic form or shape. In this case, BREW has several primitives, including triangles, rectangles, circles, and even pie slice wedges. These primitives are called such because they are very basic shapes. A number of different primitives can be combined to create more complicated scenes or various special effects.

Geometric graphics are much simpler and provide much less detail than a bitmap. So why use primitives at all? There are two main advantages to primitives: They are scalable, and they take up less space.

Primitives are drawn by providing BREW with a description of the shape that you want to display. For instance, if you want to draw a circle, tell the BREW graphics system that you would like a circle of a particular radius, centered about a specific point, and drawn in a color that you select. Because all of these parameters are controlled in code, it is possible to scale and manipulate all of these parameters in real time. For instance, if you want a circle to shrink, you merely have to provide BREW with a smaller radius value the next time you draw it. Not only does this make simple effects, such as scaling, easy to do, it also provides an easy way to deal with varying screen resolutions by simply providing different size values for different screen sizes.

How do primitives take up less space? Primitives are basically a simple description of a shape. As in the previous example, simply give a location and radius to draw a circle. A bitmap is an array of bits that determine which pixels are colored appropriately. Therefore, if we were to have a circle of radius 16, the primitive would simply have an x, y, and radius variable—a grand total of three shorts, hence 6 bytes of information. You would need a 16x16 bitmap to represent the same circle. In the case of an 8-bit image, you would need 16x16x8 bits of storage. This totals to 256 bytes. In this case, the geometric primitive is roughly 2 percent the size of the bitmap!

As you can see, geometric primitives provide a definite advantage over bitmaps when appropriate. For graphics that require more detail, bitmaps are mandatory. But in many cases, geometric primitives are suitable replacements for bitmaps, yielding huge memory savings and resolution independence.

Graphics Properties

Before we get into the details of drawing geometric primitives, you should be familiar with a few of the settings used for these operations. BREW allows for many different attributes when drawing; however, we will go over the two most common: color and fill modes.

Before you draw anything, you need to pick the appropriate color for your primitive. This is done through the simple IGRAPHICS_SetColor command. The syntax is as follows:

```
void IGRAPHICS_SetColor(IGraphics * pIGraphics, uint8 r, uint8 g, uint8 b, uint8 alpha)
```

The first argument is the typical interface pointer that we are all used to by now. Next up are the four color components that BREW uses. However, only three are actually relevant. Once direct color hardware is released, the alpha value will modulate the amount of translucency with which the color is drawn. Until then, the alpha argument is ignored.

By mixing the amount of red, green, and blue, you generate a single color. Each value has a range from 0 to 255. If combined, there are over 16 million color combinations. Of course, because at this time all BREW hardware is palette-based, this color will be mapped to one of the 216 built-in BREW colors (that is, if we are in 8-bit mode). As you can imagine, mapping a general range of 16 million colors down to 216 is quite a task. This is why some images that are not in the native BREW palette do not look very good when used in an applet.

The second attribute that warrants mention is the fill mode. It is possible to draw enclosed shapes, such as squares and circles, either filled or as an outline. By default, they are set to draw as outlines. To turn on or off shape filling, use the IGRAPHICS_SetFillMode function. It works like this:

```
void IGRAPHICS_SetFillMode(IGraphics * pIGraphics, boolean bFill)
```

After the interface pointer, a simple Boolean argument is all that is necessary. Pass in a True if you want to fill primitives and a False if not. One thing you must be aware of is that the color setting described previously only affects the outline of a shape. It is necessary to set the fill color with IGRAPHICS_SetFillColor if you want to alter the color of the filled primitive. The syntax is identical to IGRAPHICS_SetColor:

```
void IGRAPHICS_SetFillColor(IGraphics * pIGraphics, uint8 r, uint8 g, uint8 b,
    uint8 alpha)
```

If you want a primitive with one solid color, simply set the color and fill color to the same thing. Otherwise, you can have shapes with outlines of a different color than their insides. Now on to the actual shapes!

Points

To start off easy, we will cover perhaps the simplest graphics primitive of all: the point.

```
void IGRAPHICS_DrawPoint(IGraphics * pIGraphics, AEEPoint * pPoint)
```

The key argument here is the AEEPoint pointer. This points to the AEEPoint structure that is defined in the header AEEDisplay.h. In fact, all of BREW's primitive drawing operations use their own handy data structures to define the parameters of the shape. The members of the AEEPoint structure are as follows:

```
typedef struct _point
{
    int16    x, y;
}AEEPoint;
```

The point structure simply contains an x and y variable for the x and y coordinate of the point. A single-pixel point may not sound like a big deal. Luckily, BREW provides a way to control the size of the points through this function:

```
void IGRAPHICS_SetPointSize(IGraphics * pIGraphics, uint8 u8size)
```

Simply pass a value from 0 to 255 into the **u8size** argument, and you can create some fairly wide points.

Figure 9-2: A single point

Lines

Only slightly more difficult than the point is the line. The function to use here is **IGRAPHICS_DrawLine**:

```
void IGRAPHICS_DrawLine(IGraphics * pIGraphics, AEELine * pLine)
```

```
typedef struct _line {
    int16 sx, sy;      // starting point
    int16 ex, ey;      // ending point
} AEELine;
```

This is self-explanatory. The members **sx** and **sy** define the x and y pixel coordinates of the start of the line. The members **ex** and **ey** specify the x and y coordinates of the endpoint of the line.

Figure 9-3: A single line. The sx and sy variables define the starting point on the left, and the ex and ey define the endpoint on the right.

Polylines

A polyline is basically a number of connected lines, essentially connecting the dots. The function call and associated structure look like this:

```
int IGRAPHICS_DrawPolyLine(IGraphics * pIGraphics, AEEPolyline * pPolyline)
```

```
typedef struct _polyline {
    int16       len;        // Number of vertices
    AEEPoint *points;       // Array of points
} AEEPolyline;
```

The difference between this and normal shape structures is the
AEEPoint pointer. Basically, you need to allocate an array of AEEPoints
that represents the dots that you wish to connect with a line. The
pointer must address this array and deallocate when appropriate. You
then must provide the number of points in the polyline via the len
variable.

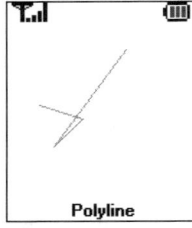

Figure 9-4: This polyline is composed of three
lines defined by four points.

Circles

Moving right along, we have a slightly more complicated primitive: the
circle. Here is an overview of the IGRAPHICS_DrawCircle function and
its associated data structure:

```
void IGRAPHICS_DrawCircle(IGraphics * pIGraphics, AEECircle * pCircle)
```

```
typedef struct _circle {
    int16 cx, cy;      // Center of the circle
    int16 r;           // Radius of the circle
} AEECircle;
```

Once again, the data structure is very straightforward. The cx and cy
members define the x and y coordinates of the circle's origin. The r
member defines the radius of the circle. As with all filled primitives,
IGRAPHICS_SetColor determines the color of the outside of the shape
and IGRAPHICS_SetFillColor defines the color of the inside.

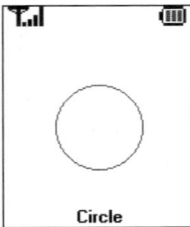

Figure 9-5: A circle

Ellipses

Vaguely related to the circle, another handy primitive is the ellipse. Known to most as an oval, an ellipse is basically an egg shape. The IGRAPHICS_DrawEllipse function and its associated data structure are used like so:

```
int IGRAPHICS_DrawEllipse(IGraphics * pIGraphics, AEEEllipse * pEllipse)

// Ellipse (axis-aligned only)
typedef struct _ellipse {
    int16 cx, cy;    // center of ellipse
    int16 wx;        // semimajor/minor axis along x-axis
    int16 wy;        // semimajor/minor axis along y-axis
                     // which is major or minor depends on the relative values of wx and wy
} AEEEllipse;
```

Figure 9-6: An ellipse

Rectangles

Another basic primitive is the rectangle. Not only useful for images, the rectangle can be used to clear portions of the display as well. The unique thing about IGRAPHICS_DrawRect is that the AEERect structure is defined in the header AEE.h instead of in AEEGraphics.h with the rest of them. Other than that, it is business as usual:

```
void IGRAPHICS_DrawRect(IGraphics * pIGraphics, AEERect * pRect)

typedef struct
{
    int16    x, y;
    int16    dx, dy;
} AEERect;
```

In the AEERect structure, x and y determine the origin of the upper left-hand corner of the rectangle, and dx and dy specify the width and height. You may have seen this used previously in the text drawing function. The AEERect structure is used in many cases where a rectangular area must be described.

Figure 9-7: A rectangle. The upper left-hand corner is defined by x and y, while the lower right-hand point is defined by dx and dy.

Triangles

We all know what a triangle is. BREW provides a simple function for drawing triangles. You simply provide three different points representing the corners of the triangle. The function works like this:

```
void IGRAPHICS_DrawTriangle(IGraphics * pIGraphics, AEETriangle * pTriangle)
```

```
typedef struct _triangle {
    int16 x0, y0;     // First point
    int16 x1, y1;     // Second point
    int16 x2, y2;     // Third point
} AEETriangle;
```

As you can see, we list each of the three points as a series of x and y coordinates inside the **AEETriangle** structure.

Figure 9-8: A triangle

A polygon is essentially the same thing but with more or an equal number of points.

Polygons

A polygon is a closed shape defined by three or more line segments. By this definition, a triangle is a polygon. However, the polygon drawing functions in BREW are traditionally used to draw shapes with more than three sides. You can draw triangles with **IGRAPHICS_DrawPolygon**, but it is really meant for more complicated shapes. A polygon is very similar to a polyline in that the structure is a series of points that are then connected to form a polygon. The function and structure look like this:

```
void IGRAPHICS_DrawPolygon(IGraphics* pIGraphics, AEEPolygon* pPolygon)
typedef struct _polygon {
```

```
    int16   len;       // Number of vertices
    AEEPoint *points;  // Array of vertices
} AEEPolygon;
```

As with the polyline, the one thing the polygon has to take into account is the fact that there are a variable number of points. Therefore, the **AEEPolygon** structure contains the number of points and then a pointer to an array of **AEEPoint** structures.

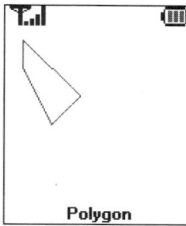

Figure 9-9: A four-point polygon

Arcs and Pies

Another very useful graphics primitive is the arc. An arc is basically a simple curve. **IGRAPHICS_DrawArc** works like this:

```
void IGRAPHICS_DrawArc(IGraphics * pIGraphics, AEEArc * pArc)

typedef struct _arc {
    int16 cx, cy;      // Center of the reference circle
    int16 r;           // Radius of the reference circle
    int16 startAngle;  // In degrees from 0 to 359
    int16 arcAngle;    // In degrees from 0 to 360
} AEEArc;
```

The arc is perhaps the most complicated primitive available. In BREW, arcs are drawn as sections of a circle. Essentially what you do is determine the location and radius of the circle and then define the starting and ending degree marking that you want to draw between. It is a far cry from the grace and usefulness of a Bezier curve or B-Spline, but hey, we will take what we can get.

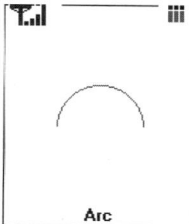

Figure 9-10: An arc

The pie primitive is essentially the same thing. In fact, the **AEEPie** structure is virtually identical to **AEEArc**. The only difference is that when a pie is drawn, it connects each endpoint of the arc to the origin of the circle. This creates a filled-in pie slice useful for charts and diagrams.

```
void IGRAPHICS_DrawPie(IGraphics * pIGraphics, AEEPie * pPie)

// Pie
typedef struct _pie {
    int16 cx, cy;     // Center of the reference circle
    int16 r;          // Radius of the reference circle
    int16 startAngle; // In degrees
    int16 arcAngle;   // In degrees
} AEEPie;
```

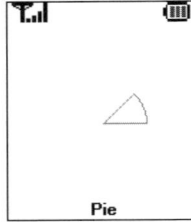 Figure 9-11: A pie

To see an example of each of these primitives in action, check out the Source\Chapter 09\geometry project in the companion files. This applet is a slide show demonstrating every primitive described in this chapter. In fact, the screen shots from this chapter are taken from this applet. Feel free to use it as a quick reference on how to draw geometric primitives with BREW.

Coordinate Systems

If you have not noticed, the coordinates for graphical primitives cover a massive range. The arguments for both the x and y coordinates are stored as 16-bit integers, essentially giving us a range of 65,536 units. How can we plot points at these ranges when most current BREW devices can barely break the 100-pixel mark in either dimension?

The answer is that the graphics primitives are drawn in a world-coordinate system. The world ranges from 0 to 65,536, and the screen of the BREW device acts as a tiny window into this much larger world. BREW allows this window to be moved around in world space, as well as modified in size and shape.

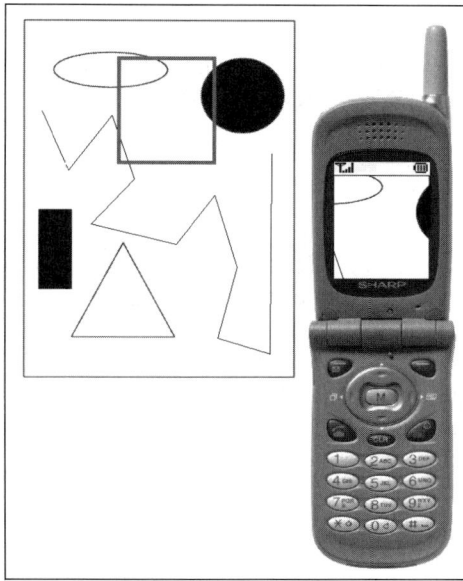

Figure 9-12: This shows the relationship between world and screen coordinates. The large picture to the left represents the geometric scene in world coordinates. The square is where the viewport is in screen coordinates. You can see how it looks on the device's screen to the right.

By default, the window, or viewport as it is called, is in the upper left-hand corner of the world. That is, the origin of the viewport is at 0,0 in world coordinates. The world coordinate origin of the viewport can be modified using the simple command IGRAPHICS_Translate:

```
boolean IGRAPHICS_Translate(IGraphics * pIGraphics, int16 x, int16 y)
```

One other attribute of the viewport is its shape. By default, the viewport is the size of the BREW device's screen. The viewport can actually be shrunk in size to present a letter-boxed or otherwise smaller view of the area. The function IGRAPHICS_SetViewport handles the viewport attributes:

```
void IGRAPHICS_SetViewport(IGraphics * pIGraphics, AEERect * pRect, uint8 nFlag)
```

Here the AEERect structure defines the size, in pixels, of the viewport rectangle. The flags field allows a series of bit flags to be combined to customize the appearance and behavior of the viewport. These flags are defined in AEEGraphics.h.

There is one other attribute of the viewport that must be discussed: clipping. If a geometric object is outside of the viewport, it is a waste of resources to draw it; this is what clip regions are for. By default, the clipping region is the size of the default viewport (that is, the size of the physical screen). However, when you shrink the size of the viewport through the IGRAPHICS_SetViewport command, it is necessary to also define a clip region with the same size using IGRAPHICS_SetClip:

```
boolean IGRAPHICS_SetClip(IGraphics * pIGraphics, AEERect * pRect, uint8 nFlag)
```

As you can see, the parameters are exactly the same, and you should pass the same parameters from IGRAPHICS_SetViewport to IGRAPHICS_SetClip.

Conclusion

In this chapter you have learned about BREW's geometric graphics functionality in depth. You have learned how to draw and manipulate a variety of geometric shapes and text. Using a combination of these techniques, it is possible to create some decent visuals for your mobile game. Future editions of BREW are bound to have more thorough graphics libraries; however, these are suitable for the simple kinds of games that the average handset can handle.

Chapter 10

GUI

Introduction

Graphical user interfaces are an important part of almost any interactive application these days. On tiny devices such as mobile phone handsets, the GUI becomes much more critical. The limited controls on the device and the miniscule amount of screen real estate means that clear, concise, and usable interfaces become all the more important.

BREW provides a fairly robust GUI package that includes all the basic elements such as text input boxes, menus, progress bars, and other GUI controls that are usually seen on desktop computers. GUI elements can be created in code or loaded out of a resource file. This chapter will explain both methods, as well as detail the common GUI components that you would need to use in a typical BREW game.

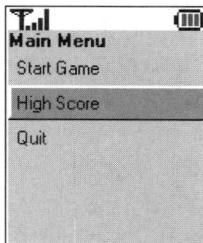

Figure 10-1: This is an extremely basic example of a game GUI. In this case, it is a simple main menu.

Creating an Object Instance

To create a GUI control and any object or interface in the BREW SDK, use the ISHELL_CreateInstance function:

```
int ISHELL_CreateInstance(IShell * pIShell, AEECLSID cls, void ** ppobj)
```

This creates an instance of a BREW object. What is an object? An *object* is another term for an interface in BREW—a collection of functions and data related to a specific purpose. We have seen interfaces quite often so far. The first argument of most functions is an interface pointer. We have not seen the actual creation of an instance previously because that was done behind the scenes. For instance, when loading a BMP image, BREW automatically creates an instance of a bitmap and sends it back to us as the returned pointer. In the case of GUI components, we have to create our own instances with this function.

As usual, the first argument is an instance. In this case, it is the Shell interface that is allocated for us upon creation of the applet. Secondly, we have the class ID of the object that we are creating. Much like our applet has a unique class ID stored in the BID file, each object or interface in BREW has a unique class ID as well. This is the key to creating different GUI controls; you simply pass the class ID of the control that you want to this function. Finally, pass a pointer to a pointer that will be modified to point to our allocated instance upon return from this function. This pointer must be cast to that of the control we need. From there, that control's interface can be accessed for our needs.

Using Static Text

The simplest GUI element in BREW is the static control. This control is a simple window for displaying large amounts of text. As explained in the previous chapter, the text drawing routines are very basic. While they support clipping areas, they do not wrap text around the boundaries. The static control solves these problems and also allows for scrolling through large amounts of text for easy reading.

Figure 10-2: A static control with some meaningless text in it

Creating a Static Control

To create a menu, use ISHELL_CreateInstance, like this:

```
IStaticCtl * pIStatic;

ISHELL_CreateInstance(pApp->a.m_pIShell, AEECLSID_STATIC, (void **)&pIStatic);
```

Now that we have our pointer to a freshly allocated static control object, we can set its dimensions and add the text. Each static control needs to have its rectangle defined. This is the area on the screen where the control resides. It is important to note that if the rectangle is not set before the text is added, some implementations of BREW will hang or crash. Set the rectangular area of the control with ISTATIC_SetRect:

```
void ISTATIC_SetRect(IStatic * pIStatic, const AEERect * prc)
```

The first argument is a pointer to the static control that we wish to modify. The second is a pointer to the same **AEERect** structure that we used in the previous chapter. By defining a rectangle in **AEERect**, we tell BREW the precise pixel coordinates that the static control should occupy.

Setting the Static Text

Now all we have to do is send some text over to display in the control. This is done via the ISTATIC_SetText function:

```
boolean ISTATIC_SetText(IStatic * pIStatic, AECHAR * pTitle, AECHAR * pText, AEEFont
    fntTitle, AEEFont fntText)
```

The first argument points to the static control that we are manipulating. Next we pass two different strings. The first, **pTitle**, is a pointer to a wide-character string used for the title. The title goes in a small blank space above the static control. Secondly, we pass in **pText**, a pointer to the wide-character string containing the text we actually want displayed in the control. Finally, we pass two font arguments: **fntTitle** and **fntText**. Obviously, **fntTitle** determines the font type for the title, while the latter determines the font type for the text. These are the same font values we used in the text drawing graphics functions.

Note: It is possible to have linefeeds in a static control for some basic text formatting. However, you must manually insert the linefeeds (the "\n" character) in the string via code. There currently is no way to specify a linefeed in a text resource.

Note: In BREW 1.0 you have no control over the timing and speed of the scrolling if you have more text than can be displayed in the static control. It may be easier for the user if you have multiple "pages" of text on different static controls that are updated on a button click.

The most annoying thing about the static control is that you cannot directly send text from a resource file to the control itself. Instead, the

ISTATIC_SetText function only takes arrays of **AECHAR**s. This means you have to load up the string from the resource into a buffer and send that to the control.

Drawing the Static Control

Much like images and other graphical elements, GUI controls have to be redrawn before **IDISPLAY_Update** is called. In the case of the static control, we use the function **ISTATIC_Redraw** to display the control on the screen:

```
boolean ISTATIC_Redraw(IStatic * pIStatic)
```

We simply pass the pointer to the static control that we want to draw, and that is it. By default, if there is more text in the control than can physically fit on the screen, it will slowly scroll through the text.

Note: Unlike most other BREW GUI components, there is no way to "hide" or otherwise deactivate a static control. When an EVT_APP_SUSPEND event is received by your applet, you are to release the static control. Otherwise, it may draw on top of whatever event is interrupting your applet. When the applet is resumed, you must manually recreate the static control and redraw it. Not doing this will almost certainly cause your program to fail the True BREW test process.

Changing the Static Control

In general, this is all you need to do with a static control. However, you may want to change the way the control looks or behaves. This is done through **ISTATIC_SetProperties**:

```
void ISTATIC_SetProperties(Istatic * pIStatic, uint32 nProperties)
```

The first argument is, of course, the pointer to our static control. The second is a concatenated field of bit flags that govern various attributes of the control. These bit flags are defined in AEEShell.h. They range from centering the text to allowing the control to handle the memory allocation of its text strings.

Note: By default, the static control allocates memory and copies the text for both the title and body text to these new buffers. When you destroy a static control, these buffers are freed. You can choose to manage your own string buffers for the title and body text by using ST_TITLEALLOC and ST_TEXTALLOC, respectively. When you do this, you must deallocate the string buffers yourself.

Using Menus

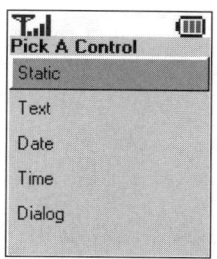

Perhaps the most useful GUI control in BREW is the menu. At its most basic level, the menu is a list of items that are selectable by the user. Simply push the control pad up or down to find the item that you wish to select and push the select button to select the item. BREW menus can be used in a variety of ways to provide different methods of displaying and interacting with menu items.

Figure 10-3: A basic menu control in action

Creating a Menu Object

The simplest form of a BREW menu control is a simple list of selectable items. Each menu item has text and an optional bitmap image associated with it. They are displayed vertically down the screen. If more menu items exist than can fit on the display, the additional items can be scrolled through by moving the joypad. In learning how to use this basic menu, you will grasp the basics of all menu controls, as they all mainly differ in visual style. To create a menu, use ISHELL_CreateInstance, like this:

```
IMenuCtl * pIMenu;

ISHELL_CreateInstance(pApp->a.m_pIShell, AEECLSID_MENUCTL, (void **)&pIMenu);
```

This assumes that **pApp** is the applet class that we defined in our project header. Inside the **AEEApplet** structure is a pointer to the shell interface that is instantiated upon applet creation. The key here is we pass the class ID of the standard menu AEECLSID_MENUCTL, defined in AEEClassIDs.h, as the second argument. This tells BREW that we want to create a menu control. Finally, we need to pass a pointer to a pointer that will be modified to address the allocated chunk of memory assigned to our Menu object.

Setting the Menu Title

Now that we have a menu, how is it used? First, we need to set the title. Above the menu is a small blank bar. This bar can remain blank or be filled with the title of the menu. The actual text can be a wide-character string or a resource ID from our resource file. Here's the function IMENUCTL_SetTitle in detail:

```
boolean IMENUCTL_SetTitle(IMenuCtl * pIMenuCtl, const char * pszResFile, uint16 wResID,
    TCHAR * pText)
```

The first argument is the pointer to our actual menu control that we created with ISHELL_CreateInstance. With the next set of arguments, we have the choice of either giving the filename of our resource file and the resource ID of the title string or passing in an actual wide-character string.

Setting Various Menu Properties

Now that we have set the title, we need to actually define the dimensions of the menu. We can define the rectangle that the entire menu occupies on the screen so that we can display other images or even GUI controls on the same screen as our menu. Use the simple IMENUCTL_SetRect function:

```
void IMENUCTL_SetRect(IMenuCtl * pIMenuCtl, const AEERect * prc)
```

The first argument points to the menu control that we want to modify. The second pointer addresses an **AEERect** structure that defines the dimensions of the menu in screen coordinates. This is the very same **AEERect** used to actually draw rectangles, as we saw in the previous chapter.

Every menu control has individual properties that are modifiable to alter the look and behavior of the control. By using **IMENUCTL_SetProperties**, we can change these:

```
void IMENUCTL_SetProperties(ImenuCtl * pIMenuCtl, uint32 dwProps)
```

The only surprise here is the **dwProps** argument. This is a series of bit flags similar in nature to the ones used for text formatting. BREW often uses bit flags as a mechanism for determining attributes of all sorts of operations. These flags are defined in AEEMenu.h. If you look in that file, you will see properties that will do everything from underline the title to arrange the menu items like a calendar. In most cases, we can do without heavy customization of controls. However, it is a handy function to know about when creating advanced interfaces.

Adding Menu Items

The meat of the menu control comes in the actual menu items. These are the individual menu items that populate the control and are selectable by the user. The function **IMENUCTL_AddItem** covers this:

```
boolean  IMENUCTL_AddItem(IMenuCtl * pIMenuCtl, const char * pszResFile, uint16 wResID,
    uint16 nItemID, AECHAR * pText, uint32 lData)
```

This is another one of those functions that allows either text straight from a variable or strings that are pulled out of a resource file as arguments. In this case, after passing the pointer to the menu control we are working with, it is possible to pass in the name of the resource file that we are dealing with and the resource ID of the string or just the string itself as a wide-character array. The next argument, nItemID, is very important. This integer is the unique value by which we will identify this menu item as being selected in the message handler. The final argument, lData, is a 32-bit wide integer to use as you please. Any sort of information from simple numerical data to a pointer can be placed in this variable to be retrieved at a later time. It comes in handy when you want to associate a more complicated data structure with each menu item. Simply cast a pointer to an integer and recast it back to a pointer to your data later when you retrieve the selected menu item. However, it is optional and can be ignored in most cases. Each successive call to IMENUCTL_AddItem will create a new entry below the previous one.

Note: You cannot add multiple menu items that have the same item ID. When you try to add an item that has the same ID as an existing entry, the call will fail.

Using the Menu Messages

Because it is possible to have multiple controls on a single screen, BREW must be notified of which one has "focus." By focus I mean which control gets input from the user. The call IMENUCTL_SetActive does the trick:

```
void IMENUCTL_SetActive( IMenuCtl * pIMenuCtl)
```

By calling this function, all keypresses and user interactions will be directed to this control. How do you process these actions? All menu selections send a special event message to our message handler: EVT_COMMAND. When EVT_COMMAND is sent to the message handler, the handler's wParam argument contains the value of nItemID that we set in IMENUCTL_AddItem.

For example, say we created a menu item with the item ID of 1:

```
boolean IMENUCTL_AddItem(pMenuCtl, "resource.bar", ITEM_STRING, 1, NULL, 0)
```

As you can see, we create an item by pulling the string with the ID ITEM_STRING out of our resource file and give it an item ID of 1 with the fourth argument. In order for the menu control to send events to our applet, we need to call the menu's own event handler upon a keypress message. Now, in our event handler, we will have some code like this:

```
boolean HandleEvent(IApplet * pi, AEEEvent eCode, uint16 wParam, uint32 dwParam)
{
    myapp_t* pApp = (myapp_t*)pi;

switch (eCode)
```

```
{
case EVT_KEY_PRESS:
if (pApp->pMenu)
IMENUCTL_HandleEvent(pApp->pMenu);
        break;

        case EVT_COMMAND:
        switch(wParam)
        {
                case 1:
                //do some special stuff here
        }
}

    return (FALSE);
}
```

As you can see in this simple message handler, upon any keypress, we send the key message to the menu's own message handler. We are assuming our own applet structure has a pointer, **pMenu**, which points to a current menu item. If an item is selected, the menu's message handler will send an **EVT_COMMAND** message to the applet. In the handler, we use **wParam** to see if it is a selection upon the menu item that we want and act accordingly. This is the basis for all menu interaction.

Drawing the Menu

Finally, as with any control, you need to draw the control before any call to **IDISPLAY_Update**. The call follows the same form as any of the other control drawing functions:

```
boolean IMENUCTL_Redraw(IMenuCtl * pIMenuCtl)
```

Different Menu Styles

The menu control itself comes in many different forms. There are several different class IDs that you can create a menu object with that will create menus with drastically different interfaces and appearances. The class IDs are as follows:

AEECLSID_MENUCTL

This is the menu we detailed in the previous example. Pretty much the standard version, it is perhaps the most common kind of menu control used with BREW. Most of the pictures in this chapter up to this point have been of standard menu controls.

AEECLSID_LISTCTL

The list control only displays one item. This entry can be cycled with the joypad to move to different items in the list. The list control is useful when screen real estate is tight, as is often the case with a mobile device.

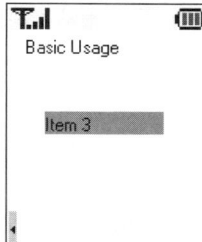

Figure 10-4: A list control from Qualcomm's menu usage example

AEECLSID_SOFTKEYCTL

The soft key control is similar to a regular menu, except it arranges each item horizontally instead of vertically.

Figure 10-5: A soft key control from Qualcomm's menu usage example

AEECLSID_ICONVIEWCTL

The icon view control uses a bitmap instead of text for each item in the menu. There is still a string associated with each menu item. However, this text is only displayed at the bottom of the screen for the currently highlighted element. This is the same control used in the emulator to browse applet icons.

Figure 10-6: An icon view control from Qualcomm's menu usage example

All of these different menu types behave in a similar fashion—items are added and removed the same way and messages are processed the same. Once you know how to use one menu type, you know them all.

Changing Menu Colors

Another way to customize the appearance of your menus is to change their color. BREW gives you a lot of flexibility in choosing the color for various properties such as the background, title text, and other elements. This is done through the function IMENUCTL_SetColors:

```
void IMENUCTL_SetColors(IMenuCtl * pIMenuCtl, AEEMenuColors * pc)
```

Obviously, the first argument is a pointer to our IMenu control interface. Next is a pointer to an AEEMenuColors structure. This structure contains the flags telling BREW which menu colors we are setting and the color values for each one. The structure is defined in AEEMenu.h like so:

```
typedef struct
{
    uint16      wMask;            // Mask of bits to pay attention to in struct...
    RGBVAL      cBack;
    RGBVAL      cText;
    RGBVAL      cSelBack;
    RGBVAL      cSelText;
    RGBVAL      cFrame;
    RGBVAL      cScrollbar;
    RGBVAL      cScrollbarFill;
    RGBVAL      cTitle;
    RGBVAL      cTitleText;
} AEEMenuColors;
```

The first member, wMask, contains bit flags telling BREW which colors we are going to change. The rest are a series of RGBVALs for each changeable item. An RGBVAL is defined in AEEDisp.h as an unsigned integer. The bit flags used in wMask are also defined in AEEDisp.h. The ones relevant to this function are described in the following table:

Table 10-1: Menu color flags

Flags	Description
MC_BACK	Unselected item background
MC_TEXT	Unselected item text
MC_SEL_BACK	Selected item background
MC_SEL_TEXT	Selected item text
MC_FRAME	Simple frame color
MC_SCROLLBAR	Scroll bar frame color

Flags	Description
MC_SCROLLBAR_FILL	Scroll bar fill color
MC_TITLE	Title background color
MC_TITLE_TEXT	Title text color

As you can see, there are quite a few different options to choose from. Be aware that not all of these options may be available; it all depends on whether the manufacturer of the phone supports them. The BREW GUI is not totally consistent, as each handset seems to have at least a few slight visual differences from each other—not to mention major functional changes such as the scroll bars on IStatic controls as discussed earlier.

In this case, we will change the title text color of a menu control. To do this, we must first set **wMask** to indicate this:

```
AEEMenuColors mc;
mc.wMask = MC_TITLE_TEXT;
```

Next, we need to set the RGB value of the **cTitleText** member to the appropriate color. In this case we will make it red:

```
mc.cTitleText = MAKE_RGB(255, 0, 0);
```

Here we use the handy macro **MAKE_RGB** to create an RGB color in the format BREW needs. **MAKE_RGB** is very simple. The macro takes three integer arguments—r, g, and b. These are 8-bit values from 0 to 255 for each corresponding color component. The result is an integer with all three components combined in a format that BREW can process. This macro can be used for any call that requires a color in the form of an integer.

Now that we have the color we want, we simply have to call **IMENUCTL_SetColors**:

```
void IMENUCTL_SetColors(IMenuCtl * pIMenuCtl, AEEMenuColors * pc)
```

After the pointer to our **IMenuCtl** interface, we then pass a pointer to the **AEEMenuColors** structure we previously set up. You might want to spend some time playing with various color schemes until you find one suitable for your applet.

Using Text Controls

The menu is great for simple user interaction. However, many times it is necessary to get more complicated input from the user. The text control gives the user the ability to type in text with the keypad for use with the applet. Of course, typing is very clumsy on a nine-digit keypad.

Therefore, text input should be kept to a minimum. However, BREW has some shortcuts, such as predictive text input, that make typing slightly less of a chore.

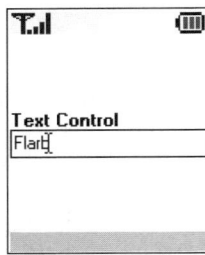

Figure 10-7: A text control with some text being entered in it

Creating a Text Control

Create a text control using the **AEECLSID_TEXTCTL** class ID:

```
ITextCtl * pIText;

ISHELL_CreateInstance(pApp->a.m_pIShell, AEECLSID_TEXTCTL, (void **)&pIText);
```

The first thing on the agenda is to set the size and location of the text control. As usual, this is done with **ITEXTCTL_SetRect**:

```
void ITEXTCTL_SetRect(ITextCtl * pITextCtl, const AEERect * prc)
```

The control will display as a window with a small frame around it for the entire span of the rectangle. When characters are typed, a cursor moves ahead as each character is deposited behind it. Hitting the clear button will delete the last character.

When more characters are typed than are visible in the control, the window will scroll ahead. It is possible through the control's properties to enable multi-line input. Therefore, if the rectangle is taller than one character of text, it is possible to enter multiple lines.

Getting the Text Out

Now that you have typed some text in, how do you read it? This is done by the **ITEXTCTL_GetText** function:

```
boolean ITEXTCTL_GetText(ITextCtl * pITextCtl, TCHAR * pBuffer, unsigned int nMaxChars)
```

The first argument is the pointer to the text control that we are using. The second is a pointer to a preallocated buffer of characters. Note that these are not wide characters. This is where the text will be copied after this call returns. The final argument, **nMaxChars**, is the size of the text buffer that we are passing to the function.

Handling Text Control Events

As with the menu control, the text control needs to be sent events from the message handler. Therefore, the text control has its own **ITEXTCTL _HandleEvent** function:

```
boolean ITEXTCTL_HandleEvent(ItextCtl * pITextCtl, AEEEvent evt, uint16 wp, uint32 dwp)
```

This is used in a similar fashion as the menu version. Put this function in your message handler to allow it to handle its own key input. Also, much like the menu control, the text control has to be set to an active state via **ITEXTCTL_SetActive** in order to receive any events:

```
void ITEXTCTL_SetActive(ItextCtl * pITextCtl, boolean bActive)
```

Changing the Text Control

The text control has a few different properties to change its interface and appearance. As with all other controls, using **ITEXTCTL_Set-Properties** allows its few attributes to be set:

```
void ITEXTCTL_SetProperties(ITextCtl * pITextCtl, uint32 dwProps)
```

The text control properties involve such attributes as multi-line editing and predictive text input.

Using Time Controls

BREW also contains controls for the convenient display and input of time. Although not necessarily a common element in game interfaces, you might find it handy in the future. This works in a similar manner to other controls.

Figure 10-8: A time control in action

Creating a Time Control

There are actually three different kinds of time controls accessible through the **ITimeCtl** interface. Each one has a different class ID, but all use the same interface type. The three kinds of controls are a simple clock display, a stopwatch display, and a countdown timer. To create a clock, we use the **AEECLSID_CLOCKCTL** class ID:

```
ITimeCtl * pITime;

ISHELL_CreateInstance(pApp->a.m_pIShell, AEECLSID_CLOCKCTL, (void **)&pITime);
```

To create a stopwatch, we use **AEECLSID_STOPWATCHCTL**:

```
ITimeCtl * pITime;

ISHELL_CreateInstance(pApp->a.m_pIShell, AEECLSID_STOPWATCHCTL, (void **)&pITime);
```

Finally, to create a countdown timer, we use **AEECLSID_COUNT-DOWNCTL**:

```
ITimeCtl * pITime;

ISHELL_CreateInstance(pApp->a.m_pIShell, AEECLSID_COUNTDOWNCTL, (void **)&pITime);
```

Setting and Getting the Time

Now that we have the control, how do we set or get the time? To set the current time of the control, use **ITIMECTL_SetTime**:

```
void ITIMECTL_SetTime(ITimeCtl * pITimeCtl, int32 tod)
```

This function sets the time and redraws the control. After passing the pointer to the time control that we are updating, pass the current time in milliseconds via the **tod** argument.

Now, if we wish to get the current time of the control, we would use **ITIMECTL_GetTime**:

```
int32 ITIMECTL_GetTime(ITimeCtl * pITimeCtl)
```

Predictably, this returns the time in milliseconds from the time control pointed to by the **pITimeCtl** argument.

Handling Time Events

A time control receives events much like a text control does in the process of accepting user input. In this case, the user can use the time control to set the current time. To handle these events, the timer has its own function, much like the rest of BREW's GUI controls:

```
boolean ITIMECTL_HandleEvent(ITimeCtl * pITimeCtl, AEEEvent evt, uint16 wp, uint32 dwp)
```

The time control also has its own standard property and release functions that work much in the same way as every other BREW GUI control.

Using Date Controls

BREW also has controls for the display and input of dates. They work in a similar fashion to the time controls in that there are a few different types. There is the standard date control that shows the month, date, and year, and then there is the date picker control, which shows a monthly calendar instead. Each control uses the same IDateCtl interface but are constructed using different class IDs.

Figure 10-9: A date control

Creating a Date Control

To construct a standard date control, use the class ID AEECLSID_DATECTL:

```
IDateCtl * pIDate;

ISHELL_CreateInstance(pApp->a.m_pIShell, AEECLSID_DATECTL, (void **)&pIDate);
```

To create a date picker that has a monthly display, use AEECLSID_DATEPICKCTL:

```
IDateCtl * pIDate;

ISHELL_CreateInstance(pApp->a.m_pIShell, AEECLSID_DATEPICKCTL, (void **)&pIDate);
```

Setting and Getting the Date

Much like the time control, we can set and get the date information out of a date control with a few simple functions. To set the date, use IDATECTL_SetDate:

```
boolean IDATECTL_SetDate(IDateCtl * pIDateCtl, unsigned int nYear, unsigned int nMonth,
    unsigned int nDay)
```

After passing a pointer to the date control that we wish to change, pass the year, month, and day in the form of integers. The year has to be specified in four digits, as in 2002. The month and day are both in two-digit format.

To get the data, use the function IDATECTL_GetDate:

```
boolean IDATECTL_GetDate(IDateCtl * pIDateCtl, unsigned int * pnYear, unsigned int *
    pnMonth, unsigned int * pnDay)
```

The argument structure is similar to that of **IDATECTL_SetDate**, except we pass pointers to integers for the month, day, and year (the integers that these pointers address are filled upon return of this function).

Handling Date Events

As with the other controls, the date control has its own message handler:

```
boolean IDATECTL_HandleEvent(IDateCtl * pIDateCtl, AEEEvent evt, uint16 wp, uint32 dwp)
```

This works like every other event handler, allowing the control to process user input.

Using Dialogs

Now that we know how to use the basic BREW GUI components, it is time to learn how to pull these out of the resource files. I find that for most purposes, simply creating and positioning the control in code is sufficient. However, using resource files to store GUI layouts makes it much easier to manage projects that have complicated interfaces. Granted, your interfaces should remain as simple as possible on such a limited device, but sometimes this is inevitable.

The resource file introduces the concept of a dialog to the BREW GUI. Previous to this, you learned about controls. The controls are the individual parts of the GUI that make an entire interface. A dialog contains multiple controls to create a complete interface. By using the Resource Editor's dialog facilities, it is possible to set the position and contents of entire screens of GUI components. Handling the events and such are the same; however, this frees the programmer from the mundane chores of creating and setting up the positions and initial states of GUI components. It also allows non-programmers to go into the resource file and alter the properties of the interface without having to recompile the project.

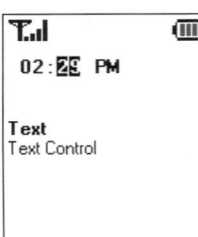

Figure 10-10: A dialog with two controls. The one at the top is a time control, and the bottom one is a text control.

Dialog Creation

You saw how to create dialogs in a resource file in Chapter 7, "Using Resources." In the examples for this section we will use a precreated dialog in the example applet Source\Chapter 10\guitest. Assuming we have a dialog ready to go in a resource file, there are a number of steps that you must take before displaying it on the screen.

The first function we need to use is ISHELL_CreateDialog:

```
int ISHELL_CreateDialog(IShell * pIShell, const char * pszResFile, uint16 wID,
    DialogInfo * pInfo)
```

After the interface pointer, pass the name of the resource file that the dialog is stored in with **pszResFile**. Then pass the resource ID as defined in the associated resource file's header via **wID**. Alternatively, we can pass a **DialogInfo** structure in the final argument, **pInfo**, to generate a dialog algorithmically instead of using entries in the resource file. A simple example of creating a dialog out of a resource file looks like this:

```
ISHELL_CreateDialog(pApp->a.m_pIShell, "resource.bar", DLG_TEST, NULL);
```

This is similar to the call we use in our example applet for this chapter. We simply provide the resource file and resource ID with no final argument.

If you insist on creating dialogs in code, you have a lot of information to provide via **DialogInfo** for the final argument. The **DialogInfo** structure looks like this, as defined in AEEShell.h:

```
typedef struct
{
    DialogInfoHead h;
    DialogItem      controls[1];
} DialogInfo;
```

The first member is a **DialogInfoHead** structure. This structure is also defined in AEEShell.h like so:

```
typedef struct
{
    uint16      wID;            // Dialog ID
    uint16      nControls;      // Number of controls
    AEERect     rc;             // Rect (-1, -1, -1, -1) is default
    uint32      dwProps;        // See IControl
    uint16      wTitle;         // Title
    uint16      wFocusID;       // ID of focus control
} DialogInfoHead;
```

All of these member variables function exactly as they do in the Resource Editor. Through this structure, you can specify the size, title, and other parameters of your custom dialog.

After the header info, the DialogInfo structure has a single element array of DialogItem structures. This is actually a variable length array, as declaring something as a one-element array is essentially the same as declaring a pointer to an arbitrary size of memory. A DialogItem is a data structure that is used to describe a GUI control that is included in the dialog. Each DialogItem structure looks like, this as defined in AEEShell.h:

```
typedef struct
{
    DialogItemHead h;
    DListItem      items[1];
} DialogItem;
```

Look familiar? The DialogItem structure works in a similar manner to DialogInfo, with a header describing the control and then an array of items for each individual element. The DialogItemHead structure is defined in AEEShell.h like this:

```
typedef struct
{
    AEECLSID   cls;        // class of the control
    uint16     wID;        // ID of the control
    uint16     nItems;     // Item count
    uint32     dwProps;    // See IControl (or specific control)
    uint16     wTextID;    // ID of initial text
    uint16     wTitleID;   // ID of initial title
    AEERect    rc;         // Rect - Relative to Dialog (-1, -1, -1, -1) is default
} DialogItemHead;
```

With this structure, you define the kind of GUI control this is by using the class ID you normally would use when creating the control classes in code. You still have to provide resource IDs of strings for title and text values. However, these can be changed later, so the entire dialog can still be generated algorithmically.

For GUI controls that list items, you need an array of items for each element in the list. For instance, with a menu control, you are going to need an item for each selection in the menu. That is what the DListItem structure is for. It is defined in AEEShell.h like this:

```
typedef struct _DListItem
{
    uint16     wID;        // ID of the item
    uint16     wTextID;    // Text
    uint16     wIconID;    // Icon ID
    uint16     pad;
    uint32     dwData;     // 32-bit data
} DListItem;
```

Here you can define the contents of each list item by specifying the resource ID of the text and any other attributes necessary. Obviously, using a resource is much more convenient; however, in some cases you may need to create dialogs on the fly by using the DialogInfo structure.

Once the dialog is created, you can get a pointer to it by calling ISHELL_GetActiveDialog:

```
IDialog * ISHELL_GetActiveDialog(IShell * pIShell)
```

This function returns a pointer to the dialog that was created via ISHELL_CreateDialog.

Dialog Event Handling

There are several messages that your applet will receive upon creation of a dialog. If these messages are not handled properly, the call to ISHELL_CreateDialog will fail and return an error code. These events are EVT_DIALOG_START, EVT_DIALOG_INIT, and EVT_DIALOG_ END.

When ISHELL_CreateDialog is called, an EVT_DIALOG_START message is immediately posted to the message handler. As you recall, the function prototype for a BREW applet message handler looks like this:

```
static boolean APPLET_HandleEvent(IApplet * pi, AEEEvent eCode, uint16 wParam, uint32
    dwParam);
```

The final two arguments, wParam and dwParam, often contain additional information to be used in conjunction with the event handler. In the case of an EVT_DIALOG_START event message (passed in via the eCode argument), the wParam argument is the ID of the dialog, as set in the resource file or DialogInfo structure, and dwParam is a pointer to the interface object of the created dialog. Because dwParam is defined as an int, this parameter has to be cast to an IDialog pointer like so:

```
IDialog * pDlg = (IDialog*)dwParam;
```

Now pDlg points to the dialog being created as a result of the ISHELL_CreateDialog call. Once you have done all of your dialog startup tasks, you must return a value of True to let BREW know this message has been successfully handled.

The second message that you must handle is EVT_DIALOG_INIT. This event is sent after EVT_DIALOG_START and is typically used to initialize dialog values, such as the contents of any text controls and such. After processing this message in the handler code, you must again return True to let BREW know that it has been dealt with. Even if you do not do anything for dialog initialization, a simple return statement in the case of EVT_DIALOG_INIT will do. The arguments for the applet message handler are used in the same way as EVT_DIALOG_START, with wParam being the dialog resource file ID and dwParam being the pointer to the IDialog interface. Of course, dwParam has to be cast to a pointer before usage.

Finally, the EVT_DIALOG_END message is sent when a dialog is closed. Here you can deallocate any memory that you may have acquired when using the given dialog. Once again, the wParam and dwParam arguments contain the dialog resource file ID and pointer to the IDialog interface, so you can tell which dialog is being shut down.

These three events manage the entire life cycle of a dialog from creation to initialization and then finally to destruction. If you do not at least return True for each of these events in your handler function, BREW will simply return an error from ISHELL_CreateDialog instead of actually creating a dialog interface.

Dialog Control Event Handling

Now that we have gone over the messages that the actual applet needs to handle in order to properly manage a dialog, we need to discuss how you manage the messages that each GUI control sends and receives.

At the most basic level, the control events for the GUI components inside a dialog are no different from those for GUI controls that you create in code. Therefore, a menu control will still send EVT_COMMAND messages to your applet's main message loop for you to handle.

As you can imagine, having a lot of different dialogs will mean that you will have to have a huge event handler function that catches all the messages generated from every control in each dialog. Conveniently, BREW allows the creation of custom message handlers for each dialog. Therefore, you can have a different message handler function for each dialog instead of having to deal with every control's messages in your main applet handler code. The amount of code is the same, but it is a bit more organized this way. This is done through the function IDIALOG_SetEventHandler:

```
void IDIALOG_SetEventHandler (IDialog * pIDialog, PFNAEEEVENT pfn, void * pUser)
```

This function takes, of course, a pointer to the dialog with which we are setting the handler in the pIDialog argument. The second argument, pfn, is a pointer to our message handler function. Finally, pUser is a pointer to any chunk of data that we wish to send to our handler. In my case, I like to send a pointer to the applet so that the handler can access global data stored in the applet structure.

The message handler function is defined with this prototype in AEEShell.h:

```
typedef boolean (*PFNAEEEVENT)(void * pUser, AEEEvent evt, uint16 w, uint32 dw);
```

If this looks familiar, it should. It is basically the same thing as the AEEHANDLER function prototype that we pass to AEEApplet_New as our applet message handler. Therefore, it works in the same way as the applet message handler, except in this case, you only need to be concerned with messages directed to the dialog. Usually, this means GUI control events, keypresses, and the like.

Control Focus

So we now know how to handle events sent by dialog controls. How do we determine which control in a dialog gets user input in the first place? We do this by setting control focus. A control that has focus is a control that is currently receiving user input. We can see which control in a dialog has focus by using **IDIALOG_GetFocus**:

```
int16 IDIALOG_GetFocus(IDialog * pIDialog, int16 wID)
```

Here we pass a pointer to the dialog that contains the control that we want to give focus to and the resource ID of the control itself.

Drawing the Dialog

Now that we know how to create a dialog and handle dialog control events, how do we draw it? Good question. There is no redraw function for a dialog as there is for individual controls. This means that you have to pull out each GUI control in the dialog and redraw it manually when you want to draw the dialog. The dialog itself has no graphical representation. Dialogs are, instead, a collection of individual GUI components that must be managed separately.

How do you get access to the dialog's GUI components? Use the function **IDIALOG_GetControl**:

```
IControl * IDIALOG_GetControl(IDialog * pIDialog, int16 wID)
```

Pass in the pointer to the dialog that you want to access with **pIDialog** and the resource ID of the control that you want. As usual, the resource ID is defined in the header file for the resource file in which the dialog is stored. The function returns a pointer to the base control class that belongs to the GUI component that you retrieve. It is up to you to cast this **IControl** pointer to a pointer to the type of control that you are accessing. For instance, if you are trying to pull a static control out of your dialog, you might do something like this:

```
IControl * pCon = IDIALOG_GetControl(pDlg, CTRL_STATIC);
IStatic * pStatic = (IStatic *)pCon;
```

Now you can use the static interface pointer to set the text or do whatever operations are allowed on a static text control. This is also how you would redraw each control contained in a dialog. Call **IDIALOG_GetControl** on each one, and then call the appropriate redraw function in each control's particular interface.

Destroying the Dialog

To release the memory associated with the currently active dialog, we need to call ISHELL_EndDialog:

```
void ISHELL_EndDialog(IShell * pIShell)
```

Notice that there is no pointer to a dialog here. You can only have one currently active dialog at a time. You must end a dialog before you create a new one. You can keep the data structures around or have multiple dialogs in a resource file, but to display the dialog, you must call ISHELL_CreateDialog. Call ISHELL_EndDialog to destroy it before creating another one. When this function is called, an EVT_DIALOG_STOP message is posted to the applet's message handler, as illustrated previously.

A Simple Example

The companion files include an example called guitest. This demonstrates the basic functionality of each control discussed in this chapter. Let's start examining the code by looking at our applet structure, as defined in guitest.h:

Listing IO-I: The applet structure

```
typedef struct myapp_s
{
    AEEApplet      a;       //applet header

    AEEDeviceInfo di;

    boolean bDialog;

    IMenuCtl    * pIMenu;
    IStatic     * pIStatic;
    ITextCtl    * pIText;
    IDateCtl    * pIDate;
    ITimeCtl    * pITime;
    IDialog     * pIDialog;

    AECHAR szTextBuf[256];

} myapp_t;
```

Here you see the standard AEEApplet structure included at the start with the DeviceInfo following it to record the handset capabilities. Next we have a Boolean that we use to determine if a dialog is currently being displayed instead of an individual control. We then have a series of pointers to GUI control interfaces, which are used to manage the individual components on the screen whether in a dialog or standalone.

Finally, we have a simple wide-character text buffer used for printing text.

Looking at the code in guitest.c, we see that the applet construction is fairly standard, with the familiar applet message handler and **CleanUp** function passed to **AEEApplet_NEW**:

Listing 10-2: Creating the applet instance

```
int AEEClsCreateInstance(AEECLSID ClsId, IShell * pIShell, IModule * po, void ** ppObj)
{
    *ppObj = NULL;

    if(ClsId == AEECLSID_GUITEST_BID )
    {
        if(AEEApplet_New(sizeof(myapp_t), ClsId, pIShell, po, (IApplet**)ppObj,
                (AEEHANDLER)GUITEST_HandleEvent, (PFNFREEAPPDATA)GUITEST_CleanUp)
                == TRUE)
        {
            return (AEE_SUCCESS);
        }
    }

    return (EFAILED);
}
```

The applet message handler is a bit of a monster, as it has to handle not only applet messages but messages from each individual control. We will take a look at it in its entirety and then break it down by message:

Listing 10-3: The message handler

```
static boolean GUITEST_HandleEvent(IApplet * pi, AEEEvent eCode, uint16 wParam,
    uint32 dwParam)
{
    ITimeCtl * pClock;
    ITextCtl * pText;
    myapp_t * pApp = (myapp_t *)pi;
    AEEApplet * pMe = &pApp->a;

    switch (eCode)
    {
        case EVT_APP_START:

            ISHELL_GetDeviceInfo (pMe->m_pIShell, &pApp->di);
            IDISPLAY_ClearScreen (pMe->m_pIDisplay);

            pApp->pIDate    = NULL;
            pApp->pIMenu    = NULL;
            pApp->pIStatic  = NULL;
            pApp->pIText    = NULL;
            pApp->pITime    = NULL;
            pApp->bDialog   = FALSE;

            GUITEST_BuildMainMenu(pApp);
            IDISPLAY_Update (pMe->m_pIDisplay);

            return(TRUE);
```

```
            break;

        case EVT_COMMAND:
        switch(wParam)
        {
            case MENUID_STATIC:
                GUITEST_BuildStaticControl(pApp);
                break;

            case MENUID_TEXT:
                GUITEST_BuildTextControl(pApp);
                break;

            case MENUID_DATE:
                GUITEST_BuildDateControl(pApp);
                break;

            case MENUID_TIME:
                GUITEST_BuildTimeControl(pApp);
                break;

            case MENUID_DIALOG:
                GUITEST_BuildDialogScreen(pApp);
                break;
        }

            return(TRUE);
            break;

        case EVT_KEY_PRESS:
            if (pApp->pIMenu != NULL)
                IMENUCTL_HandleEvent(pApp->pIMenu, EVT_KEY, wParam, 0);
            else if (pApp->pIText != NULL)
            {
                ITEXTCTL_HandleEvent(pApp->pIText, EVT_KEY, wParam, 0);
                ITEXTCTL_Redraw(pApp->pIText);
            }
            else if (pApp->pIDate != NULL)
            {
                IDATECTL_HandleEvent(pApp->pIDate, EVT_KEY, wParam, 0);
                IDATECTL_Redraw(pApp->pIDate);
            }
            else if (pApp->pITime != NULL)
            {
                ITIMECTL_HandleEvent(pApp->pITime, EVT_KEY, wParam, 0);
                ITIMECTL_Redraw(pApp->pITime);
            }

            return(TRUE);
            break;

        case EVT_DIALOG_START:

            IDISPLAY_ClearScreen(pApp->a.m_pIDisplay);
            pApp->pIDialog = (IDialog*)dwParam;
            IDIALOG_SetEventHandler(pApp->pIDialog,
                    (PFNAEEEVENT)GUITEST_DialogHandleEvent, (void*)pApp);
            pClock = (ITimeCtl*)IDIALOG_GetControl(pApp->pIDialog, CTL_CLOCK);
```

```
            pText = (ITextCtl*)IDIALOG_GetControl(pApp->pIDialog, CTL_TEXT);

            ITEXTCTL_Redraw(pText);
            ITIMECTL_Redraw(pClock);
                IDISPLAY_Update(pApp->a.m_pIDisplay);

                return(TRUE);
                break;

        case EVT_DIALOG_INIT:
                return(TRUE);
                break;

        case EVT_DIALOG_END:
                return(TRUE);
                break;
        }

    return(FALSE);
}
```

The first message that we will concern ourselves with is the **AEE_
APP_START** case in our message switch statement:

```
case EVT_APP_START:

            ISHELL_GetDeviceInfo (pMe->m_pIShell, &pApp->di);
            IDISPLAY_ClearScreen (pMe->m_pIDisplay);

            pApp->pIDate     = NULL;
            pApp->pIMenu     = NULL;
            pApp->pIStatic   = NULL;
            pApp->pIText     = NULL;
            pApp->pITime     = NULL;

            pApp->bDialog = FALSE;

            GUITEST_BuildMainMenu(pApp);
            IDISPLAY_Update (pMe->m_pIDisplay);

            return(TRUE);
                break;
```

As with many of our other applets, we initialize some variables, get the
device information, and clear the screen to start. We will use the handset screen resolution as recorded in the **DeviceInfo** structure to
properly size GUI controls later on.

The interesting function here is GUITEST_BuildMainMenu. If you
run this applet in the emulator, the first thing visible is a menu of GUI
objects. This menu itself is a basic menu control, as detailed earlier in
this chapter. The GUITEST_BuildMainMenu function sets up this menu
and displays it on the screen. In fact, each GUI control has a similar
build function associated with it. Looking at GUITEST_BuildMainMenu,
we see:

Listing IO-4: The main menu creation function

```
void GUITEST_BuildMainMenu(myapp_t* pApp)
{
    //create the menu
    ISHELL_CreateInstance(pApp->a.m_pIShell,
    AEECLSID_MENUCTL, (void **)&pApp->pIMenu);

    //set up the menu
    IMENUCTL_SetTitle(pApp->pIMenu, RES_FILE, STR_MENUTITLE, NULL);
    IMENUCTL_SetRect(pApp->pIMenu, NULL); //full-screen

    //Add in our menu items
    IMENUCTL_AddItem(pApp->pIMenu, RES_FILE, STR_MENUSTATIC,
                                        MENUID_STATIC, NULL, 0);
    IMENUCTL_AddItem(pApp->pIMenu, RES_FILE, STR_MENUTEXT,
                                        MENUID_TEXT, NULL, 0);
    IMENUCTL_AddItem(pApp->pIMenu, RES_FILE, STR_MENUDATE,
                                        MENUID_DATE, NULL, 0);
    IMENUCTL_AddItem(pApp->pIMenu, RES_FILE, STR_MENUTIME,
                                        MENUID_TIME, NULL, 0);
    IMENUCTL_AddItem(pApp->pIMenu, RES_FILE, STR_MENUDIALOG,
                                        MENUID_DIALOG, NULL, 0);

    IMENUCTL_SetActive(pApp->pIMenu,TRUE);
}
```

The first thing the function does is create an instance of the **IMenuCtl** interface with the **ISHELL_CreateInstance** call. Next, we set the title text displayed at the top of the menu by pulling it out of the resource file with **IMENUTL_SetTitle**. We also set the size of the menu control to be that of the entire screen by passing a NULL pointer to **IMENU_SetRect**. By not sending a valid pointer to an **AEERect**, BREW assumes the control is to take up the entire width and height of the handset's screen.

The biggest chunk of code in this function is the addition of the individual menu items. This is done with the familiar **IMENUCTL_AddItem** control. Each menu item has its text pulled out of the resource file using the definition in the guitest_res.h header file. The item IDs are defined in guitest.h, like this:

```
enum
{
    MENUID_STATIC = 0,
    MENUID_TEXT,
    MENUID_DATE,
    MENUID_TIME,
    MENUID_DIALOG,
};
```

Using an enumeration instead of a bunch of defines is a handy way to create consecutive values. In this case, it also makes sure that no two item IDs are the same. When adding two menu items with the same item ID, the call to **IMENUCTL_AddItem** will fail and the item with the identical value will not be added. So you must make sure that every control has unique menu IDs.

Finally, we set the menu to be the active control with **IMENUCTL_SetActive**. Now this menu has focus and is ready to accept user selections.

Getting back to the message handler, we have the **EVT_COMMAND** case that catches all menu item selection events. This part of the message handler looks like this:

```
case EVT_COMMAND:
    switch(wParam)
    {
            case MENUID_STATIC:
                GUITEST_BuildStaticControl(pApp);
                break;

            case MENUID_TEXT:
                GUITEST_BuildTextControl(pApp);
                break;

            case MENUID_DATE:
                GUITEST_BuildDateControl(pApp);
                break;

            case MENUID_TIME:
                GUITEST_BuildTimeControl(pApp);
                break;

            case MENUID_DIALOG:
                GUITEST_BuildDialogScreen(pApp);
                break;
    }

            return(TRUE);
            break;
```

When we get an **EVT_COMMAND** message, the **wParam** argument to the message handler contains the menu ID of the selected menu item that we defined when calling **IMENUCTL_AddItem**. Each one of these menu items is defined in guitest.h as detailed earlier. Each menu item calls the appropriate build function that creates their respective GUI component. We will take them in order, starting with GUITEST_BuildStaticControl:

Listing 10-5:

```
void GUITEST_BuildStaticControl(myapp_t* pApp)
{
    AEERect qrc;
    AECHAR szTitle[32];

    GUITEST_ClearGUI(pApp);

    SHELL_CreateInstance(pApp->a.m_pIShell,
```

```
        AEECLSID_STATIC, (void **)&pApp->pIStatic);

    //load up our text in a global buffer
    ISHELL_LoadResString(pApp->a.m_pIShell, RES_FILE,
        STR_STATICCONTENT, pApp->szTextBuf, sizeof(pApp->szTextBuf));

    //load up title text in our title character array
    ISHELL_LoadResString(pApp->a.m_pIShell, RES_FILE,
        STR_STATICTITLE, szTitle, sizeof(szTitle));

    //make dimensions of the control take up the entire screen
    qrc.x     = 0;
    qrc.y     = 0;
    qrc.dx    = pApp->di.cxScreen;
    qrc.dy    = pApp->di.cyScreen;

    ISTATIC_SetRect(pApp->pIStatic, &qrc); //full screen

    ISTATIC_SetText(pApp->pIStatic, szTitle,
        pApp->szTextBuf, AEE_FONT_BOLD, AEE_FONT_NORMAL);

    //draw control
    ISTATIC_Redraw(pApp->pIStatic);
}
```

This function builds a static text control that displays text stored in the resource file. First call GUITEST_ClearGUI, a function that releases all GUI objects pointed to by the pointers defined in the applet structure.

After creating an instance of the IStatic interface, set its title and contents to strings pulled out of the resource file. Unfortunately, text can only be sent to the control by sending a pointer to a character buffer. Instead of passing resource IDs, we have to load the strings out of the resource file into our character buffers to later be sent to the control as title and content text. Using ISHELL_LoadResString, pull the title text string out of the resource file and into our smaller stack-allocated array of characters, szTitle. Using that same call, put the contents text in our larger character buffer, szTextBuf, defined in the applet structure.

Next, fill the rectangle structure, qrc, with the dimensions of the screen read out of the DeviceInfo structure that we filled at the start of the applet. Pass the address of this structure to ISTATIC_SetRect, which makes the static text control take up the size of the entire screen. Our subsequent call to ISTATIC_SetText fills the title and contents of the text control with the strings that we pulled out of the resource file earlier. Finally, redraw the control with ISTATIC_Redraw.

Taking a quick look at GUITEST_ClearGUI, we see this code:

Listing 10-6: Destroying the GUI

```
void GUITEST_ClearGUI(myapp_t* pApp)
{
    ISHELL_EndDialog(pApp->a.m_pIShell);
```

```
    if (pApp->pIMenu)
    {
        IMENUCTL_Release(pApp->pIMenu);
        pApp->pIMenu = NULL;
    }

if (pApp->pIStatic)
{
    ISTATIC_Release(pApp->pIStatic);
    pApp->pIStatic = NULL;
}

if (pApp->pIText)
{
    ITEXTCTL_Release(pApp->pIText);
    pApp->pIText = NULL;
}

if (pApp->pIDate)
{
    IDATECTL_Release(pApp->pIDate);
    pApp->pIDate = NULL;
}

if (pApp->pITime)
{
    ITIMECTL_Release(pApp->pITime);
    pApp->pITime = NULL;
}
}
```

Basically, if any GUI control pointer is valid in our applet structure, release the interface and NULL it out. This ensures that the memory used up by the control is free. Call ISHELL_EndDialog to destroy any dialog that may be currently active.

Continuing through our GUI builder functions, we come across GUITEST_BuildTextControl:

Listing l0-7: Building the text entry screen

```
void GUITEST_BuildTextControl(myapp_t* pApp)
{
    AEERect qrc;

    GUITEST_ClearGUI(pApp);

    ISHELL_CreateInstance(pApp->a.m_pIShell,
        AEECLSID_TEXTCTL, (void **)&pApp->pIText);

    qrc.x     = 0;
    qrc.y     = 50;
    qrc.dx    = pApp->di.cxScreen;
    qrc.dy    = qrc.y + 15;

    ITEXTCTL_SetRect(pApp->pIText, &qrc);
    ITEXTCTL_SetTitle(pApp->pIText, RES_FILE, STR_TEXTTITLE, NULL);

    ITEXTCTL_SetActive(pApp->pIText, TRUE);
```

```
    ITEXTCTL_Redraw(pApp->pIText);
}
```

This function builds a simple text input box that allows the user to input any text she wants. Of course, begin with a call to GUITEST_ClearGUI to remove any GUI components that may already be active. Next, construct the ITextCtl interface with a call to ISHELL_CreateInstance.

Next, define the region of the screen the control will occupy. In this case, 15 pixels of height should be enough to display an entire character. The control spans the entire width of the screen by taking the device's resolution into account.

Finally, set the title of the text control from a string stored in the resource file, give the control focus by calling ITEXTCTL_SetActive, and draw it on the screen with ITEXTCTL_Redraw.

Moving right along, build the date control with GUITEST_BuildDateControl:

Listing IO-8: Building the date control

```
void GUITEST_BuildDateControl(myapp_t* pApp)
{
    AEERect qrc;

    GUITEST_ClearGUI(pApp);

    ISHELL_CreateInstance(pApp->a.m_pIShell,
        AEECLSID_DATECTL, (void **)&pApp->pIDate);

    qrc.x  = 0;
    qrc.y  = 0;
    qrc.dx = pApp->di.cxScreen;
    qrc.dy = pApp->di.cyScreen;

    IDATECTL_SetRect(pApp->pIDate, &qrc);
    IDATECTL_SetActive(pApp->pIDate, TRUE);
    IDATECTL_Redraw(pApp->pIDate);
}
```

After looking at a few of these functions, this should be largely familiar to you. Clear the previous GUI components, create the instance, set the rectangle, give the control focus, and draw it. The function to build the time control, GUITEST_BuildTimeControl, is the same:

Listing IO-9: Building the time control

```
void GUITEST_BuildTimeControl(myapp_t* pApp)
{
    AEERect qrc;

    GUITEST_ClearGUI(pApp);

    ISHELL_CreateInstance(pApp->a.m_pIShell,
            AEECLSID_CLOCKCTL, (void **)&pApp->pITime);
```

```
qrc.x  = 0;
qrc.y  = 0;
qrc.dx = pApp->di.cxScreen;
qrc.dy = pApp->di.cyScreen;

ITIMECTL_SetRect(pApp->pITime, &qrc);
ITIMECTL_SetActive(pApp->pITime, TRUE);
ITIMECTL_Redraw(pApp->pITime);
}
```

Now we come to building the dialog. The actual dialog build function is quite simple:

```
void GUITEST_BuildDialogScreen(myapp_t* pApp)
{
    GUITEST_ClearGUI(pApp);

    pApp->bDialog = TRUE;

    ISHELL_CreateDialog(pApp->a.m_pIShell, RES_FILE, DLG_TEST, NULL);
}
```

Because we are pulling the dialog out of the resource file, there is not much code to create it. However, in the obligatory handler code for **EVT_DIALOG_START**, we have some setup code. Referencing the message handler in Listing 10-3, we see the handler code for all three dialog messages:

```
case EVT_DIALOG_START:

        IDISPLAY_ClearScreen(pApp->a.m_pIDisplay);
        pApp->pIDialog = (IDialog*)dwParam;
        IDIALOG_SetEventHandler(pApp->pIDialog,
            (PFNAEEEVENT)GUITEST_DialogHandleEvent, (void*)pApp);

        pClock = (ITimeCtl*)IDIALOG_GetControl(pApp->pIDialog, CTL_CLOCK);
        pText = (ITextCtl*)IDIALOG_GetControl(pApp->pIDialog, CTL_TEXT);

        ITEXTCTL_Redraw(pText);
        ITIMECTL_Redraw(pClock);

        IDISPLAY_Update(pApp->a.m_pIDisplay);

        return(TRUE);
        break;

    case EVT_DIALOG_INIT:
        return(TRUE);
        break;

    case EVT_DIALOG_END:
        return(TRUE);
        break;
```

Do nothing other than return True in **EVT_DIALOG_INIT** and **EVT_ DIALOG_END**. However, **EVT_DIALOG_START** is another story. Here you want to clear the screen, cast the dialog pointer out of the **dwParam**

argument, and then assign a custom message handler to the dialog.
Then pull out the two controls in this dialog and redraw each one. Even
though these controls are contained inside a dialog, we still have to
draw them individually. Looking at the event handler, GUITEST_
DialogHandleEvent, we can see how to deal with them on an individual
basis:

Listing 10-10: The dialog event handler

```
boolean GUITEST_DialogHandleEvent(void * pUser, AEEEvent evt, uint16 w, uint32 dw)
{
    ITimeCtl * pClock;
    ITextCtl * pText;
    IDialog * pDialog    = NULL;
    myapp_t * pApp       = (myapp_t *)pUser;

    pClock = (ITimeCtl*)IDIALOG_GetControl(pApp->pIDialog, CTL_CLOCK);
    pText  = (ITextCtl*)IDIALOG_GetControl(pApp->pIDialog, CTL_TEXT);

    switch (evt)
    {

    case EVT_KEY_PRESS:
        ITEXTCTL_HandleEvent(pText, EVT_KEY, w, 0);
        ITEXTCTL_Redraw(pText);

        ITIMECTL_HandleEvent(pClock, EVT_KEY, w, 0);
        ITIMECTL_Redraw(pClock);

        return(TRUE);
        break;

    case EVT_COMMAND:
        switch (w)
        {
            case CTL_CLOCK:
                ITIMECTL_HandleEvent(pClock, EVT_KEY, w, 0);
                ITIMECTL_Redraw(pClock);
                break;

            case CTL_TEXT:
                ITEXTCTL_HandleEvent(pText, EVT_KEY, w, 0);
                ITEXTCTL_Redraw(pText);
                break;

        }

            return(TRUE);
            break;
    }

        return(FALSE);
}
```

This function behaves in a similar manner to the applet message handler when dealing with control events. Because this applet does nothing

special with the data inside the controls, we simply pull each component out using **IDIALOG_GetControl** and redraw them if we get an **EVT_ KEYPRESS** or **EVT_COMMAND** message. We also call the controls' message handlers so that they can process their own input.

Looking at the message handler from Listing 10-3, you will see similar code for the stand-alone controls constructed in all of those build functions:

```
case EVT_COMMAND:
    switch(wParam)
    {
        case MENUID_STATIC:
            GUITEST_BuildStaticControl(pApp);
            break;

        case MENUID_TEXT:
            GUITEST_BuildTextControl(pApp);
                break;

        case MENUID_DATE:
            GUITEST_BuildDateControl(pApp);
            break;

        case MENUID_TIME:
            GUITEST_BuildTimeControl(pApp);
            break;

        case MENUID_DIALOG:
            GUITEST_BuildDialogScreen(pApp);
            break;
    }

        return(TRUE);
        break;

    case EVT_KEY_PRESS:
        if (pApp->pIMenu != NULL)
            IMENUCTL_HandleEvent(pApp->pIMenu,
                EVT_KEY, wParam, 0);
        else if (pApp->pIText != NULL)
        {
            ITEXTCTL_HandleEvent(pApp->pIText, EVT_KEY,
                wParam, 0);
            ITEXTCTL_Redraw(pApp->pIText);
        }
        else if (pApp->pIDate != NULL)
        {
            IDATECTL_HandleEvent(pApp->pIDate, EVT_KEY,
                wParam, 0);
            IDATECTL_Redraw(pApp->pIDate);
        }
        else if (pApp->pITime != NULL)
        {
            ITIMECTL_HandleEvent(pApp->pITime, EVT_KEY,
                wParam, 0);
            ITIMECTL_Redraw(pApp->pITime);
```

```
    }
    return(TRUE);
    break;
```

Simply pass the events off to the controls and redraw them on their associated messages. The only other function to look at is the **CleanUp** function, which is very simple:

Listing IO-II: The CleanUp function

```
void GUITEST_CleanUp(myapp_t* pApp)
{
    GUITEST_ClearGUI(pApp);
}
```

A simple call to **GUITEST_ClearGUI** is all it takes to release all the memory that we have allocated in this applet. This applet does not do any complicated data manipulation or validation of controls. However, it serves as a good framework if you want to play with the different functions and options in each control type.

Conclusion

In this chapter you have learned how to use all of the basic BREW GUI components. BREW contains a wide variety of GUI controls that can create any user interface necessary for a handheld device. Dealing with control events and redrawing can be a bit of a pain. However, most GUIs for such small devices are rather simple; therefore, these issues are not very serious. By using dialogs and encapsulating your control behavior code in custom message handlers, complicated GUIs can be managed fairly well.

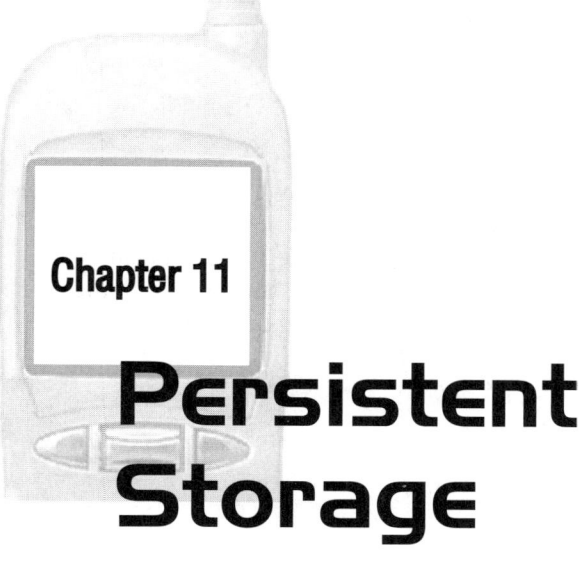

Chapter 11

Persistent Storage

Introduction

Persistent file storage is an important feature of many games. Perhaps the simplest case is writing and reading to a file to record high score data or perhaps having the levels of your game stored in a separate file that can be easily updated for new maps and creatures. BREW's flexible file I/O system allows several different kinds of persistent storage. First, there are traditional files. These can contain any kind of data—it is up to the applet to interpret it. Second, there are database files, which are simple databases that are read and written to from within BREW. Finally, there are applet preferences that are small chunks of data that belong to the applet. These are typically used to save simple user preferences and other very small chunks of data. This chapter will focus on all three, showing how you can use persistent storage for a variety of purposes.

Basic File I/O

You may think, why is this guy explaining file I/O when it has been a standard feature of C practically since its inception? The reason is BREW's restriction on Standard C Library calls. Since no Standard C Library functions are available from within a BREW applet, Qualcomm has duplicated the functionality of many common Standard C Library features in its SDK. Perhaps the biggest feature is, of course, file access.

There are two interfaces that we deal with when using files. They are the IFile and IFileMgr interfaces. IFile represents the file that you are accessing, while **IFileMgr** allows access to BREW's own file system functions for reading and writing data.

Before doing anything, you have to make sure the applet's security permissions are set to allow file access. This can be checked by loading up the applet's MIF file in the MIF Editor, clicking on the General tab, and looking at the privilege level check boxes. If the File check box is not checked, do so and save the MIF file.

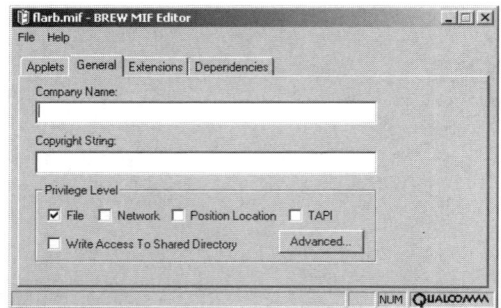

Figure 11-1: You must check the File box in order to use file I/O.

Using the File Manager Interface

Now that our applet has been granted access to the file system, we merely have to create an instance of the **IFileMgr** interface. This is done with **ISHELL_CreateInstance**, just as we did with the GUI controls. Naturally, we have to release this interface when we are done using it. The class ID for the **IFileMgr** is **AEECLSID_FILEMGR**, as defined in AEEClassIDs.h.

```
IFileMgr * pIFileMgr;
ISHELL_CreateInstance(pApp->a.m_pIShell, AEECLSID_FILEMGR,(void **)&pIFileMgr);
```

After we have created the **IFileMgr** instance, we then have the access to use its functions to manipulate files. The form of BREW's file I/O functions is very similar to the Standard C Library calls that they mimic. We will look at opening files for starters with the function **IFILEMGR_ OpenFile**:

```
IFile * IFILEMGR_OpenFile(IfileMgr * pIFileMgr, const char * pszFile, OpenFile mode)
```

After passing the pointer to our **IFileMgr** interface, pass in a filename string and a mode flag that dictates what kind of access you have to the files. These flags are detailed in Table 11-1.

Table 11-1: File mode bit flags

Flags	Description
_OFM_READ	Read access to file
_OFM_READWRITE	Read and write access
_OFM_CREATE	Create file if it does not already exist
_OFM_APPEND	Append file if it already exists

As you can see, these are essentially the same sorts of flags used in Standard C Library file functions. The function returns a pointer to the file that you just opened (in this case, a pointer to the IFile interface). By using this pointer you can then read, write, and close the file.

Note: Keep in mind that the BREW file system has a limited number of files that it can handle. Therefore, you should be prepared to gracefully handle situations where your call to create a file fails. This number varies per device. Check your handset's development documentation on the BREW Developer Extranet to see what the file system restrictions are.

The IFileMgr interface allows you full reign over the phone's file system. You can make and remove directories and create and destroy multiple files, as well as search for files and collect data about the space left in the handset's file system. These functions are all very similar to their analogous Standard C Library calls, so please refer to the BREW documentation for more information.

Using Files

Now that we have a pointer to a file from our previous open operation, we can use it to read and write data. Once again, these functions are very similar to their Standard C Library equivalents:

```
uint32 IFILE_Read(IFile * pIFile, void * pBuffer, uint32 dwCount)
```

Obviously this is equivalent to fread. In this case you simply pass the file pointer, a buffer to read your data into, and the number of bytes you wish to read from the file.

```
uint32 IFILE_Write(IFile * pIFile, PACKED const void * pBuffer, uint32 dwCount)
```

Here we have the substitute for fwrite. Again, this is very similar to its Standard C Library equivalent. Simply pass a pointer to the file, a buffer you wish to write from, and how many bytes you want to write to the file.

```
uint32 IFILE_Seek(IFile * pIFile, FileSeekType seekType, int32 move Distance)
```

Finally, we have **IFILE_Seek**, which works in a similar manner to **fseek**. Here we pass a file pointer, a seek type enumeration, and how many bytes we wish to move. The **FileSeekType** enumerations are defined in AEEFile.h, as detailed in Table 11-2.

Table 11-2: File seek types

Types	Description
_SEEK_START	Start seek from beginning of file
_SEEK_END	Start seek from end of file
_SEEK_CURRENT	Start seek from current position in file

Finally, when you are done using a file, you need to close it. In this case, closing a file means releasing its interface with **IFILE_Release**.

Sharing Files

If you want to share files between applets, BREW has a special directory set aside for just this purpose. If you use the **AEE_SHARED_DIR** definition, as defined in AEE.h, you can access a directory strictly for use with programs that need to exchange files with others. This can be used to store BMPs, databases, or just about any file you need. Simply append **AEE_SHARED_DIR** to your filename, and you can save and load files from this special directory.

As you can see, using files in BREW is virtually identical to using standard C and C++ files. Standard files are probably the best way to store persistent data in BREW, as they allow you to create your own custom file formats and data storage conventions. Simple flat files may be cumbersome if you are doing any advanced work with data structures and storage. BREW does have additional ways to store data persistently that solve these issues.

Databases

BREW also contains a very simple database system that is very handy for organizing small amounts of data. This is analogous to J2ME's RecordStore mechanism. Although RecordStores are an alternative to traditional file operations in Java, they complement the BREW file system by providing a very simple structured file format for simple databases and other collections of data. It is not as robust as something like SQL, but for the intended purposes of database functionality on a handheld device, it satisfies most needs.

Creating a Database

Much like the file system, there is a manager object that needs to be created in order to manipulate individual databases. The **IDBMGR** interface is created in the standard way:

```
IDBMGR* pIDBMgr;
ISHELL_CreateInstance(pApp->a.m_pIShell, AEECLSID_DBMGR,(void **)&pIDBMgr);
```

Now that we have created our interface, we can open or create a database in much the same way that we do a file with the function **IDBMgr_ OpenDatabase**:

```
IDatabase * IDBMgr_OpenDatabase(IDBMgr * pif, const char * pszFile, boolean bCreate)
```

As usual, the first argument is our **IDBMgr** interface. Following this parameter is the filename of the database that we want to open or create. The Boolean argument **bCreate** determines whether the call fails or creates a new database if the one specified by the filename is not found. The return value of this function is a pointer to the database opened, otherwise known as the **IDatabase** interface. Through this object, it is possible to manipulate the data inside the open database.

Destroying a Database

Sure, you can create a database with the **IDBMgr_OpenDatabase** function, but how do you delete one? This is achieved by using the **IDBMgr_Remove** function:

```
int IDBMgr_Remove(IDBMgr * pif, const char * pszFile)
```

As you can see, this function is simple. Aside from the obligatory interface pointer, simply pass along the filename and you can consider the database obliterated.

Creating a Database Record

Now what exactly is in one of these databases? A BREW database consists of a series of records. Each record has a number of different fields, determined by its creator. Create a record in an open database through the function **IDATABASE_CreateRecord**:

```
IDBRecord * IDATABASE_CreateRecord(IDatabase * pIDatabase, AEEDBField * pDBFields,
    int iNumFields)
```

After the interface pointer, pass in a pointer to an **AEEDFfield** structure. This structure describes how a field in the record is laid out. The final argument counts how many field descriptions are pointed to by **pIDatabase**. The **AEEDBField** structure looks like this:

```
typedef struct
{
    AEEDBFieldType    fType;      // Field Type
    AEEDBFieldName    fName;      // Field Name
    uint16            wDataLen;   // Data Length
    void *            pBuffer;    // Data Ptr
} AEEDBField;
```

The first two arguments are emumerations defined in the header AEEDB.h. They describe the type of data stored in this field. The second argument is an emumeration that is also defined in the same header. It describes the use of this field ranging from an e-mail address to a phone number. These settings are used by the **IDBRECORD** interface to parse the different fields in a record. The **wDataLen** argument specifies the size in bytes of this record. The **pBuffer** pointer points to the data that will be copied into this record.

One thing to remember is that the database record interface must be managed, just as the actual database interface is. That is, **IDBRECORD_Release** must be called on the open record. Otherwise, it will not be possible to close the database to which the record belongs.

Destroying a Database Record

Removing a record from a database is a simple operation. Simply use the function **IDBRECORD_Remove**:

```
int IDBRECORD_Remove(IDBRECord * pIDBRecord)
```

Predictably, this function removes the record from the database. It is not necessary to release the interface once this is called.

Finding a Database Record

Now that we are familiar with database records, how do we find them when we have more than one? The **IDATABASE** interface contains a number of different functions for searching the records inside an open database. The basic call for finding a record is **IDATABASE_GetRecordByID**:

```
IDBRecord * IDATABASE_GetRecordByID(IDatabase * pIDatabase, uint16 u16RecID)
```

After the pointer to the open database interface, pass in the ID of the record that we are looking for. The record ID is set automatically by the **IDatabase** interface when a record is created. The **IDatabase** interface makes sure that no two records have the same ID, which is why you are not allowed to directly modify this data. To find the ID of an open record, call the function **IDBRECORD_GetID**.

```
uint16 IDBRECORD_GetID(IDBRecord * pIDBRecord)
```

Simply put, this returns the ID number of the record passed to it. These ID numbers must be used again to retrieve the record via IDATABASE_ GetRecordByID.

Retrieving Field Data

Now that you have the record that you want, how do you access the data inside it? The return value from IDATABASE_GetRecordByID returns the pointer to the IDBRecord interface. Through this interface, it is possible to access the data stored in the field using the function IDBRECORD_GetField:

```
byte * IDBRECORD_GetField(IDBRecord * pIDBRecord, AEEDBFieldName * pName, AEEDBFieldType
    * pDBFieldType, uint16 * pnLen)
```

Through the IDBRECORD_GetField function, you will find the field by using the very same attributes set with the DATABASE_Create-Record's AEEDBField pointer argument. The function returns a pointer to the data stored in the desired field. The final argument is a pointer to a uint16 that will be filled with the length of the argument upon return of the function. This way, you know how long the chunk of data is to which the byte pointer returns value points.

Applet Preferences

The final way to store persistent data is with applet preferences. Each applet installed on a device is able to save small chunks of data in a space reserved for applet-specific information. This data is not stored in a traditional file but in a reserved area of the phone's memory for applet preferences, as they are called. This preference data is automatically erased when the applet it belongs to is removed or disabled. This is slightly easier than using files, but it is only recommended for saving very small amounts of applet-specific data.

Setting Preferences

First, we will discuss how to set applet preferences. Preference data is saved in the phone's persistent storage memory as a raw block of bytes. It is up to the applet to interpret this data. Therefore, a common technique is to create a custom struct for the preference data but treat it as a void pointer with the preference function calls. Setting applet preference data is done with ISHELL_SetPrefs:

```
int ISHELL_SetPrefs(IShell * pIShell, AEECLSID cls, uint16 wVer, void * pCfg, uint 16
    nSize)
```

Naturally, the first argument is the interface to the shell. The second argument is the class ID of the applet. It is possible to store preference data for any class currently installed on the phone. However, in most cases, we are simply interested in the applet that is currently running. The **wVer** argument is the version number. This way, you can make sure the preference data is current with the applet code. The void pointer, **pCfg**, points to the data that we are saving. In many cases, this is a pointer to a custom structure that we use as a void pointer. Finally, pass the size of this structure or raw data with the **nSize** parameter.

Getting Preferences

Retrieving preferences is just as easy. Use the function **ISHELL_Get-Prefs**, like so:

```
int ISHELL_GetPrefs(IShell * pIShell, AEECLSID cls, uint16 wVer, void* pCfg, uint16 nSize)
```

After the obligatory interface pointer, pass in the class ID of the preferences that we are interested in. It is possible to get preferences from other applets installed on the device by passing in different class IDs. Next, pass the version number as we set with **SetPrefs**. If the version is not the same as the preference data on the phone, an error will be returned. Next, pass a pointer to the memory that we want to fill with our preference data with **pCfg**. And finally, specify the size of this data chunk with **nSize**.

Applet preference data is a good way to store simple things such as the last score attained, the user's desired difficulty level, and other basic applet options. The space available for applet preferences is tiny. So saving large amounts of data should be reserved for the regular file I/O functions.

Conclusion

We have discussed three different ways to save persistent data with the BREW SDK. First, there are the standard file I/O functions that we are all used to from C/C++'s Standard Library. Then there is the simple database system for basic record keeping. Finally, we have the preferences data that is meant for storing tiny amounts of data related to an applet's operation. This may have seemed like a rather boring chapter, but it is important to know about persistent data and file operations if you want to keep your applet flexible and data-driven (not to mention the fact that BREW's discouragement of large static data tables makes loading in data from files a necessity).

Chapter 12

Sound and Music

Introduction

Sound effects are usually an important part of a game; however, in the mobile realm it is something of a low priority. The problem is that most handsets support sound in only very primitive form. This usually means the different ringer sounds and perhaps some basic tone generation routines, if you are lucky. Advanced handsets support MIDI and MP3 formats. Until these are commonplace, you will have to deal with rather meager sound capabilities. It is worth noting that many of the methods available in BREW's sound modules cannot be used on older handsets or even accurately represented in the emulator.

The Beep

The most basic form of sound in BREW is the beep. A *beep* is a general term for the various ringers present in the handset hardware. These ring sounds vary between handsets, so you cannot be sure that the beep operation will perform the same sound effect on any two handsets. In fact, the emulator simply plays a WAV file of a phone ringing. However, it is useful for simple notifications, such as the end of a level or the accumulation of a high score.

To perform an actual beep operation, call the function ISHELL_Beep:

```
boolean ISHELL_Beep(IShell * pIShell, BeepType nBeep, boolean bLoud)
```

The beep function is part of the virtual grab bag known as the shell interface. After the shell interface pointer, you have an enumerated type representing the kind of beep tone to play. These are defined in the AEEShell.h header, as detailed in Table 12-1.

Table 12-1: Beep tone types

Type	Function
BEEP_OFF	Turn off the beep tone
BEEP_ALERT	Alert beep
BEEP_REMINDER	Reminder beep
BEEP_MSG	Message beep (SMS or other text message)
BEEP_ERROR	Error beep
BEEP_VIBRATE_ALERT	Alert beep
BEEP_VIBRATE_REMIND	Reminder beep

The enumerations are fairly self-explanatory. However, each phone manufacturer treats these events differently. Not only that, but in many handsets, it is possible for the user to define her own tones for these events. Interestingly, two of the beeps are actually vibrate functions. However, it is also possible to have more control over the phone's vibration functions by using the ISOUND_Vibrate function. Finally, the argument bLoud tells BREW to play the sound at a higher volume.

Tones and Tone Lists

Tones are a more complicated way to express sound. Users of Nokia phones may be familiar with the concept of *ringtones*—simple jingles that you can upload to your phone to play various songs and tunes instead of a plain ring effect. Some more advanced phones have a built-in composer that allows the user to create lists of tones. These tones form a song when played. This is basically what BREW's tones are.

Note: One interesting non-game application of BREW tone lists is Moviso's Ringster. Using a BREW client, it is possible to browse a list of ringtones based on pop hits, theme songs, and other genres. Once you find the tune you like, you can download it for a fee and use it for your ring.

BREW tones are a series of built-in sounds that can be strung together in a series to create simple musical sounds. The problem is that there really is no good tool available to compose tone lists. Therefore, creating anything that even vaguely resembles music is a massive exercise in trial and error. Since the tones differ from phone to phone, this feature remains almost completely useless. For the sake of completion, it is worth a mention.

Before we play a tone or tone list, we need to construct the ISOUND interface:

```
ISOUND* pISound;
ISHELL_CreateInstance(pApp->a.m_pIShell, AEECLSID_SOUND,(void **)&pISound);
```

Now use the ISOUND_PlayTone function to, well, play a tone:

```
void ISOUND_PlayTone(ISound* pISound, AEESoundToneData toneData)
```

After the obligatory sound interface pointer, pass an AEESoundToneData structure. This structure is defined in the AEESound.h header file:

```
typedef struct _AEESoundToneData
{
    AEESoundTone    eTone;
    uint16          wDuration;
} AEESoundToneData;
```

There are two members in this structure. The wDuration argument determines the length, in milliseconds, of the sound effect. The first, eTone, is an AEESoundTone enumeration. This defines the kind of tone we are going to play. These enumerations are also defined in the AEESound.h header. If you look at these enumerations, they range from touch-tone sounds to specific frequencies that can be played in succession for rudimentary music and sound effects in combination with judicious usage of the wDuration argument.

However, to play a series of tones, you do not need to call the ISOUND_PlayTone function over and over. Instead, it is possible to use tone lists to send off a series of tones or notes to be played by BREW's sound sub-system:

```
ISOUND_PlayToneList(ISound * pISound, AEESoundToneData * pToneData, uint16 wDataLen)
```

As you can see, this is fairly similar to ISOUND_PlayTone. The big difference is you pass a pointer to a list of tones instead (that is, an array of AEESoundToneData structures). Finally, the length of this array is passed in via the wDataLen argument.

The big question here is, how do you compose tone lists that actually sound good? The BREW SDK, as of this writing, contains no tone composition tools. Therefore, it is basically an exercise of trial and error to coax anything worthwhile out of BREW's meager sound capabilities. You might be able to convert ringtone notes meant for other phones to BREW tone lists; however, since most ringtones seem to incur the bloodthirsty wrath of neighboring listeners, perhaps it is better for the game (and the safety of the player) to ignore sound altogether.

Advanced Sound Capabilities

After wallowing in the mire of beeps, tones, and tone lists, it may be quite a shock to see the next section refer to the usage of MIDI and MP3 files. Regardless, the BREW API supports these advanced sound formats. Most new handsets have fairly robust MIDI support with 16-note polyphonic sound and other features. However, we are some ways off before MP3 playback is commonplace.

Using the SoundPlayer

The playing of MIDI and MP3 files is done through the usage of the ISoundPlayer interface. This object allows the playing of the aforementioned file formats, as well as pausing, resuming, seeking, and other basic functions common to many other sound APIs. As usual, begin using the interface by creating an instance of it:

```
ISoundPlayer * pISoundPlayer;
SHELL_CreateInstance(pMe->a.m_pIShell, AEECLSID_SOUNDPLAYER, (void **)pISoundPlayer);
```

Getting the MIDI or MP3 File

Now we can access all that this interface has to offer. At the most basic level, we can load and play a MIDI file. The first thing we have to do is set the source of the sound file. This can be two different sources: a file or a buffer. Since the buffer type is not supported in the emulator, we will use a MIDI file stored in the root directory of our applet. This is done through the call ISOUNDPLAYER_Set:

```
void ISOUNDPLAYER_Set(ISoundPlayer * pISoundPlayer, AEESoundPlayerInput t, void * pData)
```

After the interface pointer, we have an enumeration that describes the kind of source from which we are getting the data. These enumerations are defined in the AEESoundPlayer.h header file. However, because the emulator only supports file sources, we can only use the SDT_FILE enumeration as input to the argument. The final argument is a pointer to the source data, which in the case of a file source is simply a string containing the name of the MIDI or MP3 file that we are using.

It is also possible to retrieve sound data from a stream. This way, you can retrieve a MIDI or MP3 file from the Internet and stream it to the handset instead of requiring the tune to be included with the applet or downloaded to memory before playing. This is done with the ISOUNDPLAYER_SetStream function:

```
void ISOUNDPLAYER_SetStream (ISoundPlayer * pISoundPlayer, IAStream * ps)
```

The second argument is an open stream that is created using the networking functions of BREW. The actual creation and use of a stream is beyond the scope of this chapter. However, it is good to know that there are alternate means of feeding data to BREW's sound engine.

Note: You may want to consider the fact that MIDI instruments are not standardized in BREW. Therefore, your MIDI song may sound completely different from one handset to the next.

Playing and Controlling the Tune

Once we have the source set, we merely have to call **ISOUNDPLAYER_ Play** to start the show:

```
void ISOUNDPLAYER_Play(ISoundPlayer * pISoundPlayer)
```

Now that we have a tune playing, we can stop it by using the **ISOUND-PLAYER_Stop** function:

```
void ISOUNDPLAYER_Stop(ISoundPlayer * pISoundPlayer)
```

We can also change the volume of the current tune by using the **ISOUNDPLAYER_SetVolume** function:

```
void ISOUNDPLAYER_SetVolume(ISoundPlayer * pISoundPlayer, uint16 wVolume)
```

The valid range of values for the **wVolume** argument range from 0 to the **AEE_MAX_VOLUME** definition given in the AEESound.h header file.

  If you want more control over the playing of the file, we can manipulate it by using BREW's simple playback control commands. First, there is the ability to pause and resume the playing tune:

```
void ISOUNDPLAYER_Pause(ISoundPlayer * pISoundPlayer)
void ISOUNDPLAYER_Resume(ISoundPlayer * pISoundPlayer)
```

Next, we can skip through the file by using the **Rewind** and **FastForward** functions:

```
void ISOUNDPLAYER_FastForward(ISoundPlayer * pISoundPlayer, uint32 dwTime)
void ISOUNDPLAYER_Rewind(ISoundPlayer *  pISoundPlayer, uint32 dwTime)
```

The key argument here is the final one, **dwTime**. This determines how long in milliseconds we fast forward or rewind from our current point in the playback of the file.

  One of the stranger functions of the **SoundPlayer** interface is the changing of the playback tempo of an MP3 or MIDI file. This is done through the **ISOUNDPLAYER_SetTempo** function:

```
void ISOUNDPLAYER_SetTempo(ISoundPlayer * pISoundPlayer, uint32 dwTempoFactor)
```

After the interface pointer, pass the **dwTempoFactor** argument. This is the percentage of the standard tempo that we want the tune to play back at. For instance, passing a value of 5 will slow down the playback to 5 percent of its default tempo. The acceptable values range from 1 to 500 percent of the normal tempo. The documentation claims that this opera-tion does not work in the emulator but does on hardware.

Another rather esoteric ISOUNDPLAYER function is the ability to alter the pitch of the currently playing file. This is done through the function ISOUNDPLAYER_SetTune:

```
void ISOUNDPLAYER_SetTune(ISoundPlayer * pISoundPlayer, uint8 nStep)
```

The nStep argument modifies the tune in what the documentation calls "half-step increments." The values range from –12 to 12, which repre-sents a tune range from one octave lower than normal playback to one octave higher. Of course, for those of us who are not musicians, this is probably gibberish. In essence, you can make the song sound like it is sung by a chorus of tree sloths or a group of wild chipmunks by altering the octave range of the music.

Making Sounds Interactive

Playing sounds is useful, but many times the sound needs to have more direct control over the program. For instance, it may be necessary to synchronize the start of a level with the end of an introduction tune. Or perhaps it might be convenient to remove an enemy from the game only after its explosion sound effect has finished playing. For cases such as these, BREW has several callback mechanisms. A *callback* is a function that is called at the end of specific events. Callbacks are commonly used in the SDK for all sorts of non-blocking function calls, including sound. Sound callbacks are used via the convenient, yet somewhat confusing, ISOUND_RegisterNotify function:

```
void ISOUND_RegisterNotify(ISound * pISound, PFNSOUNDSTATUS pfn, const void * pUser)
```

The first argument is the interface pointer. An important thing to remember about this interface pointer is that the callback function is registered for only the sound that belongs to this interface. Therefore, you must register callbacks individually on any sound for which you wish status notification.

The second argument, pfn, is a function pointer. This points to the function that will be called when various sound events occur. You can think of this callback command as a kind of HandleEvent for the sound. BREW will send various messages indicating volume changes, sound playback status, and other events to this function. It is up to this call-back to process these messages as it sees fit. Finally, the argument pUser is a pointer to an arbitrary chunk of data. This pointer will be sent as an argument to the callback.

Getting back to the callback function, the PFNSOUNDSTATUS function prototype is defined in AEESound.h like so:

```
typedef void (*PFNSOUNDSTATUS)
(
    void*          pUser,
    AEESoundCmd    eCBType,
    AEESoundStatus eSPStatus,
    uint32         dwParam
);
```

The pUser argument is a pointer to the data specified in the final argument of ISOUND_RegisterNotify. Probably the most common usage of this is to pass the applet structure pointer back to the callback. This way, your callback function can access all of the applet's global data.

The eCBType argument specifies which kind of sound command is being sent. As of this revision of BREW, there are only two command types, as defined in AEESound.h: AEE_SOUND_STATUS_CB and AEE_SOUND_VOLUME_CB. AEE_SOUND_STATUS_CB commands are messages that signify the starting and stopping of sound playback including tones, MIDI, MP3s, and just about any sound type that BREW supports. Obviously, an AEE_SOUND_VOLUME_CB command is used for any change in volume of the BREW sound system.

The eSPStatus argument is of the AEESoundStatus type. This enumeration is defined in the AEESound header file, as detailed in Table 12-2.

Table 12-2: Sound status types

Status Types	Description
AEE_SOUND_UNKNOWN	Unknown status
AEE_SOUND_SUCCESS	Sound has successfully played.
AEE_SOUND_PLAY_DONE	Sound is done playing.
AEE_SOUND_FAILURE	Sound has failed playing.
AEE_SOUND_LAST	This is the last sound to be played (as in a tone list).

These values reflect the status of the command being sent. For instance, if a volume change in your sound has failed for some reason, you will get an eCBType of AEE_SOUND_VOLUME_CB with an eSPStatus argument value of AEE_SOUND_FAILURE.

Finally, there is the dwParam argument. Although this is defined as an unsigned integer, it is actually a pointer to an AEESoundCmdData structure. If the callback message does not need this pointer, the value is NULL. Otherwise, if must be cast to a pointer of AEESoundCmd-Data, which is a union defined in the AEESound header, like so:

```
typedef union
{
    uint16 wVolume;
```

```
    uint16 wPlayIndex;
} AEESoundCmdData;
```

The first member, wVolume, is used in any **AEE_SOUND_VOLUME_ CB** command type. This way upon a volume change, it is possible to see where the current volume of the sound is set. The second member, **wPlayIndex**, is the number of the tone played in a tone list. Every single tone in a tone list will trigger a callback. Of course, this is a union, so you can only have either or. This way, it is possible to know how far the playback is in the tone list for any special processing necessary.

Simple Examples

This chapter has a few simple example applets associated with it. Source\Chapter 12\sound1 in the companion files demonstrates playing a tone list with a callback function, and Source\Chapter 12\sound2 shows how to play an MP3, also with a callback function.

A Tone List Example

We will start by looking at the tone list sample project, sound1. Looking at the applet structure defined in sound1.h:

Listing 12-1: The applet structure

```
typedef struct myapp_s
{
    AEEApplet       a;        //applet header
    AEEDeviceInfo di;

    ISound *        pISound;

    AEESoundToneData * pToneData;
    uint16      nTones;
    uint16      nCurrentTone;

} myapp_t;
```

Here we have the standard pairing of the mandatory **AEEApplet** structure and **AEEDeviceInfo**. Next, we have a pointer to our ISound interface, **pISound**. We also have a pointer to our eventual dynamically allocated array of tones, **pToneData**. Finally, we have **nTones**, which counts how many tones are in our tone list, and **nCurrentTone**, which will be used to mark which tone we are currently playing.

Once again, we have a standard **AEE_AppletNew** call in AEECls-CreateInstance:

Listing 12-2: Creating the applet object

```
int AEEClsCreateInstance(AEECLSID ClsId, IShell * pIShell, IModule * po, void ** ppObj)
{
```

```
    *ppObj = NULL;
    if(ClsId == AEECLSID_SOUND1_BID )
    {
        if(AEEApplet_New(sizeof(myapp_t), ClsId,
            pIShell,po,(IApplet**)ppObj,
            (AEEHANDLER)Sound1_HandleEvent,
            (PFNFREEAPPDATA)Sound1_Cleanup)  == TRUE)
        {
            return(AEE_SUCCESS);
        }
    }
        return(EFAILED);
}
```

As you can see, we pass our message handler Sound1_HandleEvent and the cleanup function Sound1_Cleanup to AEEApplet_New. Looking at Sound1_HandleEvent, we see the bulk of the applet's code:

Listing 12-3: The event handler

```
static boolean Sound1_HandleEvent(IApplet * pi, AEEEvent eCode, uint16 wParam,
    uint32 dwParam)
{
    myapp_t * pApp      = (myapp_t*)pi;
    AEEApplet * pMe = &pApp->a;
    AECHAR szBuf[] = {'N', 'o', 't', 'e', ':', ' ', '0', '\0'};

    switch (eCode)
    {
        case EVT_APP_START:

            pApp->nCurrentTone = 0;

            //clear the screen and display status
            IDISPLAY_ClearScreen(pApp->a.m_pIDisplay);
            IDISPLAY_DrawText(pApp->a.m_pIDisplay,
                AEE_FONT_NORMAL, szBuf, -1, 0, 0, NULL,
                IDF_ALIGN_CENTER | IDF_ALIGN_MIDDLE);
            IDISPLAY_Update(pApp->a.m_pIDisplay);

            //create the sound interface
            ISHELL_CreateInstance(pApp->a.m_pIShell,
                AEECLSID_SOUND, (void **)&pApp->pISound);

            //create our music
            Sound1_CreateTones(pApp);

            //register our callback
            ISOUND_RegisterNotify(pApp->pISound,
                Sound1_Callback, (void *)pApp);

            //play our music
            ISOUND_SetVolume(pApp->pISound, AEE_MAX_VOLUME);
            ISOUND_PlayToneList(pApp->pISound,
                pApp->pToneData, pApp->nTones);

            return(TRUE);
            break;
    }
```

```
    return(FALSE);
}
```

Here, the only message we handle is **EVT_APP_START.** In our handler, clear the screen first and display some text to notify the user that we are on the first note of the tone list. Create the instance of **ISound** so that we can start playing our sounds. Next, call the function **Sound1_ CreateTones.** This function builds our simple tone list. Then, register our callback function, **Sound1_Callback,** via the call to **ISOUND_Regis-terNotify. Sound1_Callback** will be called every time the sound system generates a message related to the playing of our tone list. Finally, set the volume of the sounds to the maximum level with **ISOUND_Set-Volume** and start playing our tone with **ISOUND_PlayToneList.** We will now look at how we create the tone list data in **Sound1_CreateTones:**

Listing 12-4: Creating the tones

```
void Sound1_CreateTones(myapp_t * pApp)
{
    pApp->nTones = 5;

    pApp->pToneData  = (AEESoundToneData*)MALLOC(pApp->nTones
                                        * sizeof(AEESoundToneData));

    //now hardcode in some tone values.  In a real-world example, you
    //would probably read these out of a file.

    pApp->pToneData[0].eTone = AEE_TONE_RING_A;
    pApp->pToneData[0].wDuration = 500;

    pApp->pToneData[1].eTone = AEE_TONE_RING_C;
    pApp->pToneData[1].wDuration = 1000;

    pApp->pToneData[2].eTone = AEE_TONE_RING_A;
    pApp->pToneData[2].wDuration = 1000;

    pApp->pToneData[3].eTone = AEE_TONE_RING_C;
    pApp->pToneData[3].wDuration = 500;

    pApp->pToneData[4].eTone = AEE_TONE_RING_C;
    pApp->pToneData[4].wDuration = 500;
}
```

We are going to make a tone list that is five tones long. Therefore, set the **nTones** variable, which is declared in our applet header, to 5. Next we must allocate enough space to hold five tones in the tone list with a call to **MALLOC,** multiplying the size of an individual tone structure by the number of tones we are using. Subsequently, manually enter in the tone data for each of the five tones. For each tone, set the **eTone** variable to one of the predetermined note definitions given in AEESound.h. Also set **wDuration** to the number of milliseconds that we want each tone to play for. The result is an array of five tones, each with its own note and duration. The pointer to this dynamically allocated array of tones and the **nTones** variable representing the number of tones in the array are both sent as arguments to **ISOUND_PlayToneList** to start the playing of our sounds.

Now, moving on to the callback function, **Sound1_Callback**:

Listing 12-5: Callback function

```
void Sound1_Callback(void * pUser, AEESoundCmd eCBType, AEESoundStatus eStatus,
    uint32 dwParam)
{
    AECHAR szBuf[16];
    myapp_t* pApp = (myapp_t*)pUser;
    AEESoundCmdData * pData = (AEESoundCmdData *) dwParam;
    AECHAR szFormat[] = {'N', 'o', 't', 'e', ':', ' ', '%', 'd', '\0'};

    if (!pApp)
        return;

    switch (eStatus)
    {

        case AEE_SOUND_PLAY_DONE:

                ++pApp->nCurrentTone;

                //if we have played all the tones, stop
                if (pApp->nCurrentTone >= pApp->nTones)
                {
                    STR_TO_WSTR("End.", szBuf, sizeof(szBuf));
                    ISOUND_StopTone(pApp->pISound);
                }
                else
                    WSPRINTF(szBuf, sizeof(szBuf), szFormat, pApp->nCurrentTone);

                IDISPLAY_ClearScreen(pApp->a.m_pIDisplay);
                IDISPLAY_DrawText(pApp->a.m_pIDisplay, AEE_FONT_NORMAL,
                    szBuf, -1, 0, 0, NULL, IDF_ALIGN_CENTER | IDF_ALIGN_MIDDLE);
                IDISPLAY_Update(pApp->a.m_pIDisplay);
                break;
    }
}
```

Basically, a sound callback is essentially the same as a message handler; it just takes different arguments, and the messages it handles are all sound related. We allow the callback to access the main applet structure by passing a pointer to the applet through the **pUser** pointer. This pointer is actually set in the final argument of **ISOUND_RegisterNotify**. This way, the sound callback can access our global data and other applet variables.

There are a number of different messages that BREW's sound system sends to the callback. The only one we handle is **AEE_SOUND_PLAY_DONE**. In a tone list, this message is sent to the callback every time one tone in the list has finished playing. Therefore, in the handler

code for this event, we update the display to show which note we are
currently playing.

In the handler, you can see that the first thing we do is increment
the current tone number. If we have hit the end of the tone list, we display the word "End" on the screen and stop the playing of the tone with
ISOUND_StopTone. Otherwise, the string we display is set to show the
current value of nCurrentTone. If we do not stop the tone, it will loop
back to the beginning when the end of the tone list is hit.

Finally, the CleanUp function deallocates the memory that we
acquired for the tone list:

Listing 12-6: Cleaning up the applet

```
void Sound1_Cleanup(myapp_t* pApp)
{
    //stop sound and release object
    if (pApp->pISound)
    {
        ISOUND_StopTone (pApp->pISound);
        ISOUND_StopVibrate (pApp->pISound);
        ISOUND_Release (pApp->pISound);
        pApp->pISound = NULL;
    }

    //release our allocated storage for tone data
    if (pApp->pToneData)
    {
        FREE(pApp->pToneData);
        pApp->pToneData = NULL;
    }
}
```

In addition to freeing the memory, we also release the ISound interface
to release all resources that we acquired when initializing this applet.

An MP3 Example

Now we will show you how to play an MP3 in the example sound2.
Looking at the applet structure defined in sound2.h, we see:

Listing 12-7: The applet structure

```
typedef struct myapp_s
{
    AEEApplet    a; //applet header
    ISoundPlayer * pISoundPlayer;

} myapp_t;
```

This one is very simple. Aside from the obligatory AEEApplet structure,
we have a pointer to the ISoundPlayer interface, which is how we will
play the MP3 file. Next, we have the standard applet construction code
in sound2.c:

Listing 12-8: Creating the applet instance

```
int AEEClsCreateInstance(AEECLSID ClsId,IShell * pIShell,IModule * po,void ** ppObj)
{
    *ppObj = NULL;

    if(ClsId == AEECLSID_SOUND2_BID )
    {
        if(AEEApplet_New(sizeof(myapp_t), ClsId, pIShell,po,(IApplet**)ppObj,
            (AEEHANDLER)Sound2_HandleEvent,
            (PFNFREEAPPDATA)Sound2_Cleanup) == TRUE)
        {
            return (AEE_SUCCESS);
        }
    }

    return (EFAILED);
}
```

Looking at the event handler, Sound2_HandleEvent, we see:

Listing 12-9: The event handler

```
static boolean Sound2_HandleEvent(IApplet * pi, AEEEvent eCode, uint16 wParam, uint32
    dwParam)
{
    myapp_t* pApp = (myapp_t*)pi;
    AECHAR szTitle[] = {'T', 'h', 'e', ' ', 'G', 'u', 'i', 'd', 'o', ' ', 'S', 'h', 'u',
                        'f', 'f', 'l', 'e', '\0'};
    AECHAR szArtist[] = {'b', 'y', ' ', 'R', 'i', 'c', 'o', ' ', 'B', 'a', 'r', 'r',
                        '\0'};

    switch (eCode)
    {
        case EVT_APP_START:

            //clear screen and display track information
            IDISPLAY_ClearScreen(pApp->a.m_pIDisplay);
            IDISPLAY_DrawText(pApp->a.m_pIDisplay, AEE_FONT_NORMAL,
                szTitle, -1, 0, 25, NULL, IDF_ALIGN_CENTER);
            IDISPLAY_DrawText(pApp->a.m_pIDisplay, AEE_FONT_NORMAL,
                szArtist, -1, 0, 45, NULL, IDF_ALIGN_CENTER);
            IDISPLAY_Update(pApp->a.m_pIDisplay);

            ISHELL_CreateInstance(pApp->a.m_pIShell,
                AEECLSID_SOUNDPLAYER, (void **)&pApp->pISoundPlayer);

            //set the file we want to play
            ISOUNDPLAYER_Set(pApp->pISoundPlayer, SDT_FILE,
                "The_Guido_Shuffle.mp3");

            //register our callback function
            ISOUNDPLAYER_RegisterNotify(pApp->pISoundPlayer,
                Sound2_Callback, (void *)pApp);

            //play the file
            ISOUNDPLAYER_Play (pApp->pISoundPlayer);

            return TRUE;
                break;
```

```
    }

    return FALSE;
}
```

Because the playing of an MP3 is automatic, all we need to do is set up the sound system in the **EVT_APP_START** handler and let it go. One of the first things we do is create a few strings to note the artist and title of this MP3. It is possible through the binary MP3 file to find the ID3 tag, but for now we will just hard-code the values.

After drawing the text in the **EVT_APP_START** handler, create the ISoundPlayer interface with a call to **ISHELL_CreateInstance**. Next, tell ISoundPlayer which file to play by calling **ISOUNDPLAYER_Set**. This file is located in the same directory as the applet, so no path data is required—just the filename.

Set up the callback function with **ISOUNDPLAYER_RegisterNotify**. This works in much the same way as **ISOUND_RegisterNotify**, which we saw in the previous example. Set the callback to be the function **Sound2_Callback**. After this, we then start playing the MP3 with a simple call to **ISOUNDPLAYER_Play**.

Like the ISOUND callback in the previous example, our Sound2_Callback function gets called in response to a variety of events related to the currently playing MP3 or MIDI file. The function looks like this:

Listing 12-10: The callback function

```
void Sound2_Callback(void * pUser, AEESoundPlayerCmd eCBType, AEESoundPlayerStatus
    eStatus, uint32 dwParam)
{
    myapp_t* pApp = (myapp_t*)pUser;
    AEESoundPlayerCmdData * pData = (AEESoundPlayerCmdData *) dwParam;

    switch (eStatus)
    {
        case AEE_SOUNDPLAYER_DONE:

        //upon finishing tune, play again
        ISOUNDPLAYER_Play(pApp->pISoundPlayer);
        break;
    }
}
```

The only message we choose to process here is **AEE_SOUNDPLAYER_DONE**. This message is sent to the callback when the MP3 has finished playing. Therefore, we simply call **ISOUNDPLAYER_Play** when this message is sent to start the file playing again. This way, the MP3 loops at the end.

The only function left to detail is our cleanup code:

Listing 12-11: The Cleanup function

```
void Sound2_Cleanup(myapp_t* pApp)
{
    //stop and release tune
    ISOUNDPLAYER_Stop(pApp->pISoundPlayer);
    ISOUNDPLAYER_Release(pApp->pISoundPlayer);
}
```

Here we stop playing the MP3 and release the interface to ISound-Player. That is all we need to do in order to deallocate all of our resources.

Conclusion

In this chapter you have learned the "ins and outs" of BREW's sound system. Of course, the two major issues are the fact that the emulator does not support all of these functions, and many current handsets have primitive, if any, sound capabilities. As BREW handsets become more common, and the relentless pace of progress makes hardware cheaper and faster, phones with robust sound facilities will be more common. Indeed, major content providers, such as MP3.com, have announced streaming music services for BREW phones. In Japan, many non-BREW handsets already have the ability to stream karaoke tunes. It is only a matter of time before sound becomes a major factor.

Chapter 13

Miscellaneous Topics

Introduction

Until now, all of BREW's functionality has fit into neatly categorized chapters. Here we will focus on a few odds and ends that are necessary for game and application development with BREW but fit under no specific category.

Strings and String Manipulation

Because there is no Standard C Library in BREW, string manipulation is done through a set of BREW helper macros designed to emulate the functionality with which we are familiar. Also, because BREW uses wide-character strings for its text display functions, there are a few functions included for converting between normal and wide-character strings.

Introduction to Wide-Character Strings

The BREW API uses wide-character strings for its text display functions and GUI objects. What I mean by wide character is using 16 bits per character instead of 8 bits. In normal single-byte strings, only 8 bits are used to represent each character. A string in C is really just an array of these single-byte values. Each character in a string is actually a number from 0 to 255. Because a char is 8 bits of information, only 256 different characters can be represented. The American Standard Code for Information Interchange standard, or ASCII, is what we use to represent English characters, numbers, symbols, and special unprintable control characters that signify the end of a file and other events. Together with the Extended ASCII standard, there are 256 different characters that are represented by a given char value. This works fine for English, where we only have 26 letters, but for languages such as Japanese and Chinese that have literally thousands of symbols, a new standard is required.

There are different standards used to represent non-English characters, and many of them require the use of wide characters. These are 16-bit values that replace the standard 8-bit char as the elements in a string. So a wide-character string is really an array of these 16-bit values. Each 16-bit value can represent 65,536 different values. That's more than enough characters to represent any language's alphabet. In BREW, the wide-character type is known as the **AECHAR**. If you look where it is defined in AEE.h, it looks like this:

```
typedef uint16 AECHAR;
```

The type uint16 is actually defined in AEEComdef.h as:

```
typedef  unsigned short uint16;   /* Unsigned 16 bit value */
```

As you can see, it's really just a short, which is a 16-bit value.

If you are reading this book, you are most likely interested in using English with your BREW applet. You may deal with foreign languages once you are done with the English version, if at all. So, how do you use these wide-character strings to represent plain English strings? It is definitely a little more awkward than standard ASCII, but it is not that much of a leap.

Making a Wide-Character String

As you saw in the Hello World applet described earlier in this book, you can create a wide-character string like so:

```
AECHAR szBuf[] = {'H','e','l','l','o',' ','W','o', 'r', 'l', 'd', '\0'};
```

What this actually does is create an array of **AECHAR**s initialized to the values that you set on the right. Notice that the NULL termination character is included at the end. If you do not include this string, operations will blow past the end of the array into unclaimed memory because they do not know where the string actually ends.

Another way to create a wide-character string is to pull the string out of the resource. This involves using the function **ISHELL_Load-ResString()**:

```
Int ISHELL_LoadResString(Ishell * pIShell, const char * pszResFile, int16 nResID,
    AECHAR * pBuff, int nSize)
```

The first argument is the pointer to the Shell interface. Next, pass a character string containing the filename of the resource file that you are getting the string out of. After this, pass the resource ID, as defined in the associated header via nResID. Finally, pass a pointer to an **AECHAR** buffer with pBuff and the size of this buffer in bytes with nSize.

Using strings in resource files makes your project much easier to localize. A non-programming translator could simply go through your resource file and translate each string. Regardless, it makes your code much cleaner and avoids taking up valuable stack space with declared arrays of wide characters.

String Conversion

Now this szBuf string is ready to pass as a string to any function in the SDK that uses wide strings. What if you have plain ASCII strings that you need to convert? This situation can arise if you are reading data out of a file, or perhaps you are trying to save memory by keeping your strings single-byte until absolutely necessary. In this case, BREW provides several helper functions to convert from ASCII to wide and vice versa:

```
AECHAR * STR_TO_WSTR(char * pszIn, AECHAR * pDest, int nSize)
```

The first argument, pszIn, is the ASCII string that we want to convert. The second argument, pDest, is the destination buffer where we want the converted string to be copied. Finally, the nSize argument specifies how long the ASCII string is in characters. Although it seems redundant, the function returns a pointer to the destination string if the conversion is successful.

You also have the converse of this function with:

```
char * WSTR_TO_STR(AECHAR * pIn, char * pszDest, int nSize)
```

Here, the first argument, pIn, is the wide-character string. The pszDest argument is a pointer to the destination string (in this case, an array of chars). Finally, pass the length of this string in the nSize argument.

STDLIB String Function Equivalents

Aside from the conversion functions, most of the string functions in the BREW SDK are identical to the Standard Library functions that they replace. The following tables list the BREW string functions and the standard lib functions for which they stand in. Also, remember that string functions starting with **WSTR** instead of **STR** use wide-strings. Tables 13-1 and 13-2 show the ASCII and wide-character equivalent functions, respectively.

Table 13-1: ASCII functions and their stdlib equivalents

ASCII Function	stdlib Equivalent
int ATOI(const char * src)	atoi
int SPRINTF(char * buffer, const char * format [, argument] ...)	sprintf
char * STRCAT(char * dest, const char * src)	strcat
char * STRCHR(const char * string, int c)	strchr
int STRCMP(const char * str1, const char * str2)	strcmp
char * STRCPY(char * deste, const char * src)	strcpy
int STRLEN(const char * str)	strlen
char * STRNCPY(char * strDest, const char * strSource, size_t count)	strncpy
char * STRRCHR(const char * string, int c)	strrchr
unsigned long STRTOUL(const char * nptr, char * * endptr, int base)	--

Table 13-2: Wide-character functions and their stdlib equivalents

Wide-character Function	stdlib Equivalent
int WSPRINTF(AECHAR * pDest, int nSize, AECHAR * pFormat, ...)	sprintf
AECHAR * WSTRCAT(AECHAR * pDest, AECHAR * pSrc)	strcat
AECHAR * WSTRCHR(AECHAR * s1, AECHAR c)	strchr
int WSTRCMP(AECHAR * s1, AECHAR * s2)	strcmp
AECHAR * WSTRCPY(AECHAR * pDest, AECHAR * pSrc)	strcpy
AECHAR * WSTRDUP(AECHAR * pIn)	strdup
int WSTRLEN(AECHAR * str)	strlen
AECHAR * WSTRRCHR(AECHAR * s1, AECHAR c)	strrchr

The wide-character functions behave the same as their standard library counterparts, except they take slightly different arguments and work with, obviously, wide strings. As you can see, some functions are emulated for wide characters only. Any string processing functions not here that are missing from the standard library will have to be recreated from scratch. Luckily, BREW has covered most of the bases.

Advanced String Manipulation

BREW also contains additional string functionality. For instance, if you need to convert wide strings to all upper- or lowercase, use the following pair of functions:

```
void WSTRLOWER(AECHAR * pszDest)
void WSTRUPPER(AECHAR * pszDest)
```

In wide-character strings, it is possible to merge ASCII characters to use their own 8-bit half of a wide character. Therefore, if you have a character value that is in the range of 0 to 127, you can place two of them in a single 16-bit wide character. The function **WSTRCOMPRESS** takes an entire wide-string and merges valid characters to reduce the memory footprint:

```
WSTRCOMPRESS(const AECHAR * pSrc, int nLen, byte * pDest, int nSize)
```

In a case like this, the string's length and size are not necessarily the same. Because two characters can be occupying one **AECHAR**, you cannot simply multiply the string length by 2 to get the size in bytes of a wide-character string. It is necessary to have a function that can calculate the byte size of the string that is completely different from the string length:

```
int WSTRSIZE(AECHAR * p)
```

As you can see, you pass the wide string as an argument and it returns the size in bytes. By using this and the compression function, wide strings do not have to be a major memory hog. This is important when dealing with handsets that can have as little as 60K of available RAM.

Wide String Type Conversions

There are also a few functions to convert basic data types to wide-character strings:

```
AECHAR * WWRITELONGEX(AECHAR * pszBuf, long n, int nPad, int * pnRemaining)
```

In this case, we can convert a long to a wide string. The first argument, pszBuf, is a pointer to our wide string to which we want the result to be copied. The second argument, n, is the long we are converting. The nPad argument specifies how many zeroes to add to the beginning of the number. If you have 3 for nPad and the number is only two digits (say, 23), the resultant string will read "023." The final argument, nRemaining, specifies the size of the pszBuf to which we are copying. The function also returns a pointer to the converted string.

```
boolean FLOAT_TO_WSTR(double v, AECHAR * psz, int nSize)
```

This simple function converts a float to a wide string. The first argument, v, is the floating-point number that we wish to convert. The second argument is the wide-character string to which we want to copy the value. Finally, nSize is the length of the wide-character buffer.

Memory Management

Until now we have not dealt with allocating and deallocating memory. However, you must be wondering, if we cannot use the Standard C Library, how can we use malloc and free? Luckily, the BREW SDK has equivalent functions for these operations.

Memory Allocation

Allocating memory has two functions, one of which should be familiar:

```
void * MALLOC(dword dwSize)
```

This behaves exactly as malloc does in the Standard C Library. You can also use the function CREATEOBJ, which is just an alias for MALLOC. Naturally, realloc is also available:

```
void * REALLOC(void * pSrc, uint32 dwSize)
```

Memory Deallocation

To release memory that you have allocated with MALLOC, use FREE:

```
void FREE(void * po)
```

This behaves as free does in the Standard C Library. It is also aliased by FREEOBJ; however, both FREE and FREEOBJ do the same thing. Remember that when you use WSTRDUP, as illustrated in the previous section, you must free the memory occupied by the duplicated string when you are done with it.

There is also a second kind of memory deallocation function that has a special purpose:

```
void SYSFREE(void * pBuff)
```

SYSFREE is used to free memory allocated via the CONVERTBMP macro. If you recall from Chapter 8, "Bitmap Graphics" CONVERTBMP converts a BMP image to a device-dependent bitmap. One of the arguments to CONVERTBMP is the address of a Boolean. If this Boolean is set by CONVERTBMP, the pointer to the device-dependent bitmap must be deallocated with SYSFREE when you are done with it. This will be covered in depth in Chapter 18, "Running on Hardware."

Timers

Timing is an important part of game programming. This is especially true with action games. Until now, our applets have only reacted to incoming messages. With no main loop, how can you do something as simple as moving an object across the screen? The answer is timers.

Setting a Timer

Timers are a way for BREW to call a given function at a specified interval. You can create a function and then tell BREW to call it later using **ISHELL_SetTimer**:

```
int ISHELL_SetTimer(IShell * pIShell, int32 dwMSecs, PFNNOTIFY pfn, void * pUser)
```

This function is part of the virtual grab bag that is **IShell**. After the interface pointer, pass the number of milliseconds from this current point in time that we want the function to execute via the **dwMSecs** argument. Then pass a pointer to the callback function that we wish to call at that time with **pfn**. In AEE.h the **PFNNOTIFY** function prototype is defined like so:

```
typedef void (*PFNNOTIFY)(void * pUser);
```

As you can see, it is a simple function that receives one void pointer as an argument. That brings us to the final argument in **ISHELL_SetTimer**, **pUser**. This argument is a void pointer that is passed as the singular argument to our **PFNNOTIFY** function. I usually cast my applet pointer to void and use that. That way, the callback function can access the applet's globals through the applet pointer.

Once our time is reached and BREW calls the function pointed to by **pfn**, the timer is killed. That means we have to reset the timer in the callback if we want it to be a continuous event. Therefore, if we want a timer loop that moves a character across the screen every 15 milliseconds, we might call **ISHELL_SetTimer**, like this:

```
int ISHELL_SetTimer(pApplet->pIShell, 15, UpdateFunction, (void*)pApplet)
```

Here we assume that we have a global pointer, **pApplet**, that also contains our shell pointer. Our **UpdateFunction** callback would look something like this:

```
void UpdateFunction(void * pData)
{
    myapp_t* pApplet = (myapp_t*)pData;

    MoveCharacterAround(pApplet);

    ISHELL_SetTimer(pApp->pIShell, 15, UpdateFunction, pData);
}
```

This callback calls the fictitious MoveCharacterAround function that presumably moves an image across the screen. Finally, call ISHELL_ SetTimer again, setting it up to call this very same function in another 15 milliseconds.

Canceling a Timer

Sometimes, you need to stop a timer's operation immediately. This is often in response to a suspend event where the applet must be paused and any currently running timers have to be halted. This is done through the function ISHELL_CancelTimer:

```
int ISHELL_CancelTimer(IShell * pIShell, PFNNOTIFY pfn, void * pUser)
```

Here we pass the interface pointer, which is followed by the pointer to our callback function. BREW will cancel all functions that have that function pointer as their callback. If pfn is NULL, it will use the pUser pointer. In this case, BREW will cancel all timers that are using the same pUser pointer.

The trick is to resume these timers once the applet gets a resume message. You will have to find a way to save the state of the applet and create all the valid timers once the applet resumes. This also involves calculating how much time was left until the next tick of the timer and setting up the timers with this adjusted millisecond value.

Telling Time

This brings us to our next issue, which is finding out the time. There are several helper functions to do this. The first of these is GET_ SECONDS:

```
uint32 GET_SECONDS()
```

This function returns the number of seconds since January 6, 1980, 00:00:00 GMT. I am not sure how useful this is, but it does have its purpose, I guess.

```
uint32 GET_TIMEMS()
```

Here we get the time of day in milliseconds. This is useful if you wish to display the current time of day in your applet. Perhaps a more informative way to get the date is by using GET_JULIANDATE:

```
void GET_JULIANDATE(uint32 dwSecs, JulianType * pDate)
```

You must pass the result of GET_TIMEMS as the first argument so that it can convert this value to the current date according to the Julian calendar. It puts this date information in a JulianType structure as referenced by the final argument, pDate. The JulianType is defined in AEEShell.h as:

```
typedef struct
{
    uint16  wYear;
    uint16  wMonth;
    uint16  wDay;
    uint16  wHour;
    uint16  wMinute;
    uint16  wSecond;
    uint16  wWeekDay;
} JulianType;
```

This structure is rather self-explanatory. From this structure, you can get the calendar and time information useful for displaying the day, date, and current time in your applet.

```
uint32 GET_UPTIMEMS()
```

Finally, we have **GET_UPTIMEMS**. This gets the number of milliseconds since the device was turned on. This is the closest thing BREW has to something like **GetTickCount**, if you are familiar with Win32 programming.

Random Numbers

Another useful tool in game programming is random number generation. Without the use of the Standard C Library's rand function, how do we create random integers? BREW has a helper function, **GET_RAND**, for this purpose:

```
void GET_RAND(byte * pDest, int nSize)
```

This works in a fairly different manner from the rand function with which we are all familiar. In this case, pass a pointer to a memory buffer in the pDest argument and then the size of this buffer with nSize. GET_RAND fills this chunk of memory with random bits. So, if you wanted to get a random integer value, you might do something like this:

```
int value;
GET_RAND((byte*)(&value), sizeof(int));
```

Sure, it is a bit of a hack, but that is how you have to do it. Unfortunately, BREW provides no way to seed this random number generator. Instead, it is automatically seeded by the local time on the device. On actual hardware, the performance of this function is rather unpredictable. Sometimes you will get a genuine array of random bits, and other times you will simply get a bunch of zeroes. For the final product, it may be worth investigating creating your own random number function. At least this is true when using version 1.0 of the SDK.

Note: An interesting alternative to the GET_RAND function is the Mersenne Twister algorithm. However, it does require a fairly hefty pre-loaded table. If you can spare the RAM, it may be worth investigating. Go to www.personal.engin.umich.edu/~wagnerr/ MersenneTwister.html for more information on its implementation.

Conclusion

This chapter has been a confusing jumble of seemingly unrelated information, but all of the concepts presented here will come in handy later down the line. BREW has plenty more hidden functions and handy data structures that you can only find by poring over the documentation and SDK example code. I suggest scanning through the documentation to see if there are any handy functions that you might want to earmark for future use.

Chapter 14

Putting It All Together

Introduction

If you have followed this book through the previous chapters, you have learned about most of the BREW components necessary for developing a game. In this chapter we will develop a very simple game to illustrate how everything works together. This is not a great game by any means. However, by studying the development of a complete and finished title, you can build upon this code or use the ideas presented to create your own masterpiece.

The Mobile Game Development Process

I have seen many different authors write books about how games should be designed or developed. The truth is, if you ask a dozen different developers, you will most likely get a dozen different answers. There seems to be no one simple solution to developing a game. Depending on the type of game that you are making, the people on the team, and many other factors, the process necessary to take your project from concept to completion varies wildly.

Developing games for mobile devices also has its own unique development process. Because the games are often very simple, the development process is much like that of the early days of gaming. This means that often only one programmer and artist are necessary to put together a commercial-quality title.

The average major console or PC title can have a development time of 12 to 24 months. With a mobile game, it is often reduced to one to three months. Sometimes, it is a matter of mere weeks. Games become more complicated once you enter the multiplayer realm, with server and network protocol development. However, for simple single-player games, you can bang out a great title in an astoundingly short period of time.

As I said before, there is no one way to develop a game. The process I describe in this chapter is not necessarily one that you should follow. It is just a method cobbled together from my own personal experience and industry observations. With mobile game development being so quick and informal, it is really up for debate whether you actually need a totally organized process in the first place. Regardless, I will detail a few major phases of the development process.

Design

Ideas come from anywhere. Although there are a lot of "industry visionaries" running around doing interviews and giving lectures at development conferences, just about anyone can come up with a good game concept. The real value in a design is the implementation. If you cannot figure out a way to develop your game concept and finish it, the idea is worthless. Whether your game is fun or not is up to the players to decide. By the time you finish a game, it is nearly impossible to have an objective opinion on the quality of the gameplay. Often, the soldiers in the trenches developing the game may have a completely different opinion of the quality of their product than the general gaming public. Occasionally, I have assumed the game I was working on was headed for disaster, when in reality the public loved it once the title was released. The fickle tastes of the public are hard to define. Also, when you are too close to a game's development, it is almost impossible to be objective about the game's quality or lack thereof. Luckily, the brief development time of the average mobile product greatly reduces the significance of this issue.

Catering to the Platform

With that said, creating a design for a mobile game is totally unique from that of the average multimillion dollar console title. As a programmer, you are in the unique position to take the technical challenges and limitations into account when designing your game. Probably one of the greatest dangers in game development is for the design to be overly ambitious and just downright impossible given the technical limitations of the target platform and resources available to the developers. In this case, you must be very cautious of the tiny amount of available memory, limited graphics display capabilities, lack of storage space, and restrictive input schemes of most BREW handsets. These issues will eventually be resolved as the relentless pace of progress creates more impressive phone hardware, but you still have to keep in mind the existing user base of older phones. For quite some time, you will have to target the most basic of BREW hardware with your design if you are interested in appealing to the widest possible audience.

With technical issues out of the way, what about actual game concepts? What kinds of ideas are suitable for mobile games? There are several key issues to take into account when developing for this venue. The first is length of play. If you want to play a game for long periods of time, you most likely would rather be in the comfort of your own home, sitting in front of your console or PC rather than hunched over your wireless phone fumbling around with the keypad and squinting at a tiny LCD screen. Because of the nature of the device, chances are the only idle time a player has with her phone is brief. Therefore, games must be designed to be played in short bursts, (for instance, while waiting for the bus or in line at a movie theater).

This is not to say the game cannot be designed for long-term play. You can have a game with long-term goals that may require many hours of play to achieve. However, the game must still be able to be successfully played in short bursts and interrupted at any time. It is highly annoying to lose your game because you got a phone call while in the middle of a climactic battle. Allow the user to pause and resume the game at any time with little or no effect on the game's progress.

Catering to Your Audience

If you are looking at this from a purely commercial perspective, you have to also take into account your audience's tastes. The audience for mobile games is not necessarily the same as that of console or PC games. People do not buy phones to play games, although this is slowly changing. I do believe that once mobile phone handsets become closer in quality to the hardware present on Nintendo's Game Boy Advance and other more modern consoles, the desire for traditional games of normal lengths and play styles will override the so-called "mobile" audience considerations. For now, keep it simple.

Therefore, duplicating what is successful on consoles and PCs, which appeal to a totally different market, does not necessarily equate to a successful mobile game. Right now, we are in the pioneering days of mobile gaming. Experimentation with all sorts of crazy and wild ideas not viable for PC titles with millions of dollars at risk may yield equally bizarre success stories.

Once this market matures, we will be able to see a pattern in mobile gaming tastes. Yet, even existing patterns can be wrong. Just look at how the entire PC gaming industry was turned on its head by the massive success of a "niche" market title such as Deer Hunter. In an era where every game seemed to be appealing to the bloodlust of your average teenage male, Sunspot Interactive's low-budget hunting simulator went on to rack up sales higher than many high-profile so-called "A" titles. Who am I to say how you should design and develop a game? In the end, it is entirely up to you. You can only observe industry facts and the wisdom of others and determine if it makes sense in what you are trying to accomplish. Ultimately, the players will determine if they like it or not.

Development

The term "game development" encompasses a lot more than just programming. Design and programming are just two of the overall elements involved in creating a completed game. Granted, with mobile gaming, that is about 99 percent of it. But in traditional game development, there are management concerns, work flow, tool creation, and a host of other issues that are encompassed in the general term "game development."

Asset Creation

In the game industry, an *asset* is defined as any element of the game that is plugged into the engine for the creation of content. Usually, you will hear people refer to "sound assets" or "art assets." They are usually talking about the sound (music, dialogue, and sound effects) or artwork (sprites, 3D models, and textures). The creation of these assets involves many different tools, like 3D modeling packages such as 3D Studio Max or Maya, bitmap drawing programs including Photoshop or Paint Shop Pro, sound editors like SoundForge or CoolEdit, and a host of other tools used by artists, musicians, and others to create graphics, models, music, animations, and just about any other content used to make a game.

Most of the graphics in mobile games are bitmaps and geometric 2D shapes. In the case of bitmaps, there are a number of programs available for creating images. It is irrelevant to the programmer, as most of them output standard BMPs that can be read by BREW. In the case of 2D geometric graphics, there are currently no tools for creating such content for use with BREW. Perhaps a way to import data from structured drawing programs like CorelDRAW® will appear in the future. Also, as the hardware becomes more powerful and 3D graphics are a possibility, the use of high-end 3D modeling packages will become commonplace in the development of mobile games.

Source Control

Another important factor in game development is source control. A source control system, such as Microsoft's Visual Source Safe or the open-source solution CVS, is one which manages changes coming from different developers to the same code. Often, this means only one programmer can modify a section of code at a time, and the source control application makes sure this is so. Although primarily a tool used when multiple software engineers are working on a project, source control can also serve some use for single developers as well. This is because most source control applications keep a log of all changes in each source file over the course of the entire project. You can then easily track changes and see exactly when problematic lines of code were added into any file. Checking the differences between different versions of the source code is invaluable when fixing hard-to-find bugs. Often, if you can look at the differences between the last working version of the source and the current version, it is quite easy to tell what the problem is.

The very same source control software can also be used to manage asset creation. If you have a project where multiple artists are modifying the same graphics assets, a source control tool can manage this potential logistical nightmare quite easily. Because artwork is usually stored in binary files, a source control program cannot usually detect the differences in two versions of the same art asset. However, it can at least keep a log and backup of any changes in a given piece of artwork over the course of the project.

Testing

Once you have your game up and running, you have to make sure it is bug-free. Well, bug-free is perhaps a little extreme. It seems that most programs these days ship with many bugs. Some bugs are worse than others. In these cases, it must be decided which bugs are "show-stoppers" (ones that the game cannot ship without fixing, such as a crash) and which do not significantly impact the player's experience. Under the pressure of time and limited resources, a publisher often has to make the decision as to which bugs the game can ship with and which are worth spending more time to fix.

In the PC realm, this decision is much easier, as patching the game after release is commonplace. In fact, many game developers announce the creation of a patch before the first version of the game even hits the shelves! On a console such as the PlayStation 2 or Xbox, standards are far more stringent. It is not possible to update the game later, despite the emergence of hard drives on some modern consoles. The game has to be totally stable before shipping, as you cannot go back and correct your mistakes with a patch. Mobile games are more like consoles in that patching is really not an option. Also, carriers will demand that your product be stable before offering them to their customers to purchase.

Although you as the programmer can test the game yourself, there is a need for external testing to bang on the game for a bit and see if it breaks. These third-party testers can be anyone from a software quality assurance firm to your little brother. By bringing in an impartial third party to play with your game, you can see if the game is used in a way you never thought of. This process may reveal bugs that would otherwise have gone unnoticed and even give you a window into how much fun the game actually is.

In the case of BREW, your applet must go through a stringent testing process before being offered to carriers. This testing process costs money. Every time the testers find a bug, the applet needs to be tested again. Therefore, it makes economic sense to make sure you catch as many of your own bugs as possible before sending it off to be certified as "True BREW." Otherwise, you will have to pay for another testing run every time you fix your bugs. This can definitely add up if you have to resubmit the program several times.

Attack of the Flarb

Now that I have gone through the overview of how a game is developed, we will create a simple game in this chapter: Attack of the Flarb. This is a mobile version of a really bad arcade action game that I created as a Java applet back in 1995. I am even using the same graphics that I created for that fine classic of web gaming. This is not going to turn any heads at the next Electronic Entertainment Exhibition, but at least it is a simple enough game to use as an example of a completely functional applet and to bring together the lessons learned in previous chapters. The source for this project is found in the Source\Chapter 14\flarb folder in the companion files.

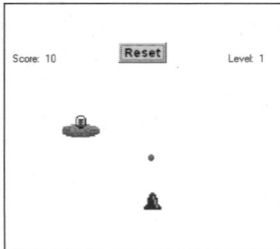

Figure 14-1: The original "classic," Attack of the Flarb, as a Java applet in 1995

In Attack of the Flarb, you move the ship at the bottom of the screen (known as the GULLET-1) and fire projectiles at the invading Flarb ship. The Flarb spacecraft travels across the screen from the left to the right, descending down a level when it reaches the edge and beginning back on the left side of the screen. When the bullet collides with the ship, the Flarb starts again at the top and descends at a faster pace. If the Flarb reaches the bottom of the screen, the game is over; their death grip on planet Earth is complete! Eventually, the relentless pace of the advancing Flarb will make success impossible and humanity will die. But such is our lot in life. Hey, maybe Attack of the Flarb is actually some kind of existential metaphor for life. Well, perhaps not. Either way, it is an undeniably cheesy attempt at some late '70s gameplay on a mobile phone.

As I have said, this game is not a prime candidate for quality; however, it exhibits a few of the design principles illustrated earlier. First, it adheres to the limited interface of the mobile phone by having the controls all one-handed. All gameplay is controlled by the joypad; therefore all you really need is your thumb. Also, it properly accepts the suspend and resume events, so the game pauses when an incoming call, text message, or other event is received. The actual game is an exceedingly simple arcade challenge, which lends itself to short bursts of play. However, the ability to save the high score gives the player a long-term goal in trying to beat her personal best.

The Elements of the Code

The main programming concepts behind this game are displaying moving bitmaps, basic collision detection, and simple file I/O for keeping track of high scores, not to mention handling various system events, such as suspending and resuming the applet. Looking at the cast of characters, you can see all of the bitmaps used in the game. Each one of these is stored in the associated resource file flarb.bid. You can open it with the BREW Resource Editor and see what else is in there. Aside from the graphics, we also store the strings used in the game and the interface in the resource file.

This game is designed to be used with the 7GP_256Color device profile. This simulates a device with 8-bit color depth and a resolution of 120x160. I have yet to see a BREW handset with these exact specifications, but if you would like to see this on a real handset, it should be no problem to modify the graphics to fit on a smaller screen. In fact, I try to use the screen resolution as described in the **AEEDeviceInfo** structure to make it somewhat compatible with most phone screen sizes.

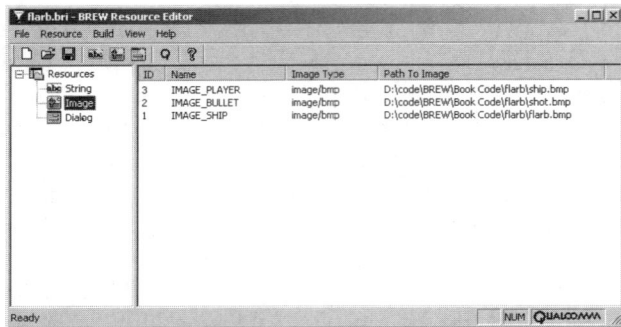

Figure 14-2: The Resource Editor with Flarb's BRI file open

We also created a BID file that contains another handmade BREW class ID number for this applet. If this were to be distributed commercially, we would need to generate a legitimate value from the BREW developer web site. For our purposes, it allows us to test the applet in the emulator as long as the class ID is not used by any other applets.

Header Definitions and Structures

With these two files out of the way, we have the source code to examine. We have one header and a single source file, as with most of the examples in this book. In the header we have a few enumerations and definitions that we need to look at:

Listing 14-1: Menu ID codes

```
enum
{
    MENUID_START    = 0,
    MENUID_SCORES   = 1,
    MENUID_QUIT     = 2,
};
```

Here we have an enumeration that defines the menu ID codes that we use in the menu control. By using an enumeration, we make sure that we do not have any two menu items with the same ID. This can wreak havoc with our menu event parsing code, and adding two menu items with the same ID will fail out. Therefore, it is a good idea to use enumerations instead of straight definitions if you want to make sure that all your values are unique.

Listing 14-2: The game modes

```
enum
{
    GAME_MODE_NONE = 0,
    GAME_MODE_MENU = 1,
    GAME_MODE_PAUSE = 2,
    GAME_MODE_OVER = 3,
```

```
        GAME_MODE_PLAY = 4,
        GAME_MODE_SCORES = 5,
        GAME_MODE_INSTRUCTIONS = 6,
};
```

Here we define the different game modes as enumerations. The game itself can be in a number of different states. For instance, when displaying the main menu, the game is set to the menu state. Most of the code checks to see the current state of the game before it does anything. This way, we know when to display the menu, draw the game graphics, and such.

I use this convention in pretty much all of my games.

Listing 14-3: Sprite types

```
enum
{
    SPRITE_PLAYER = 0,
    SPRITE_BULLET,
    SPRITE_SHIP,
};
```

Here we have a simple enumeration to index into our array of images used for sprite graphics. Our sprite structure looks like this:

Listing 14-4: The sprite structure

```
typedef struct sprite_s
{
    unsigned char nX;
    unsigned char nY;

    unsigned char nWidth;
    unsigned char nHeight;

    unsigned char nImage;

} sprite_t;
```

The first two members, nX and nY, are used to represent the x and y coordinates of the upper left-hand corner of the image in screen coordinates. We also have an nWidth and nHeight member to store the dimensions of the graphic for use with collision detection. Finally, we have nImage, which is used to index into the array of images loaded from the resource file. Whenever we draw a sprite, we use this value to pull the proper bitmap from our array.

The Applet Structure

Speaking of the array of bitmaps, this and many other global data structures are stored in our custom applet structure:

Listing 14-5: The applet structure

```
typedef struct myapp_s
{
    AEEApplet      a;        //applet header
    AEEDeviceInfo di;

    IMenuCtl    * pIMenu;
    IStatic     * pIStatic;

    boolean bPaused;
    boolean bSuspended;
    boolean bBullet;
    boolean bGameOver;

    int nScore;
    int nSpeed;
    int nHighScore;

    int nGameMode;

    void*      pRawImagePtrs[MAX_IMAGES];
    AEEBmp     pBMPImagePtrs[MAX_IMAGES];
    byte       pAllocFlags[MAX_IMAGES];

    AECHAR     szBuf[MAX_STRING_LENGTH];

    sprite_t    sprBullet;
    sprite_t    sprPlayer;
    sprite_t    sprAlien;

} myapp_t;
```

As is the standard with any BREW applet, we must first provide the
AEEApplet structure as our first member. Next, we have the customary
AEEDeviceInfo structure used for determining the capabilities of the
handset. Following this is a series of Booleans used to determine a few
different status events. For instance, if the applet is paused, the
bPaused Boolean is set to True. If there is currently a bullet in play,
bBullet is set to True. You get the idea. Next up, we have a few integers
to determine the current score, the speed of the alien Flarb spaceship,
and the current high score to beat. We also have an integer that represents the game mode. The nGameMode variable can be set to any one
of the game mode enumerations in Listing 14-2. The value of this member is checked to determine which mode the game is in.

Following these simple variables we have a few pointers and custom
data structures to illustrate. First, we have an array of void pointers
called pRawImagePtrs. This array will store pointers to our device-
dependent bitmaps used for drawing on the screen. Next, pBMPImage-
Ptrs contains pointers to our loaded bitmaps before we convert them to
device-dependent bitmaps. We have so few images that memory is not
really an issue, but in an actual commercial product, you would most

likely only keep pointers to the device-dependent bitmaps and not store the loaded BMPs.

Applet Creation

Looking at the source code to the applet in flarb.c, we see the standard applet construction routine common to all BREW applets:

Listing 14-6: Creating the applet instance

```
int AEEClsCreateInstance(AEECLSID ClsId, IShell * pIShell, IModule * po, void ** ppObj)
{
    *ppObj = NULL;

    if(ClsId == AEECLSID_FLARB_BID )
    {
        if(AEEApplet_New(sizeof(myapp_t), ClsId, pIShell, po, (IApplet**)ppObj,
            AEEHANDLER)Flarb_HandleEvent, (PFNFREEAPPDATA)Flarb_Cleanup)
            == TRUE)
        {
                return(AEE_SUCCESS);
        }
    }

    return(EFAILED);
}
```

There is nothing out of the ordinary here. We simply construct our applet with **AEEApplet_New** and pass it a pointer to the event handler function, **Flarb_HandleEvent**, and our cleanup function, **Flarb_Cleanup**.

The Message Handler

The message handler looks like this:

Listing 14-7: The event handler

```
static boolean Flarb_HandleEvent(IApplet * pi, AEEEvent eCode, uint16 wParam, uint32
    dwParam)
{
    FileInfo fileInfo;
    IFileMgr  *  pIFileMgr = NULL;
    IFile     *  pIFile    = NULL;
    myapp_t   *  pApp      = (myapp_t*)pi;
    AEEApplet *  pMe       = &pApp->a;

    switch (eCode)
    {
        case EVT_APP_START:

        //initialize variables
        pApp->bPaused     = FALSE;
        pApp->bSuspended  = FALSE;
        pApp->pIMenu      = NULL;
        pApp->pIStatic    = NULL;
        pApp->nGameMode   = GAME_MODE_NONE;
```

```
        pApp->pIMenu = NULL;
        pApp->pIStatic = NULL;

        //open our high score file
        ISHELL_CreateInstance(pApp->a.m_pIShell,
            AEECLSID_FILEMGR,(void **)&pIFileMgr);
        pIFile = IFILEMGR_OpenFile(pIFileMgr, "highscore.txt", _OFM_READ);

        //if we cannot find the file, create it
        if (!pIFile)
        {
int nZero = 0;

 pIFile = IFILEMGR_OpenFile(pIFileMgr,
            "highscore.txt", _OFM_CREATE);
            //write a zero score entry, and rewind
            IFILE_Write(pIFile, &nZero, sizeof(int));
            IFILE_Seek(pIFile, _SEEK_START, 0);
        }

            //read in our high score if we have one. If not, make the
            //high score 0
            IFILE_GetInfo(pIFile, &fileInfo);

            if (fileInfo.dwSize == sizeof(int))
            {
                IFILE_Read(pIFile, &pApp->nHighScore, sizeof(int));
            }
            else
            {
                int nTemp = 0;

                IFILE_Write(pIFile, &nTemp, sizeof(int));
                pApp->nHighScore = 0;
            }

            //release our file resources
            IFILE_Release(pIFile);
                IFILEMGR_Release(pIFileMgr);

                //get the device info (screen resolution, etc.)
                ISHELL_GetDeviceInfo (pMe->m_pIShell, &pApp->di);

                //set background color to white
                IDISPLAY_SetColor(pApp->a.m_pIDisplay,
                    CLR_USER_BACKGROUND, MAKE_RGB(0xFF, 0xFF, 0xFF));

                //clear out our data arrays
                MEMSET(pApp->pAllocFlags, 0, sizeof(char) *
                        MAX_IMAGES);
                MEMSET(pApp->pBMPImagePtrs, 0, sizeof(int) *
                        MAX_IMAGES);
                MEMSET(pApp->pRawImagePtrs, 0, sizeof(int) *
                        MAX_IMAGES);

                Flarb_LoadResources(pApp);

                Flarb_SetGameMode(pApp, GAME_MODE_MENU);
                 return(TRUE);
```

```
                                        break;

                        case EVT_APP_SUSPEND:
                                Flarb_SuspendGame(pApp);
                                break;

        case EVT_APP_RESUME:
                        Flarb_ResumeGame(pApp);
                        return(TRUE);
                        break;

                case EVT_KEY_PRESS: //input from keypad
                        if (pApp->pIMenu != NULL)
                        {
                                IMENUCTL_HandleEvent(pApp->pIMenu, EVT_KEY, wParam, 0);
                                return(TRUE);
                        }
                        else if (pApp->pIStatic != NULL)
                        {
                                Flarb_SetGameMode(pApp, GAME_MODE_MENU);
                                return(TRUE);
                        }
                                else if (pApp->nGameMode == GAME_MODE_PLAY)
                                        Flarb_GameInput(pApp, wParam);
                        break;

                case EVT_COMMAND:   //Menu GUI commands
                        switch(wParam)
                        {

                                case MENUID_START:
                                        Flarb_SetGameMode(pApp, GAME_MODE_PLAY);
                                        return(TRUE);
                                        break;

                                case MENUID_SCORES:
                                        Flarb_SetGameMode(pApp, GAME_MODE_SCORES);
                                        return(TRUE);
                                        break;

                                case MENUID_QUIT:
                                        ISHELL_CloseApplet(pApp->a.m_pIShell, FALSE);
                                        return(TRUE);
                                        break;
                        }
                        break;
        }
            return(FALSE);
}
```

At the start of this function, we declare a few variables. The first, fileInfo, is of a FileInfo structure. This is used to retrieve the size of the high score file. The pointer to the IFileMgr interface, pIFileMgr, is used to read the score file. Of course, the IFile interface pointer, pIFile, will point to the score file that we want to read once it is opened or created. Finally, we have pointers to our applet structure, pApp, and to the BREW AEEApplet structure, pMe.

Starting the Applet

Now we will move on to the first event handler, the EVT_APP_START case in Listing 14-7. Here, we initialize all of our global applet variables and proceed to open the high score file. In this case, we are saving our high score in the file highscore.txt with IFILEMGR_OpenFile. As you can see, we pass the OFM_CREATE and OFM_READ bit flags to the call, which means we will create the file if it does not exist. If the file cannot be opened for some reason, release the IFileMgr interface and close the applet with a simple call to ISHELL_CloseApplet. Why would this call fail if it creates the file if it is not found? Each phone has a limit to how many files can exist in the handset's file system. In some rare case it is possible to hit this limit, and thus the attempt to create a non-existent file will fail.

Next, if we have a high score file present, read the first integer in this file into our nHighScore member. If not, clear the high score and write the value of 0 to the file. This creates a proper cleared high score file with the high score set to 0 points.

After releasing the file interfaces, get the device information with a call to ISHELL_GetDeviceInfo, and then set the background color to white with a call to IDISPLAY_SetColor. Finally, clear out our global arrays declared in our applet structure. We cannot guarantee that these memory areas will be initialized upon applet construction.

Loading Images

Next, load the applet resources with a call to our own function, Flarb_LoadResources:

Listing 14-8: Loading the image resources

```
void Flarb_LoadResources(myapp_t* pApp)
{
    //Load each image and fill sprite structure accordingly
    Flarb_LoadImage(pApp, IMAGE_SHIP, SPRITE_SHIP, &pApp->sprAlien);
    Flarb_LoadImage(pApp, IMAGE_PLAYER, SPRITE_PLAYER, &pApp->sprPlayer);
    Flarb_LoadImage(pApp, IMAGE_BULLET, SPRITE_BULLET, &pApp->sprBullet);
}
```

The Flarb_LoadResources function is a simple one, which calls Flarb_LoadImage for every bitmap that we want to use. Next, we will look at Flarb_LoadImage.

Listing 14-9: Loading the actual bitmap

```
void* Flarb_LoadImage(myapp_t* pApp, int nResID, int nIndex, sprite_t* pSprite)
{
    AEEBmp pbmSource;
    AEEImageInfo imageInfo;
    byte* pDataBytes;
```

```
boolean bVal = FALSE;

//pull the BMP from the resource file
pApp->pBMPImagePtrs[nIndex] = pbmSource = ISHELL_LoadResData(pApp->a.m_pIShell,
        RES_FILE, (short)nResID, RESTYPE_IMAGE);

if (pApp->pBMPImagePtrs[nIndex] == NULL)
    return(NULL);

pDataBytes = (byte *)pbmSource + *((byte *)pbmSource);
pApp->pRawImagePtrs[nIndex] = CONVERTBMP(pDataBytes, &imageInfo, &bVal);

//if we have allocated memory in the conversion, signify
if (bVal)
{
    pApp->pAllocFlags[nIndex] = 1;
}
else
    pApp->pAllocFlags[nIndex] = 0;

//if we want to retain the sprite info, fill it in
if (pSprite)
{
    pSprite->nWidth = (unsigned char)imageInfo.cx;
    pSprite->nHeight = (unsigned char)imageInfo.cy;
    pSprite->nImage = nIndex;
}

return(pApp->pRawImagePtrs[nIndex]);
}
```

This function does several things. First, look at the arguments:

```
void* Flarb_LoadImage(myapp_t* pApp, int nResID, int nIndex, sprite_t* pSprite)
```

This function takes a pointer to the applet. Next, pass the resource ID of the bitmap that we want to load. This bitmap ID is defined in the header file generated by BREW's Resource Editor, flarb_res.h. Next, pass the index in our sprite image array where we want this image to be loaded. Finally, pass the address of a sprite structure that we want to fill with information about the image. The function returns a pointer to the device-dependent bitmap, which is just an entry in the array declared in myapp_t.

Looking at the function itself, start with a few variable declarations. The first variable, pbmSource, is of type AEEBmp. AEEBmp is actually just defined as a void pointer in the AEEDisp.h header file. In the end, it serves as a pointer to raw BMP image data. Next, we have an AEEImageInfo structure, imageInfo, used to collect information about the loaded BMP, such as the width and height. We then have a pointer of type byte, which will be used as another pointer into our raw BMP data. Finally, we have the Boolean bVal, which is used to determine if the CONVERTBMP macro has allocated memory or not.

After the variable declarations, the first thing we do is load the bitmap out of the resource file with a call to ISHELL_LoadResData. The resultant IImage pointer is stored in our pBMPImagePtrs array indexed by the nIndex argument. If for some reason this pointer is NULL, return from the function with a NULL value.

The next few lines of code may be a bit confusing:

```
pDataBytes = (byte *)pbmSource + *((byte *)pbmSource);
pApp->pRawImagePtrs[nIndex] = CONVERTBMP(pDataBytes, &imageInfo, &bVal);
```

When a BMP is pulled from a resource file, the first part of the file contains some header information. The first byte of the file is a number that tells how long this header is in bytes. Hence, we skip ahead by the number stored in the first byte of pbmSource and set this new position to the pDataBytes pointer. The pDataBytes pointer now points to the actual BMP data that can then be converted with the CONVERTBMP macro, as we have seen in previous chapters.

Next, set the flag to 0 or 1, depending on the value of our Boolean bVal, which is modified by CONVERTBMP if the macro allocates memory in the process. You will find in the emulator that bVal is always set to False. However, on a real handset, CONVERTBMP will often allocate memory, and thus set bVal to True.

Finally, if a valid pointer to a sprite structure has been passed to the function, set a few of its members according to the information set in imageInfo by the CONVERTBMP macro. Finally, return a pointer to this new image at the end of the function.

Game Modes

Getting back to the message handler, call our Flarb_SetGameMode function. This function is used frequently to set the game into different states. In this case, we are setting it to the main menu state.

Listing 14-10: Setting the game mode

```
void Flarb_SetGameMode(myapp_t* pApp, int nMode)
{
    //Kill timer clear gui for all state changes
    ISHELL_CancelTimer(pApp->a.m_pIShell, NULL, NULL);
    Flarb_ClearGUI(pApp);

    switch (nMode)
    {

        case GAME_MODE_MENU:
            Flarb_BuildMainMenu(pApp);
            break;

        case GAME_MODE_PLAY:
            pApp->nScore = 0;
            pApp->nSpeed = START_SPEED;
```

```
            Flarb_SetupGame(pApp);
            break;

        case GAME_MODE_SCORES:
            Flarb_BuildScoreScreen(pApp);
            break;
    }

        //This is now our current game mode
        pApp->nGameMode = nMode;
}
```

This function simply does a little housekeeping by canceling any current timers and clearing the GUI components, much like our example applet in Chapter 10. After calling the appropriate mode function, set the game mode variable and end the function. Looking at Flarb_ClearGUI, you see some code that you should be familiar with by now:

Listing 14-11: Destroying the GUI

```
void Flarb_ClearGUI(myapp_t* pApp)
{
    //Release all valid GUI controls
    if (pApp->pIMenu)
    {
        IMENUCTL_Release(pApp->pIMenu);
        pApp->pIMenu = NULL;
    }

    if (pApp->pIStatic)
    {
        ISTATIC_Release(pApp->pIStatic);
        pApp->pIStatic = NULL;
    }
}
```

This is basically the same code we used in Chapter 10. We see if any of the GUI object pointers are valid and release their interfaces if so. This way, we deallocate any memory allocated in the creation of these GUI controls, as well as remove them from the screen. Therefore, use this function upon applet close to clean up memory—it is not just a cosmetic function.

The Main Menu

Getting back to the Flarb_SetGameMode function, take a look at Flarb_BuildMainMenu.

Listing 14-12: Building the menu screen

```
void Flarb_BuildMainMenu(myapp_t* pApp)
{
    //create the menu
    ISHELL_CreateInstance(pApp->a.m_pIShell,
```

```
          AEECLSID_MENUCTL, (void **)&pApp->pIMenu);

     //set up the menu
     IMENUCTL_SetTitle(pApp->pIMenu, RES_FILE, STR_MENUTITLE, NULL);
     IMENUCTL_SetRect(pApp->pIMenu, NULL); //full-screen

     //Add in our menu items
     IMENUCTL_AddItem(pApp->pIMenu, RES_FILE,
         STR_MENUSTART, MENUID_START, NULL, 0);
     IMENUCTL_AddItem(pApp->pIMenu, RES_FILE,
         STR_MENUSCORES, MENUID_SCORES, NULL, 0);
     IMENUCTL_AddItem(pApp->pIMenu, RES_FILE,
         STR_MENUQUIT, MENUID_QUIT, NULL, 0);

     IMENUCTL_SetActive(pApp->pIMenu,TRUE);
}
```

Once again, we see a function similar to those used in the GUI example. Here we create a menu control with a call to ISHELL_CreateInstance. Next, set up the menu's title and screen dimensions. Finally, add in each menu element and set this menu control as the active GUI component. Each call to IMENUCTL_AddItem passes along a menu item ID, as defined in the header file, along with a resource ID for the string used in each menu item.

Now, looking at our message handler again, we see how we handle the message sent by this menu control:

```
case EVT_COMMAND:      //Menu GUI commands
    switch(wParam)
    {
        case MENUID_START:
            Flarb_SetGameMode(pApp, GAME_MODE_PLAY);
            return(TRUE);
            break;

        case MENUID_SCORES:
            Flarb_SetGameMode(pApp, GAME_MODE_SCORES);
            return(TRUE);
            break;

        case MENUID_QUIT:
            ISHELL_CloseApplet(pApp->a.m_pIShell, FALSE);
            return(TRUE);
            break;
    }
    break;
```

We simply call Flarb_SetGameMode to the mode appropriate for each menu selection. The function we call does all of the work. As seen in Listing 14-10, calling Flarb_SetGameMode with a game mode of GAME_MODE_PLAY starts the game with a call to Flarb_SetupGame.

Listing 14-13: Setting up the game

```
void Flarb_SetupGame(myapp_t* pApp)
{
    pApp->bGameOver = FALSE;
    pApp->bBullet = FALSE;

    //the player starts at the bottom, centered
    pApp->sprPlayer.nX = pApp->di.cxScreen / 2 - (pApp->sprPlayer.nWidth / 2);
    pApp->sprPlayer.nY = pApp->di.cyScreen - pApp->sprPlayer.nHeight;

    //the alien starts near the top
    pApp->sprAlien.nX = 0;
    pApp->sprAlien.nY = pApp->sprAlien.nHeight;

    //Set the main game loop timer up
    ISHELL_SetTimer(pApp->a.m_pIShell, TIMER_PERIOD,
        (PFNNOTIFY)Flarb_TimerCallback, pApp);
}
```

The Timer

This function is used to reset the status of the game, not only for when you start a new game but when you successfully destroy the Flarb and the ship returns to the top of the screen. This is why before we call this function in our Flarb_SetGameMode function, you can see in Listing 14-10 that we reset the score and speed of the enemy ship. The function Flarb_SetupGame handles the rest of the tasks, including setting the player and enemy positions and starting the timer. You can see that we take the width of the sprites and the size of the screen into account when positioning the sprites. This will ensure that the game at least looks somewhat normal when running on devices with differing resolutions. At the conclusion of this function, you can see that we start the timer with a call to ISHELL_SetTimer. The callback function, Flarb_TimerCallback, handles the main loop tasks of the game.

Listing 14-14: The timer callback

```
void Flarb_TimerCallback(myapp_t* pApp)
{
    if (pApp->bPaused)
        return;

    if (pApp->bSuspended)
        return;

    if (pApp->bGameOver)
        return;

    //move objects
    Flarb_MoveEnemy(pApp);
    Flarb_MoveBullet(pApp);

    //perform collision check if we have a bullet in play
    if (pApp->bBullet)
```

```
        Flarb_DoCollision(pApp);

    //redraw screen
    Flarb_DrawScreen(pApp);

    //Set the timer up again—time for another go
    ISHELL_SetTimer(pApp->a.m_pIShell, TIMER_PERIOD,
        (PFNNOTIFY)Flarb_TimerCallback, pApp);
}
```

This is the heart of the game. Every time this callback is fired off, if the game is not over or paused, we move the objects in the game, check for collisions, redraw the screen, and then start the process all over again with a call to ISHELL_SetTimer.

Enemy Movement

The first function that we call to move our on-screen objects is Flarb_MoveEnemy.

Listing 14-15: Enemy movement

```
void Flarb_MoveEnemy(myapp_t* pApp)
{
    pApp->sprAlien.nX += pApp->nSpeed;

    //If we've gone off the edge, move the alien down a notch
    if (pApp->sprAlien.nX >= (pApp->sprAlien.nWidth + pApp->di.cxScreen))
    {
        pApp->sprAlien.nX = 0;
        pApp->sprAlien.nY += pApp->sprAlien.nHeight;
    }

    //If the alien has hit the 'bottom' of the screen,then the game is over
    if (pApp->sprAlien.nY >= (pApp->di.cyScreen - pApp->sprAlien.nHeight - 10))
    {
        pApp->bGameOver = TRUE;

        if (pApp->nScore > pApp->nHighScore)
        {
            IFileMgr * pIFileMgr = NULL;
            IFile * pIFile = NULL;

            //if we have achieved a high score, write it out to the file
            ISHELL_CreateInstance(pApp->a.m_pIShell,
                AEECLSID_FILEMGR,(void **)&pIFileMgr);
            pIFile = IFILEMGR_OpenFile(pIFileMgr, "highscore.txt", _OFM_READ);

            if (!pIFile)
            {
                int nErr = IFILEMGR_GetLastError(pIFileMgr);
                    IFILEMGR_Release(pIFileMgr);
                ISHELL_CloseApplet(pApp->a.m_pIShell, FALSE);
            }

            pApp->nHighScore = pApp->nScore;
```

```
        IFILE_Write(pIFile, &pApp->nHighScore, sizeof(int));

        //release our file resources
        IFILEMGR_Release(pIFileMgr);
        IFILE_Release(pIFile);
      }
    }
}
```

The general purpose of this function is to move the enemy ship and detect if it has either hit the edge of the screen or ended the game by reaching the bottom. We do this by accessing the **sprAlien** member of our applet structure. This contains all of the position and size information about our sprite, some of which was set upon loading the image resource.

The first thing we do in this function is move the alien sprite over to the right by adding the **nSpeed** member to its x coordinate. Detect if the left edge of the ship has passed the screen, and if so, move the alien down one notch and move it all the way back to the left edge of the screen.

Next, check if the y coordinate is such that the alien has hit the bottom of the screen. If so, the game is over. Therefore, set the **bGameOver** value to True and see if we have reached a high score. If a high score has been achieved, write the score value out to the file. By setting the **bGameOver** variable to True, we are telling the applet that we are in the game over state. This information is used by the drawing and timer functions to pause the game and display "Game Over" text.

Getting back to the timer callback function, move the bullet after the enemy if there is a bullet in play. This simple function moves the bullet up the screen and checks if it is off the edge of the screen.

Listing 14-16: Moving the bullet

```
void Flarb_MoveBullet(myapp_t* pApp)
{
    pApp->sprBullet.nY -= 2;

    //If the bullet has passed the top of the screen, remove it from play
    if (pApp->sprBullet.nY <= 5)
    {
        pApp->bBullet = FALSE;
    }
}
```

As you can see, we simply subtract from the bullet sprite's y value to move it up the screen. If the bullet has gone near the top of the screen, remove it from play and set our bullet status Boolean, **bBullet**, to False.

Collision Detection

If a bullet is in play, the next operation is to check for collisions between the bullet and the invading Flarb spaceship. Do this through the function Flarb_DoCollision.

Listing 14-17: Collision detection

```
void Flarb_DoCollision(myapp_t* pApp)
{
    //check to see if bullet rectangle intersects space ship
    //rectangle

    //this is based off of a handy algorithm found at:
    //http://www.gamedev.net/reference/articles/article735.asp

    short left1, left2;
    short right1, right2;
    short top1, top2;
    short bottom1, bottom2;

    left1 = pApp->sprBullet.nX;
    left2 = pApp->sprAlien.nX;
    right1 = pApp->sprBullet.nX + pApp->sprBullet.nWidth;
    right2 = pApp->sprAlien.nX + pApp->sprAlien.nWidth;
    top1 = pApp->sprBullet.nY;
    top2 = pApp->sprAlien.nY;
    bottom1 = pApp->sprBullet.nY + pApp->sprBullet.nHeight;
    bottom2 = pApp->sprAlien.nY + pApp->sprAlien.nHeight;

    //do we not have a collision?
    if (bottom1 < top2)
        return;

    if (top1 > bottom2)
        return;

    if (right1 < left2)
        return;

    if (left1 > right2)
        return;

    //otherwise, we have a collision!

    //increase our score
    pApp->nScore += 100;

    //hide the bullet
    pApp->bBullet = FALSE;

    //place alien back at the top
    pApp->sprAlien.nX = 0;
    pApp->sprAlien.nY = pApp->sprAlien.nHeight;

    //increase the alien's movement speed
    pApp->nSpeed += 2;
```

```
    if (pApp->nSpeed > MAX_SPEED)
        pApp->nSpeed = MAX_SPEED;
}
```

This function is based off of a handy code snippet found on the enormously useful game programming site www.gamedev.net. This function checks to see if the rectangle of the ship intersects the rectangle of the bullet. The dimensions of the rectangle are taken from the image dimensions, as stored in the relevant sprite structures. If there is a collision, add 100 points to the score, move the alien ship back to the top of the screen, and increase the movement speed of the Flarb. This way, the game becomes harder with every successful hit. Eventually, the Flarb will reach its maximum speed to which the nSpeed variable is capped. You will most likely be dead by the time that happens.

Drawing the Screen

Before setting the timer off again, the last thing the timer callback function does is draw the screen via a call to Flarb_DrawScreen.

Listing 14-18: Drawing the screen

```
void Flarb_DrawScreen(myapp_t* pApp)
{
    AECHAR szTempBuf[] = {'%','d','\0'};

    //clear background
    IDISPLAY_ClearScreen(pApp->a.m_pIDisplay);

    //draw score and number of lives
    WSPRINTF(pApp->szBuf, (MAX_STRING_LENGTH * sizeof(AECHAR)),
        szTempBuf, pApp->nScore);
    IDISPLAY_DrawText(pApp->a.m_pIDisplay, AEE_FONT_NORMAL,
        pApp->szBuf, -1, 5, 0, NULL, IDF_TEXT_TRANSPARENT);

    //draw player
    IDISPLAY_BitBlt(pApp->a.m_pIDisplay, pApp->sprPlayer.nX,
        pApp->sprPlayer.nY, -1, -1, pApp->pRawImagePtrs[pApp->sprPlayer.nImage],
        0, 0, AEE_RO_COPY);

    //draw alien
    IDISPLAY_BitBlt(pApp->a.m_pIDisplay, pApp->sprAlien.nX,
        pApp->sprAlien.nY, -1, -1, pApp->pRawImagePtrs[pApp->sprAlien.nImage],
        0, 0, AEE_RO_COPY);

    if (pApp->bBullet)
        IDISPLAY_BitBlt(pApp->a.m_pIDisplay, pApp->sprBullet.nX,
            pApp->sprBullet.nY, -1, -1, pApp->pRawImagePtrs[pApp->sprBullet.nImage],
            0, 0, AEE_RO_COPY);

    //draw game over text
    if (pApp->bGameOver)
    {
        ISHELL_LoadResString(pApp->a.m_pIShell, RES_FILE,
            STR_GAMEOVER, pApp->szBuf, sizeof(pApp->szBuf));
```

```
        IDISPLAY_DrawText(pApp->a.m_pIDisplay, AEE_FONT_NORMAL,
            pApp->szBuf, -1, 0, 0, NULL,
            IDF_TEXT_TRANSPARENT | IDF_ALIGN_CENTER | IDF_ALIGN_MIDDLE);
    }

    //refresh screen
    IDISPLAY_Update(pApp->a.m_pIDisplay);
}
```

This function performs the basic tasks of drawing the graphics of the game. The first thing we do is create a string, which is used for the **WSPRINTF** macro to convert the current score to a printable wide-character string. Of course, also clear the screen with a call to **IDISPLAY_ClearScreen** before the drawing begins.

The first thing we draw on the screen is the score. Use the **WSPRINTF** macro to convert the score to a text string, as previously described. Then draw this text at the top of the screen with a call to **IDISPLAY_DrawText**. Next, draw the player and enemy ship with a call to **IDISPLAY_BitBlt**. Looking at the player drawing call, we see the following code:

```
IDISPLAY_BitBlt(pApp->a.m_pIDisplay, pApp->sprPlayer.nX, pApp->sprPlayer.nY, -1, -1,
    pApp->pRawImagePtrs[pApp->sprPlayer.nImage], 0, 0, AEE_RO_COPY);
```

For each sprite, pass in the coordinates as well as the index into our array of bitmaps, as stored in the sprite structure. Using the **AEE_RO_COPY** mode is a simple pixel copy, drawing the pixels of the image on the background with no fancy processing. If a bullet is in play, make the same kind of call to draw the bullet sprite. Also, if the game is over, as signified by the **bGameOver** variable set in the **Flarb_Move-Enemy** function, draw the "Game Over" text as well. Finally, call **IDISPLAY_Update** to execute the drawing operations that we have called, hence updating the screen with the new graphics.

Player Input

Now we have seen how we move the enemy ship and bullets, as well as calculate collisions and update the screen. But how do we accept player input and move the mighty GULLET-1 ship around the screen? We do this in response to keypress events as accepted in the message handler. Looking at the handler code, we see:

```
case EVT_KEY_PRESS: //input from keypad
    if (pApp->pIMenu != NULL)
    {
        IMENUCTL_HandleEvent(pApp->pIMenu, EVT_KEY, wParam, 0);
        return(TRUE);
    }
    else if (pApp->pIStatic != NULL)
    {
        Flarb_SetGameMode(pApp, GAME_MODE_MENU);    //go back to the main menu
```

```
            return(TRUE);
        }
    else if (pApp->nGameMode == GAME_MODE_PLAY)
        Flarb_GameInput(pApp, wParam);
    break;
```

If a GUI control is up, pass the key event to the appropriate control. However, if the game is actually playing, call **Flarb_GameInput** to handle moving the ship around.

Listing 14-19: Parsing input

```
void Flarb_GameInput(myapp_t* pApp, uint16 wParam)
{
    unsigned char nTempPos = pApp->sprPlayer.nX;;

    //if the game is over, return to the main menu
    //instead of controlling the player
    if (pApp->bGameOver)
    {
        Flarb_SetGameMode(pApp, GAME_MODE_MENU);
        return;
    }

    //move the ship or fire a bullet based on key input
    switch (wParam)
    {
        case AVK_LEFT:
        pApp->sprPlayer.nX -=1;

        //check if we've hit the edge
        if (pApp->sprPlayer.nX < pApp->sprPlayer.nWidth)
            pApp->sprPlayer.nX = nTempPos;
        break;

        case AVK_RIGHT:
            pApp->sprPlayer.nX += 1;

            //check if we've hit the edge
            if (pApp->sprPlayer.nX > (pApp->di.cxScreen - pApp->sprPlayer.nWidth))
                    pApp->sprPlayer.nX = nTempPos;
            break;

        case AVK_SELECT:
            if (!pApp->bBullet)
            {
                //launch bullet
                pApp->sprBullet.nX = (pApp->sprPlayer.nX +
                    (pApp->sprPlayer.nWidth / 2) - (pApp->sprBullet.nWidth / 2));
                pApp->sprBullet.nY = pApp->sprPlayer.nY - pApp->sprPlayer.nHeight +
                    15;

                pApp->bBullet = TRUE;
            }
            break;
    }
}
```

First, check if the game is over. If so, return to the main menu. That way, when the game is over, any keypress will essentially reset the applet. Then look at what type of key message we are receiving through the **wParam** argument. In response to the left and right keys, add or subtract from the ship's position and then check if its edge is off the side of the screen. If so, move the ship back to its previous position so it does not slide off the screen.

If the select key is pressed, release a bullet (that is, if a bullet is not currently in play). We only allow one bullet on the screen at once to at least give the Flarb a fighting chance. Therefore, if there is no bullet currently in play, we place the bullet sprite just above the center of the player's ship and set the Boolean **bBullet** to True. This way, the applet knows to move and check collision against this bullet with the next call to the timer's callback function.

High Score Screen

The only other major game mode is the high score display screen. When the high score option is selected from the main menu, the message handler sets the game mode to **GAME_MODE_SCORES**, which in turn calls **Flarb_BuildScoreScreen**.

Listing 14-20: Building the high score display

```
void Flarb_BuildScoreScreen(myapp_t* pApp)
{
    AEERect qrc;
    AECHAR szTitle[32];
    AECHAR szTempBuf[] = {'%','d','\0'};

    Flarb_ClearGUI(pApp);

    qrc.x = qrc.y = 0;
    qrc.dx = pApp->di.cxScreen;
    qrc.dy = pApp->di.cyScreen;

    ISHELL_CreateInstance(pApp->a.m_pIShell, AEECLSID_STATIC,
        (void **)&pApp->pIStatic);

    //load title text
    ISHELL_LoadResString(pApp->a.m_pIShell, RES_FILE,
        STR_SCORETITLE, szTitle, sizeof(szTitle));

    //convert high score to text
    WSPRINTF(pApp->szBuf, (MAX_STRING_LENGTH * sizeof(AECHAR)),
        szTempBuf, pApp->nHighScore);

    ISTATIC_SetRect(pApp->pIStatic, &qrc);
    ISTATIC_SetText(pApp->pIStatic, szTitle, pApp->szBuf,
        AEE_FONT_BOLD, AEE_FONT_NORMAL);
```

```
ISTATIC_Redraw(pApp->pIStatic);
IDISPLAY_Update(pApp->a.m_pIDisplay);
}
```

This function takes the high score value, builds a static text control, and fills it with the high score converted to a wide-character string. This is essentially the same code that you saw in the GUI example, except we use **WSPRINTF** to convert the high score integer to a string.

Suspend and Resume

The only two events left to handle in this applet are **EVT_APP_SUS-PEND** and **EVT_APP_RESUME**. Looking at these two events in the message handler we see the following:

```
case EVT_APP_SUSPEND:
        Flarb_SuspendGame(pApp);
        return(TRUE)
        break;

case EVT_APP_RESUME:
        Flarb_ResumeGame(pApp);
        return(TRUE);
        break;
```

In response to a suspend event we call the function Flarb_Suspend-Game:

Listing 14-21: Suspending the game

```
void Flarb_SuspendGame(myapp_t* pApp)
{
    if (pApp->pIMenu)
    {
            IMENUCTL_SetActive(pApp->pIMenu,FALSE);
    }
    else if (pApp->pIStatic)
    {
            ISTATIC_Release(pApp->pIStatic);
            pApp->pIStatic = NULL;
    }

    ISHELL_CancelTimer(pApp->a.m_pIShell, NULL, NULL);
}
```

All suspend events must cancel any asynchronous activity such as timers. Otherwise, they will run in the background while the process that interrupted the applet executes. In addition, we must deactivate or otherwise destroy currently active GUI controls. If we do not, active controls may draw on top of the screen when the interrupting event is displaying. In the case of an active menu, we simply call **IMENUCTL_SetActive** and pass it an argument of False to deactivate the control. Unfortunately, **IStatics** have no such function, and thus must be destroyed only to be recreated upon resumption of the applet.

Note: Pause and resume events are no joke. Improper handling of these events is one of the top bugs found in the True BREW certification process.

To resume the game, we make a call to Flarb_ResumeGame:

Listing 14-22: Resuming the game

```
void Flarb_ResumeGame(myapp_t* pApp)
{
    if (pApp->pIMenu)
    {
            IMENUCTL_SetActive(pApp->pIMenu,TRUE);
            IMENUCTL_Redraw(pApp->pIMenu);
    }
    else if (pApp->nGameMode == GAME_MODE_SCORES)
    {
            Flarb_BuildScoreScreen(pApp);
            ISTATIC_Redraw(pApp->pIStatic);
    }
    else if (pApp->nGameMode == GAME_MODE_PLAY)
    {
            Flarb_DrawScreen(pApp);

            //set up timer
            ISHELL_SetTimer(pApp->a.m_pIShell, TIMER_PERIOD,
                            (PFNNOTIFY)Flarb_TimerCallback, pApp);
    }
}
```

In our resume function, we must first determine if we have any active GUI controls and redraw or rebuild them. If we have a valid pointer to a menu, we simply activate it and redraw it. The only menu in the game is the main menu, so it is a given that we are in the main menu if this is the case. If the game is in the score display mode, we must rebuild the high score display static control and redraw it. Otherwise, we are in the midst of gameplay and must redraw the screen and reactivate the main game loop timer. If you really want to seriously test your suspend and resume events and do not have the patience to use real hardware, use the BREW 2.0 emulator. This emulator includes a menu option for TAPI functionality that allows you to simulate the conditions of receiving an incoming call. The BREW 2.0 emulator includes a number of other debugging features that may prove useful, even for straight 1.0 code.

Cleanup

The final function to examine is our cleanup function, which is passed as an argument to **AEEApp_New** upon creation of this applet.

Listing 14-23: Cleaning up

```
void Flarb_Cleanup(myapp_t* pApp)
{
    int i;

    //remove GUI components, this will release
    //memory that they have allocated
    Flarb_ClearGUI(pApp);

    for (i = 0; i < MAX_IMAGES; i++)
    {
        //release all bitmap resources
        if (pApp->pBMPImagePtrs[i])
        {
            ISHELL_FreeResData(pApp->a.m_pIShell, pApp->pBMPImagePtrs[i]);
                pApp->pBMPImagePtrs[i] = NULL;
        }

        //if we have allocated memory in the conversion, free that as well
        if (pApp->pAllocFlags[i])
        {
            if (pApp->pRawImagePtrs[i])
            {
                SYSFREE(pApp->pRawImagePtrs[i]);
                pApp->pRawImagePtrs[i] = NULL;
                pApp->pAllocFlags[i] = 0;
            }
        }
    }
}
```

What we do in this function is call Flarb_ClearGUI to clear any memory currently used by allocated GUI components. We then proceed to go down our arrays of loaded images and free them. We march down the array of pAllocFlags to see if any of these images have had memory allocated upon the CONVERTBMP conversion process. If so, we call SYSFREE on the pointer to the device-dependent bitmap and clear out its flag value. As explained in Chapter 8, you will typically have to use the SYSFREE method when running this code on an actual handset. The emulator does not allocate memory in the conversion process.

Figure 14-3: Attack of the Flarb running in the BREW emulator

Conclusion

So, now you have a completed game. Granted, this is a really basic applet. Perhaps the biggest remaining issue is that we have not tried to compile this game for a real BREW handset. Even if we did, you will find that it is not optimized for real handset use. We will discuss optimizations and running applets on hardware in Chapter 19.

Also, the game itself is very simple. You can use this framework to add any game enhancements that you want. Try adding different enemies, weapons, and power-ups. The possibilities are endless. I hope that the end result of this chapter is a simple game framework that gives you an extremely basic applet structure to use for most any game.

You will find that this not only barely scratches the surface of BREW game programming, but it also hardly touches on game programming in general. I suggest reading books on general game development and attempt to apply them to BREW. For instance, elements of AI, graphics techniques, collision, and other basic game programming techniques are directly applicable to BREW game programming. You just have to make sure that you can apply these techniques in an efficient and compact manner.

Chapter 15

BREW Extensions

Introduction

Many game programming books tend to focus on their own custom libraries or building a game library that contains many basic game functions that can be reused between applications. Now that we have covered many of BREW's features suitable for game development, it is time that you learned how to create your own libraries. In this chapter we focus on how to create an extension class, essentially the BREW equivalent of a Windows DLL. Although not a full-fledged game library, we will walk through the process of creating an extension class for use with the simple Hello World applet. After this exercise, you will be ready to create your own custom classes with whatever functionality you wish to abstract.

What Is an Extension?

Through conventional C/C++ programming, you are probably familiar with the concept of libraries. A *library* is a compiled chunk of code from which functions can be copied into your application via the linking process. A library usually contains code that is useful for many different programs. For instance, the C language usually comes with its own Standard Library, which contains many useful functions for printing, I/O, and such.

The problem is that the code from the libraries ends up being copied into your application's compiled binary. That is, if many different applications use the same library, they will most likely have redundant code copied into them via the link process.

This is why in Windows we have the dynamic-link library, or DLL. A *DLL* acts as a library in that it is a compiled chunk of code for use with other applications. The difference is that a DLL is not statically linked like a traditional library. Therefore, the code is not copied into the application's binary by the linker at compile time. Instead, the linking occurs at run time. Whenever an application wants to access a function from a DLL, it accesses the code inside the DLL itself. Therefore, if multiple applications are using code from the same DLL, they are all accessing a single copy of the DLL itself instead of code copied into their own binaries. This not only saves space, but it allows you to update the library without recompiling the applications that use it.

Almost every OS has its own version of the DLL, or shared object as they are commonly known. BREW has the same concept with extension classes. It is possible to create your own class with functions that can be accessed by any number of BREW applets. This is essentially what every BREW system class, such as IGraphics, IDisplay, and IShell, is. Extension classes can be a little tricky to create but are very easy to use.

Creating the Extension

For this example we will create a simple extension class that prints the text "Hello World" to the display. Therefore, instead of printing the text in the applet, our extension class, using a version of Hello World, will call a function inside our extension class instead of displaying the text itself. We will start with the extension class found in the Source\Chapter 15\extclass folder in the companion files.

The Code

The first thing we have to do is come up with a BREW class ID for the extension class. If you look at the extclass.bid file, you will see this:

Listing 15-1: The BID file

```
#ifndef EXTCLASS_BID
#define EXTCLASS_BID

#define AEECLSID_EXTCLASS_CLS     0x01F1F011

#endif //EXTCLASS_BID
```

As you can see, the only real difference here is that we named the definition AEECLSID_EXTCLASS_CLS instead of the typical AEECLSID_ EXTCLASS_BID. You can use whichever notation you want; however, the CLS suffix is used with public interfaces such as this. If you look at how you create any BREW interface, all of their class IDs end with CLS. The class ID is very important in this case because it is how you will identify the class when you create the interface. As with any commercial BREW applet, these extension class IDs must be generated via the BREW web site for a final commercial product.

The extclass.h header file contains not only the function definitions for the extension but also the interface and a few other conventions necessary for the creation of an extension class.

Listing 15-2: The extension class header file

```
/*
 * extclass.h
 *
 */

#ifndef __EXTCLASS_H__
#define __EXTCLASS_H__

#include "AEE.h"
#include "AEEDisp.h"
#include "AEEModGen.h"
#include "AEEAppGen.h"
#include "AEEDisp.h"
#include "AEEClassIDs.h"
#include "AEEStdLib.h"
#include "AEEImage.h"
#include "AEEShell.h"

#include "AEEMenu.h"
#include "AEEFile.h"
#include "AEEText.h"

typedef struct _IExtClass IExtClass;

QINTERFACE(IExtClass)
{
    DECLARE_IBASE(IExtClass)
    void      (*DrawHelloWorld)(IExtClass * po);
};

#define IEXTCLASS_AddRef(p)            GET_PVTBL(p,IExtClass)->AddRef(p)
#define IEXTCLASS_Release(p)           GET_PVTBL(p,IExtClass)->Release(p)
#define IEXTCLASS_DrawHelloWorld(p)    GET_PVTBL(p,IExtClass)->DrawHelloWorld(p)

typedef struct extclass_s
{
    // Declare VTable
    DECLARE_VTBL(IExtClass)

    // Class member variables
    uint32      m_nRefs;
    IShell *    m_pIShell;
    IDisplay *  m_pIDisplay;
    IModule *   m_pIModule;

} extclass_t;

#endif
```

The first thing you see in the header is the definition of our interface structure. Although the naming convention clashes with my own, you have to use an underscore in front of the structure name for use with BREW's own custom macros used for creating extension classes. Therefore, we give our interface structure the name _IExtClass and typedef it with the more usable IExtClass.

Next we have the interface definition. You can have any number of internal functions in your extension class, but only expose a few to be used by other applets. These exposed functions are the interface. Through the use of BREW's QINTERFACE macro, you can define the pointers to functions that you wish to be made available to other users. The first line after the usage of QINTERFACE is the macro DECLARE_IBASE. This macro inserts a list of standard functions in our interface. Here is the definition of this macro in AEE.h:

```
#define DECLARE_IBASE(iname) \
  uint32  (*AddRef)        (iname*);\
  uint32  (*Release)       (iname*);
```

You see that it adds the **AddRef** and **Release** functions to the interface. These two functions are mandatory for every BREW object. If you look through the BREW documentation, you will see an **AddRef** and **Release** function in every available interface. These two functions are critical to managing the life cycle of a BREW object. If you are familiar with C++, just think of these functions as pure virtuals from some abstract base class from which all BREW objects are derived.

After the DECLARE_IBASE macro, list out a single function that we wish to expose through the interface:

```
void     (*DrawHelloWorld)(IExtClass * po);
```

After defining the interface, we must also create macros that access all of the functions in our interface, including the ones declared by DECLARE_IBASE. In our case, we only have three functions to define in this way:

```
#define IEXTCLASS_AddRef(p) \ GET_PVTBL(p,IExtClass)->AddRef(p)

#define IEXTCLASS_Release(p) \ GET_PVTBL(p,IExtClass)->Release(p)

#define IEXTCLASS_DrawHelloWorld(p) \
GET_PVTBL(p,IExtClass)->DrawHelloWorld(p)
```

These macros access the internal function table of the interface (passed in via the **p** parameter) in order to call the given function. You see these same macro definitions in any header file for the standard BREW classes.

After these macros, we define the class structure. This is similar to the applet structures we are all familiar with by now. Because this is not an applet, we do not have to include the **AEEApplet** structure first. Instead, we must include the function table that we defined previously with another BREW macro:

```
typedef struct extclass_s
{
    // Declare VTable
    DECLARE_VTBL(IExtClass)

    // Class member variables
    uint32      m_nRefs;
    IShell *    m_pIShell;
    IDisplay *  m_pIDisplay;
    IModule *   m_pIModule;

} extclass_t;
```

The macro **DECLARE_VTBL** essentially inserts our previously defined function pointers at the top of the structure. The rest of the structure is filled out with global member variables used by our extension class internally.

Looking at the source in extclass.c, we see a familiar **AEEClsCreateInstance** function:

Listing 15-3: Creating the applet instance

```
int AEEClsCreateInstance(AEECLSID ClsId,IShell * pIShell,IModule * po,void ** ppObj)
{
    *ppObj = NULL;

    if( ClsId == AEECLSID_EXTCLASS_CLS )
    {
        if( ExtClass_New(sizeof(extclass_t), pIShell, po,
            (IModule **)ppObj) == SUCCESS )
                return AEE_SUCCESS;
    }
    return EFAILED;
}
```

The one difference here is that we are calling our own custom **ExtClass_New** function instead of the familiar **AEEApplet_New**. This is because we are creating our own custom class, not an applet. Therefore, we must call our own custom function to create an instance. The **ExtClass_New** function handles all of the basic housekeeping duties of an extension class:

Listing 15-4: Constructing the class

```
int ExtClass_New(int16 nSize, IShell *pIShell, IModule* pIModule, IModule ** ppMod)
{
    extclass_t *        pMe = NULL;
    VTBL(IExtClass) *   modFuncs;

    if( !ppMod || !pIShell || !pIModule )
        return EFAILED;

    *ppMod = NULL;

    // Allocate memory for the ExtensionCls object
    if( nSize < sizeof(extclass_t) )
        nSize += sizeof(extclass_t);

    if( (pMe = (extclass_t *)MALLOC(nSize + sizeof(IExtClassVtbl))) == NULL )
        return ENOMEMORY;

    // Allocate the vtbl and assign each function pointer
    // to the correct function address
    modFuncs = (IExtClassVtbl *)((byte *)pMe + nSize);

    modFuncs->AddRef          = ExtClass_AddRef;
    modFuncs->Release         = ExtClass_Release;
    modFuncs->DrawHelloWorld  = ExtClass_DrawHelloWorld;

    //initialize the vtable
    INIT_VTBL(pMe, IModule, *modFuncs);

    //We've got one reference since this class is allocated
    pMe->m_nRefs = 1;

    // initialize our internal member variables
    pMe->m_pIShell   = pIShell;
    pMe->m_pIModule  = pIModule;

    //Add References to the interfaces we're using
    ISHELL_AddRef(pIShell);
    IMODULE_AddRef(pIModule);

    //we need to get access to the screen
    if( ISHELL_CreateInstance(pIShell, AEECLSID_DISPLAY,
        (void **)&pMe->m_pIDisplay) != SUCCESS )
    return EFAILED;

    *ppMod = (IModule*)pMe;

    return AEE_SUCCESS;
}
```

This function allocates storage for the class, links up the interface function pointers with the correct functions, and does any other variable and interface initialization necessary.

As you can see, there are three different return codes sent from this function: **EFAILED**, **ENOMEMORY**, and **AEE_SUCCESS**. In order to adhere to the standards of other interfaces, we need to properly signify success or failure of class creation.

After getting the interface function pointers with **GET_VTBL**, first see if the pointers passed as arguments are valid. If not, return **EFAILED**. Secondly, check to see if the **nSize** argument specifies enough memory to hold this class. If not, augment the size. Now let's look at how we allocate memory for this class:

```
if( (pMe = (extclass_t *)MALLOC(nSize + sizeof(IExtClassVtbl))) == NULL )
        return ENOMEMORY;
```

As you can see, we not only allocate enough memory for the size of the class itself but also for its function table. If this fails for some reason, return **ENOMEMORY**. Finally, at the conclusion of the function, return **AEE_SUCCESS** to signify successful class creation.

After this, we properly assign all of the function pointers in the function table to their appropriate functions:

```
modFuncs = (IExtClassVtbl *)((byte *)pMe + nSize);

modFuncs->AddRef            = ExtClass_AddRef;
modFuncs->Release           = ExtClass_Release;
modFuncs->DrawHelloWorld    = ExtClass_DrawHelloWorld;
```

The first thing we do is assign the **modFuncs** pointer to the additional chunk of memory that we allocated for the class itself. We do this by advancing the pointer **nSize** amount of bytes. Since we allocated **nSize** plus the size of the function table, the space underneath the memory occupied by the class is where we put our function table. As you can see, we assign each entry in the function table to the appropriate function. The first two are the standard implementations of **AddRef** and **Release** that we will define later in the source. Lastly, assign our custom **DrawHelloWorld** function pointer to the **ExtClass_DrawHelloWorld** function that is defined in this source file.

Next initialize the **vtbl** with the following macro:

```
INIT_VTBL(pMe, IModule, *modFuncs);
```

This basically sets the **vtbl** pointer to that space below the class where we put our function table. That way, every time **GET_VTBL** is called (as in our macros in extclass.h), it will refer to the pointer that we just initialized.

After the function table setup, set our internal variables. The first thing to do is set the reference count:

```
pMe->m_nRefs  = 1;
```

The **m_nRefs** variable counts how many applets are referencing this object. Because this is the creation function and the first instance, there is only one reference. You will see that in the **Release** function we only destroy this class if m_nRefs is equal to 0. Otherwise, there are still other programs accessing the object, and thus it cannot be destroyed without potentially crashing those programs.

Finally, we initialize the rest of our variables. This includes adding references to any interfaces that we are using and assigning the module pointer to the address of the class. This is the pointer that will be returned to the calling application as the interface to this class. At the conclusion of the function, we return AEE_SUCCESS, indicating that the class has been created without a hitch.

Now look at the implementation of the standard AddRef and Release functions. First, look at AddRef:

Listing I5-5: Managing references

```
static uint32 ExtClass_AddRef(IExtClass * po)
{
    return (++((extclass_t *)po)->m_nRefs);
}
```

As you can see, we simply increase the m_nRefs variable. This lets the class know that there are other applets accessing it, which comes into play with ExtClass_Release:

Listing I5-6: Destroying the class

```
static uint32 ExtClass_Release(IExtClass * po)
{
    extclass_t *     pMe = (extclass_t *)po;

    //Decrease the number of references.  If we still
    //have some references to this object, return
    //and do not free resources
    if(-pMe->m_nRefs != 0)
        return pMe->m_nRefs;

    //Release display
    if(pMe->m_pIDisplay)
            IDISPLAY_Release(pMe->m_pIDisplay);

    // Release interfaces
    ISHELL_Release(pMe->m_pIShell);
    IMODULE_Release(pMe->m_pIModule);
    //Free the object
    FREE_VTBL(pMe, IModule);
    FREE(pMe);

    return 0;
}
```

The first thing to do is decrease the reference count, as there is now one less program accessing this class. If there are still more references left, we return. Otherwise, we go on to deallocate any storage and release all the interfaces that this class uses. Note that there is a special call to free the **vtbl** separately from the class itself:

```
FREE_VTBL(pMe, IModule);
FREE(pMe);
```

Once these two chunks of memory are free, the class has been destroyed and all resources are returned to the OS. The only thing left to look at is the actual **DrawHelloWorld** function. This is basically the same thing as Hello World from our previous example:

Listing 15-7: Drawing the text

```
void ExtClass_DrawHelloWorld(IExtClass * po)
{
    //simply draw "hello world" string as we do in the early examples

    extclass_t *    pMe = (extclass_t *) po;
    AECHAR    szBuf[] = {'H','e','l','l','o',' ', 'W', 'o', 'r', 'l', 'd', '\0'};

    if(!pMe || !pMe->m_pIDisplay)
        return;

    IDISPLAY_DrawText(pMe->m_pIDisplay,
            AEE_FONT_BOLD, szBuf, -1, 0, 0, 0, IDF_ALIGN_CENTER | IDF_ALIGN_MIDDLE);
}
```

In this case, make a special check just to ensure that we actually have a valid pointer to the display and our class. After that, simply draw the text as we have seen many times before.

The MIF File

Even though this class is not to be used as an applet, a MIF file is still required. If you open up extclass.mif, you will see that there is no applet definition. This is because this is not an applet designed to be visible and selectable from the phone's menu. Instead, it is supposed to be a hidden object only used by other programs.

If you switch to the Extensions tab, you will see that we have listed the BREW class ID of our extension class here. This lets BREW know that this object is for use with other programs. If you do not have this extension defined, the call to create our extension class from another program will fail every time.

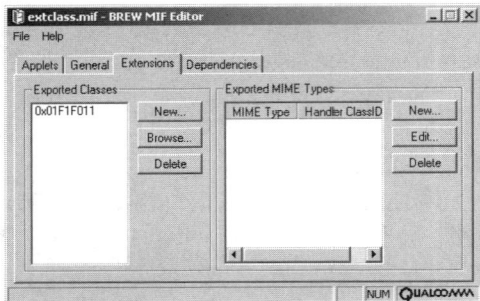

Figure 15-1: The MIF file for our extension class. Note the exported BID value.

Using the Extension

Now that we have seen how to create an extension class, we will now see how to use one. You should already know this because you have created and used instances of classes before by using any of the standard BREW interfaces. A call to ISHELL_CreateInstance using the BREW class ID of our extension is all that is necessary. The example applet in Source\Chapter 15\exthello illustrates the usage of our extension to create another Hello World style applet. Looking at the header, exthello.h, we see the standard applet structure definition:

Listing I5-8: The applet structure

```
typedef struct myapp_s
{
    AEEApplet    a;    //applet header
} myapp_t;
```

Because all of the core functionality is offloaded into an extension class, we really only need the bare-bones definition of an applet here. Looking at the source in exthello.c, we see the standard applet creation sequence:

Listing I5-9: Creating the applet

```
int AEEClsCreateInstance(AEECLSID ClsId, IShell * pIShell, IModule * po, void ** ppObj)
{
    *ppObj = NULL;

    //create applet as usual
    if( ClsId == AEECLSID_EXTHELLO_BID )
    {
        if(AEEApplet_New(sizeof(myapp_t), ClsId, pIShell, po, (IApplet**)ppObj,
            (AEEHANDLER)ExtHello_HandleEvent, NULL) == TRUE)
        {
            return(AEE_SUCCESS);
        }
    }

    return(EFAILED);
}
```

Obviously, we pass a pointer to our message handler, ExtHello_Handle-Event. This function is where all the work is done:

Listing 15-10: The message handler

```
static boolean ExtHello_HandleEvent(IApplet * pi, AEEEvent eCode, uint16 wParam,
    uint32 dwParam)
{
        IExtClass * pIExtClass;
        myapp_t* pApp = (myapp_t*)pi;
        AEEApplet * pMe = &pApp->a;

        switch (eCode)
        {

                case EVT_APP_START:

                //clear screen (default color is white)
                IDISPLAY_ClearScreen(pMe->m_pIDisplay);

                //draw the text with our function in the extension class
                if(ISHELL_CreateInstance(pMe->m_pIShell, AEECLSID_EXTCLASS_CLS,
                    (void **)&pIExtClass) )
                return (FALSE);

                IEXTCLASS_DrawHelloWorld(pIExtClass);
                IEXTCLASS_Release(pIExtClass);

                //update screen
                IDISPLAY_Update(pMe->m_pIDisplay);

                //we've successfully handled this message
                return(TRUE);
                break;
        }
    return(FALSE);
}
```

As you can see, in the **EVT_APP_START** handler, we create an instance of our extension with a call to ISHELL_CreateInstance. The only difference here from any other call to this function is that we pass our own class ID in the definition **AEECLSID_EXTCLASS_CLS**. This is included from the BID file of our extension class.

Next, use the macros defined in our extension class's header to access functions in the interface. Make a call to IEXTCLASS_Draw-HelloWorld to draw the string out to our display. Next, call IEXTCLASS_Release to deallocate this interface. Because there is only one reference to it, a single call to IEXTCLASS_Release will destroy the class instead of simply decreasing the reference count.

The MIF File

Before you run this program, you have to declare which external classes you are using in the MIF file. If you look at exthello.mif and flip to the Dependencies tab, you will see that we have listed the BREW class ID of our extension class. This is used by BREW to know which classes your applet needs in order to run. Depending on the version of BREW available, it is possible for the OS to automatically download and install the required extension objects if you do not currently have them on your handset.

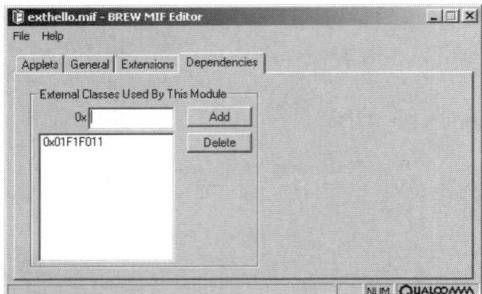

Figure 15-2: The MIF file for the applet using the extension class. Note the dependency on the BID value for the extension.

The Final Product

Now that we have put everything together, the results should look familiar, as shown in Figure 15-3.

This example is relatively useless; however, it explains how applets and extension classes work together. It is possible to make your own game library encompassing sprite drawing, collision, and other common functions in one handy extension class. Then, all of your games can access this "engine" extension class instead of including redundant code in each applet. The initial download will include the game and the engine, but any subsequent games that use the extension will benefit from a smaller download, as they only have to contain the game logic in the binary. The engine functions are in the extension class that the user has already downloaded.

Figure 15-3: Hello World using an extension class

Conclusion

In this chapter you have seen how to create a reusable extension class. Although we do not make a functional game library in this chapter, you now have the knowledge to create custom classes for your own uses. An interesting exercise may be to take the image loading, collision detecting, and sprite drawing functions that we have used in previous examples and create your own extension out of them.

Chapter 16

Tile Graphics

Introduction

Now that we have discussed most of the details of the BREW API and created a basic game, it is time to focus on a few basic game programming concepts. This chapter will focus on tile graphics, a useful way to draw and store backgrounds. We will discuss how to create tile graphic backgrounds and create a simple BREW applet that displays them.

What Are Tile Graphics?

Game graphics are typically categorized as either sprite or background graphics. *Sprite graphics* are usually the moving characters in the game. This would include the Ghost Monsters in Pac-Man or your ship in Asteroids. *Backgrounds* are typically defined as the graphics behind the sprite characters that they travel over and interact with. For instance, the maze in Pac-Man or the dungeon in Legend of Zelda would be considered backgrounds.

Figure 16-1: GameVIL's BoomBoom is a prime example of sprites and backgrounds. Note the contrast of foreground elements with the sky and trees in the background.

If you look at most any background in one of these games, you will notice that parts of the background seem to repeat. For instance, you may see the same door graphic on each wall or the same floor pattern repeated over the screen. This is because instead of storing backgrounds as one large image, the backgrounds are broken up into repeating elements, or tiles, that can be pieced together in various ways to create a variety of backgrounds in much less space.

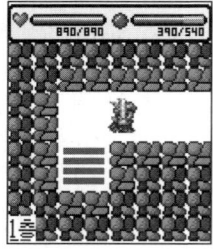

Figure 16-2: GameVIL's Last Warrior uses tile graphics for its backgrounds. Note the repeating patterns used for the dungeon walls.

Instead of using up a lot of memory to store a bunch of background images, a small set of tiles and some instructions to tell the code where to place those tiles on the screen are used to represent background graphics. This storage mechanism not only saves vast amounts of memory, but it has other advantages when referring to the background for collision, artificial intelligence, and other game-specific tasks.

So, what exactly is a tile? A *tile* is basically a square bitmap. Tiles are then arranged in "tile maps" to make an entire picture or background image. For example, let's say you have the following two tiles:

Figure 16-3: The two tiles with which we draw a background

These two tiles will be known as 0 and 1. A tile map using these two images is a simple two-dimensional array, like so:

```
char map[][] = {{0, 0, 0, 0},
                {1, 1, 1, 1},
                {0, 0, 0,  0},
                {1, 1, 1, 1}};
```

If we were to draw this tile map on the screen, we would get an image like this:

Figure 16-4: The tile background represented by our array

This does not look like much, but I am sure you can start to see the possibilities. For instance, we could make an empty room that looks like this:

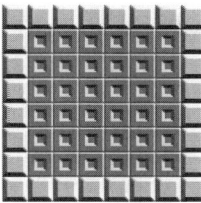

Figure 16-5: An empty room scene made of tiles

It is possible to save this background as a single large bitmap. This would involve an allocation of about 16 KB to store this 128x128 8-bit image. However, the image is really composed of two tiles: the maze wall and the floor. If it uses tiles instead and saves the maze as a tile map, we have huge memory savings. The two tiles will end up being 16x16 8-bit BMPs. Both images together consume approximately 2 KB.

In order to make a 128x128 pixel scene out of 16x16 tiles, we need an 8x8 tile map. This is because 8 multiplied by 16 pixels is 128. Therefore, with an 8x8 tile map we would reproduce our original 128x128 image. If our tile map is an 8x8 array of single-byte entries, the map itself would consume 64 bytes. To reproduce the scene, our tile map has to look like this:

```
char map[][] = { {1, 1, 1, 1, 1, 1, 1, 1},
                 {1, 0, 0, 0, 0, 0, 0, 1},
                 {1, 0, 0, 0, 0, 0, 0, 1},
                 {1, 0, 0, 0, 0, 0, 0, 1},
                 {1, 0, 0, 0, 0, 0, 0, 1},
                 {1, 0, 0, 0, 0, 0, 0, 1},
                 {1, 0, 0, 0, 0, 0, 0, 1},
                 {1, 1, 1, 1, 1, 1, 1, 1}};
```

With the two tiles and map data together, we have to use slightly over 2 KB to reproduce an image that originally was 16 KB. With memory at such a premium on mobile devices, tile maps are invaluable for most games.

Another great feature of tile maps is that they are an internal representation of the scene that is easily used in code to read and store information about the game world. For instance, in this case we know a tile entry of 1 is a wall. Therefore, we can perform collision detection against the walls of the maze by detecting if our game character is attempting to enter a tile that has an index of 1.

If you want to get more complicated, you can create a small tile structure and make the tile map an array of these structures instead of simple bytes. For instance, you might need a Boolean member variable in your tile structure to determine if there is a gold piece on a given tile. If this Boolean is set to True, when the tile is drawn, you also draw a piece of gold on top. If the character picks up the gold, you then clear out the Boolean. Then, the next time you draw the screen, the gold will be missing from that tile. A simple tile structure might look something like this:

```
typedef struct tile_s
{
    char nIndex;
    boolean bGold;
} tile_t;
```

The tile map would be a two-dimensional array of these instead:

```
tile_t map[8][8];
```

Obviously, this structure is rather inflexible. You can create your own kind of structure that allows the specification of more than just the presence of gold. I like to include a short or integer that I use to set a number of different bit flags per tile. You can even do this without a tile structure and just use shorts for the tile index with the first byte as the tile index and the second byte as special-purpose bit flags. It all depends on how large your tile maps are going to be and therefore how much memory you have to conserve. The larger the maps, the smaller the tile structure needs to be in order to make sure it fits in the average handset's miniscule amount of heap space.

Larger Worlds

Another advantage of tile maps is the ability to create a large world and make the display a smaller viewport into it. For instance, the Sharp Z-800 color phone has a 128x144 screen resolution. If we use small 8x8 tiles, we can display about 15x18 tiles on the screen at once. However, we can create a tile map much larger than this—as large as heap memory will afford. Therefore, we can make mazes and worlds that span multiple screens.

The relationship of the viewport to the larger map is on the same idea as the world coordinate/viewport concept discussed in Chapter 9, "Text and Geometric Graphics." As you move the viewport around the map, you see different sections. Using the maze example again, we can have it so that if your character walks off the side of the screen, we move the viewport over one screen width to show the next section of the map. Or, you can slide the viewport around to create a smooth scrolling effect. Unfortunately, most BREW handsets do not currently have the drawing speed to accomplish smooth scrolling. Therefore, I tend to do the former "flick-screen" approach to large map navigation.

This way, a full screen redraw is only necessary when the character crosses a screen border.

Modifying Worlds

Making your environment interactive is another way to make a game deeper and more interesting. For instance, you could have a bridge that needs to be blown up in order to prevent the enemies from advancing. Or you could have a secret wall that is revealed with the flick of a switch.

Because we are storing our worlds in the form of tile maps, it is easy to do this. All we need to do is modify the indices of the tiles that we want to change and redraw the screen. For instance, if you want a secret door to appear in a wall after a certain event, you just modify the tile indices in the wall to those of the secret door and redraw. The possibilities are endless.

Creating Tile Maps

If you have very small maps, then perhaps creating arrays by hand will work. But with BREW's restrictions against static data, even this is not a good idea. What you really need is a tile map editor so that you and anyone else working on your game can create content. In the old days, everyone used to write their own tile editor for their games. In many cases, people still do. However, there are a number of commercial and freely available tile editors that are quite powerful. Some of them also allow you to configure the output so that you can include any special game-specific data in your maps that the editor normally does not keep track of.

Mappy

The tile editor I use most often is Robin Burrow's Mappy, which is included in the companion files. Mappy can output a large file format known as FMP with many data fields and multiple tile map layers. However, it also has a simple binary file with a tile index array like we used at the beginning of this chapter. Mappy also has a few different playback utilities that allow you to view your maps outside of the editor with various environments and graphics libraries, such as Allegro, OpenGL, Java, or DirectX. Also, if you really need some extra features added, the C source code is provided, which allows you to create your own custom versions of Mappy suited for your game.

Figure 16-6: Mappy's main screen

When you first load up Mappy, you will see the Map Editor window on the left and the Block Editor window on the right. The Block Editor window is really a tile window with some extra benefits. The first thing you need to do is create a new map by clicking on File | New Map.

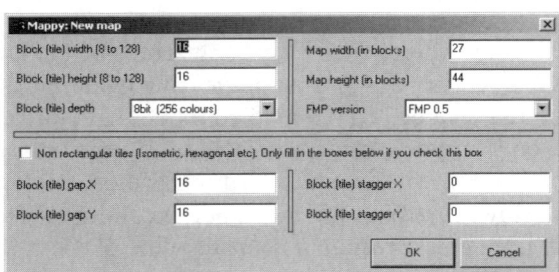

Figure 16-7: Mappy's New Map dialog

Here you see the New Map dialog. In the first quadrant, you see your block settings. Here you determine the pixel width and height of the tiles that you will be using, as well as the bit depth.

To the right of these settings are the map attributes. These include the width and height of the map in tiles. You can resize the map later so you are not locked into the settings that you make in this dialog. You can also determine the type of FMP file that you wish to output when saving the map. This is to maintain backward compatibility with game engines that use Mappy's older FMP 0.5 format. In this case, it does not matter since we are just going to use the raw bitmap output for our example code in this chapter.

At the bottom of the dialog are settings for isometric tiles. *Isometric tiles* are tiles that are shown from a slightly skewed overhead perspective. Instead of square-shaped tiles, they look like little diamonds. This provides a much more detailed perspective, which has the ability to represent height in a 2D environment. For the sake of this book, we are only using basic 2D tiles, so you can ignore these settings. For the maps in this chapter, use the settings shown in Figure 16-7.

Once we have a blank map with the appropriate settings, we need to import our tiles. The tiles are stored as a long BMP with each tile juxtaposed so that Mappy can cut them up into individual blocks. Our tile BMP file, tilegfx.bmp, can be found in the folder Source\Chapter 16\tiles in the companion files. It is the same graphic shown in Figure 16-3.

To import the tiles, click File|Import.

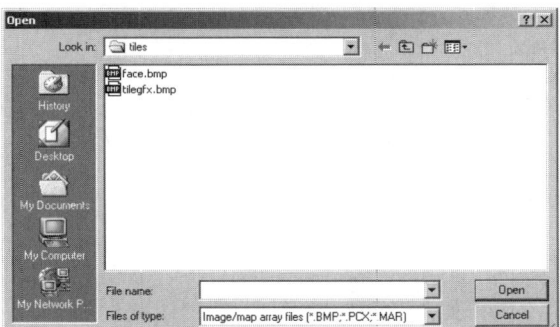

Figure 16-8: The BMP import dialog

Simply pick the BMP that you wish to use for the tiles. In this case, use tilegfx.bmp found in the companion files. Mappy will then ask if you want to import them as NEW graphics:

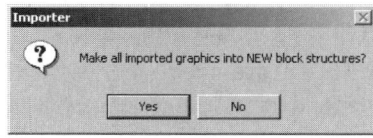

Figure 16-9: Mappy asks if you want to overwrite the existing tiles with the ones loaded from the BMP.

In this case, it does not matter. All tiles will be new since we do not have any to begin with. Click No only if you want to append the current map tiles with new ones from another BMP. Click Yes and you will see your tiles in the Block Editor with the map consisting entirely of the blank entry. If you want to import different titles and overwrite the existing ones, you can click No in the previous dialog and it will overwrite the current set of tile graphics with the ones in the BMP that you are loading. If you click No, it will add them to the end of the list of tiles, essentially appending your tile collection with new graphics.

Now that you have tiles, you can simply click on one in the Block Editor and paint with it in the Map Editor. However, if you want to see the FMP file we used to create the map for the example code, simply load the mazemap.fmp file from the project directory in the companion files.

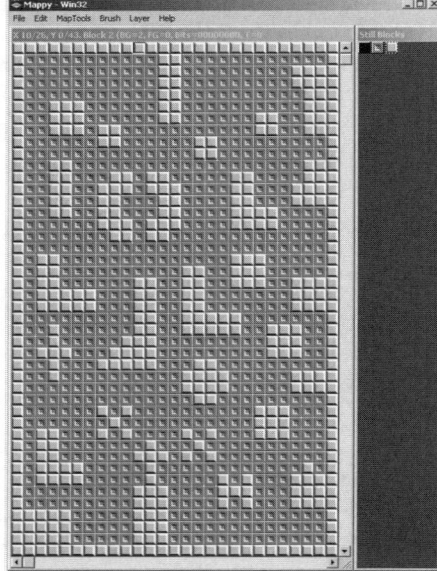

Figure 16-10: The entire map loaded and visible in Mappy

Looking at the Mappy interface, you can see a window into our larger map on the left and the tile palette on the right. If you want to edit the map, simply click on a tile block on the right, and then paint with it in the Map Editor window. You can then save your modified map and use that instead.

Our sample code will not read FMP files directly. These are too large and have many things included that we are not concerned with for our purposes. However, for more advanced games, converting the FMP files into a more compact binary format with a custom command-line tool may be a possibility. Instead, we will use Mappy's raw MAP file format.

The MAP format is a simple binary file that has the width and height of the map in tiles at the beginning followed by a raw array of bytes representing the tile indices. Luckily, Mappy allows you to customize the output format of the MAP file in its own INI file. My custom Mappy.ini file looks like this:

```
; Be sure you know what you are doing editing this file,
; comments start with a ';' at the very start of the line,
; all other lines must have an equals sign in for values
width = 640
height = 480
grid = 1
zoom = 1
```

```
picklayer = 1
transred = 255
transgreen = 0
transblue = 255
trans8bit = 0
apmode = "640*480 8bpp ?hz"
importskip = 1
csvadjust = -1
maptype = "LW4H4A1-1"
mapdefw = 100
mapdefh = 100
mapdefbw = 32
mapdefbh = 32
mapdefBMP = "nodefault.bmp"
mapstaggerx = 0
mapstaggery = 0
mapclickmask = 0
```

As you can see, there are a variety of settings that you can configure in this file. The one we are interested in is **maptype**. The following string, "LW4H4A1–1," determines how Mappy will format the raw binary MAP file.

The first character, "L," determines the endian format of the data. The only thing you need to be concerned about is that L means we are using the little endian format that all PCs use. It just so happens that the ARM processor reads bytes in the same order as CPUs found in PCs. Therefore, use the same settings for native machine-readable maps on a real BREW handset.

The next character, "W," determines that the first group of bytes in the file will represent the width of the map in tiles. The number 4 following the letter says that the width will be stored as four bytes (an integer). You can probably guess the character pairing "H4" means the next four bytes will store the height in tiles of the map.

The "A" character means the next chunk of data will be the map array. The "1" following the "A" tells Mappy that the tile indices will be stored in one byte. That means a tile entry in the map array can have any value from 0 to 255. If you need more tiles, you can choose to expand the byte size of the map array elements to any size that you would like. For most mobile games, 255 different values is sufficient.

The final part of the string, "–1," tells Mappy to subtract one from each tile index. If you look in the block array on the right, you will see a "blank" tile at the start of the tile array. Mappy creates a default blank tile when importing tile BMPs. Therefore, there are actually three tiles when we really only imported two. To fix this offset, you must subtract one from each tile value so that the first tile really is referenced as 0 and not 1. As a result, we ignore the blank tile completely. So, the string "LW4H4A1–1" means the binary MAP file is stored in little endian format, beginning with four bytes representing the width, four bytes representing the height, and then an array of zero-relative, single-byte entries containing the tile map data.

To export a MAP file, you simply need to select the MAP file format when saving the file. Mappy will warn you that you are discarding important map information by doing this. As long as you save a corresponding FMP file for each MAP file you export, you will easily be able to recreate this data. You can also load MAP files in directly, just as long as you remember the BMP file that you used for the tile graphics.

What extra information is stored in an FMP? Mappy allows you to save attribute bits on tiles and have multiple parallax layers of tile maps and all sorts of advanced features. These are all useful for more complicated games, which is why you might look into reading FMP files directly or converting them to a slightly more complicated binary format for your games.

Now that we have seen the basics of Mappy usage, we will illustrate a simple example using the map that you saw in the mazemap.FMP file previously. Although this is not a complete game, you can use this code to build your own expansive tile world-based applet.

A Simple Example

We are going to create a simple maze applet where you can move a character through a large tile map that spans several different screens. This code will serve as a solid foundation for any large tile-world games that you want to make. In fact, the map loading and drawing functionality is definite material for a reusable engine extension class.

Looking at a screen shot of our applet in action, we see a simple smiley face character in a maze of simple wall and floor tiles, as we saw previously in the Mappy tutorial.

Figure 16-11: The tiles example in action

In this applet, you move the smiley face through the maze. There is basic collision detection so that you cannot move through walls, and if you move the character off the screen, the view will change to the new section of the map to where you moved. The result is an expansive world that you can explore. Well, it is a world consisting entirely of green walls and brown floor tiles, but you get the idea.

The Resource File

If you open tiles.bri in the Resource Editor, you will see that we are including two bitmaps in this applet. One is a wide strip of tiles for our background graphics. The next is a simple smiley face image used for our player sprite. These are both 8-bit.

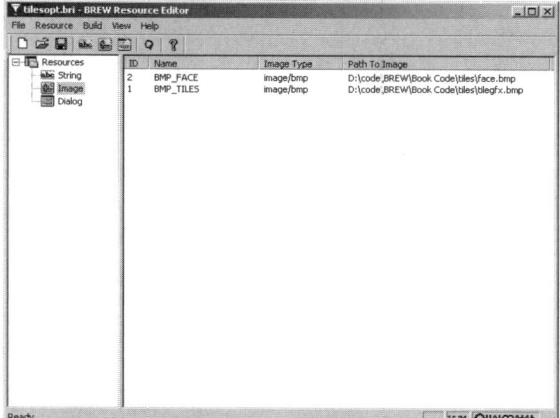

Figure 16-12: The resource file loaded in the editor

The Code

Looking at the header file, tiles.h, we see the basic structure definitions, including the applet itself. It also contains some handy defines that we will use so that we can easily modify this applet to work with different resolutions and tile sets.

Listing 16-1: The applet header

```
/*
 * tiles.h
 *
 */

#ifndef __TILES_H__
#define __TILES_H__

#include "AEE.h"
#include "AEEDisp.h"
#include "AEEModGen.h"
#include "AEEAppGen.h"
#include "AEEDisp.h"
#include "AEEClassIDs.h"
#include "AEEStdLib.h"
#include "AEEImage.h"
```

```
#include "AEEShell.h"

#include "AEEMenu.h"                  // Menu Services
#include "AEEFile.h"                  // AEEFile Services
#include "AEEText.h"

#include "tiles_res.h"

#define MAX_IMAGES       2

#define TILE_WIDTH       16
#define TILE_HEIGHT      16

#define OFFSET_X         6
#define OFFSET_Y         8

enum
{
    IMAGE_TILES = 0,
    IMAGE_PLAYER
};

typedef struct tile_s
{
    unsigned char nIndex;     // Index into bitmap array
    unsigned char nFlags;     // Tile properties

} tile_t;

typedef struct sprite_s
{
    unsigned int nX;
    unsigned int nY;

    unsigned char nWidth;
    unsigned char nHeight;

    unsigned char nImage;

} sprite_t;

typedef struct myapp_s
{
    AEEApplet      a;
    AEEDeviceInfo di;

    int      nMapWidth;
    int      nMapHeight;
    int      nTileX;
    int      nTileY;
    int      nNumTilesX;
    int      nNumTilesY;

    tile_t * pMapArray;

    void*    pRawImagePtrs[MAX_IMAGES];
    AEEBmp   pBMPImagePtrs[MAX_IMAGES];
    byte     pAllocFlags[MAX_IMAGES];
```

```
    sprite_t      playerSpr;

    IGraphics * pIGraphics;

} myapp_t;

static boolean Tiles_HandleEvent(IApplet * pi, AEEEvent eCode, uint16 wParam,
    uint32 dwParam);

boolean Tiles_ReadMap(myapp_t* pApp, char* pszFile);
void Tiles_DrawMap(myapp_t* pApp, int nX, int nY);
void Tiles_LoadResources(myapp_t* pApp);
void* Tiles_LoadImage(myapp_t* pApp, int nResID, int nIndex, sprite_t* pSprite);
void Tiles_Cleanup(myapp_t* pApp);
void Tiles_DrawScreen(myapp_t* pApp);
void Tiles_MovePlayer(myapp_t* pApp, int nX, int nY);
void Tiles_HandleInput(myapp_t* pApp, uint32 nKeyCode);

#endif
```

After the standard assortment of header includes, include the resource file header, tiles_res.h. We are storing the tile and player graphics in a resource file. Sure, there are only two images in this applet, but it is good to get used to the concept of using resource files. It makes asset management much easier and is just a good habit to get used to.

Following the header includes, we have a number of definitions:

```
#define MAX_IMAGES      2

#define TILE_WIDTH      16
#define TILE_HEIGHT     16

#define OFFSET_X        6
#define OFFSET_Y        8
```

The first, MAX_IMAGES, is used to size our image array. We know that we are only going to have two BMPs in this applet: the tile strip and the smiley face. Next, we define the width and height of the tiles in pixels. We will use these values when we have to draw our tiles across the screen. Finally, we have two offsets. This applet is designed for the default emulator device file 7GP_256Color.qsc, which is a fictional 8-bit phone with a resolution of 155x191. Since this is not a multiple of 16 in either direction, we must have a small border around the screen in order to center the tile map. These offsets will be added to all graphics drawn on the screen to properly center the image. If you want to fill larger portions of the screen, you can play with different tile sizes. I have ended up using weird dimensions, like 13x11, to suit the screens of different devices.

Next up is our enumeration of sprite images. As mentioned previously, we only have two, so we only need two enumerations. We will use these values to index into our array of loaded bitmaps. This may be overkill since we only have one sprite, but in the interest of future expansion, we will continue to use this method. When you go on to create games with many sprites, a similar enumeration scheme is a good way to keep track of images in your game.

Now we come to our tile structure. In this demonstration applet, we do not actually need one since we are just displaying images. But in a real game, you might want to store different attributes with your tiles, as we illustrated in the gold piece example earlier. Therefore, we have created a little tile structure that has the tile index and another small member for storing bit flags that we may use later.

```
typedef struct tile_s
{
    unsigned char nIndex;
    unsigned char nFlags;

} tile_t;
```

When we discuss optimizations in Chapter 19, you will see this field used. For now, we will simply use the tile index value. Keep in mind that you should keep this structure as small as possible. With large maps, many of these structures are stored in memory. An easy way to run out of heap space is to make a bloated tile structure.

Next we have our sprite structure. This is almost the same as the one we used in the Flarb example game. However, we have changed the coordinate values to be integers instead of bytes as previously used.

```
typedef struct sprite_s
{
    unsigned int nX;
    unsigned int nY;

    unsigned char nWidth;
    unsigned char nHeight;

    unsigned char nImage;

} sprite_t;
```

The reason we use integers instead of bytes is because we need to represent position values that are larger than the screen. Much like BREW's geometric graphics system, we have world and screen coordinates. The screen has pixel positions in a 155x191 range. However, the map spans many screens. This means that the screen is only a window into a much larger range of position values. This larger range of position values is the world coordinate system. The x and y coordinates of the sprite are in world coordinates, which span values much larger than the limit of 255 provided by byte-sized variables.

Other than that modification, this structure remains the same. We have the width and height values that are taken from the BMP's dimensions. We also have the image index member, n**Image**, which is an index into our array of images. This value will be set to one of the enumerations that we made earlier, which indexes our array of loaded BMPs.

Now we come to our applet structure. This is fairly standard looking if you have gone through the Flarb example game:

```
typedef struct myapp_s
{
    AEEApplet      a;
    AEEDeviceInfo di;

    int      nMapWidth;
    int      nMapHeight;
    int      nTileX;
    int      nTileY;
    int      nNumTilesX;
    int      nNumTilesY;

    tile_t * pMapArray;

    void*    pRawImagePtrs[MAX_IMAGES];
    AEEBmp   pBMPImagePtrs[MAX_IMAGES];
    byte     pAllocFlags[MAX_IMAGES];

    sprite_t playerSpr;

    IGraphics * pIGraphics;

} myapp_t;
```

Of course, we begin with the **AEEApplet** structure. Next we have **di**, which is our **AEEDeviceInfo** structure used to store information about the device. The **nMapWidth** and **nMapHeight** members are used to record the size of the entire map in tiles for both width and height dimensions. The **nTileX** and **nTileY** members are used to tell where the origin of the screen is in tile coordinates. We could use actual world coordinates, but since we are just flipping entire pages, we can simply use the tile numbers. For smooth scrolling, you would need to use pixel coordinates instead. **nNumTilesX** and **nNumTilesY** determine how many tiles can be displayed in the x and y dimensions at the current resolution. These values will vary, depending on the screen resolution recorded in the **AEEDeviceInfo** structure.

Following these integers is a pointer to a **tile_t** structure. This is the one-dimensional array representing the entire map of tiles. Why one dimensional? Allocating a large single-dimensional array and putting our tile values into it is simple. You can easily reference x and y tile coordinates with simple pointer math.

Next we have our customary arrays of loaded BMPs, device-dependent bitmaps, and allocation flags. Rounding out the header file are the function declarations that we will go into as we examine the actual code.

The code begins with another standard class allocation function:

Listing 16-2: Creating the applet instance

```
int AEEClsCreateInstance(AEECLSID ClsId, IShell * pIShell, IModule * po, void ** ppObj)
{
    *ppObj = NULL;

    if( ClsId == AEECLSID_TILES_BID )
    {
        if(AEEApplet_New(sizeof(myapp_t), ClsId, pIShell, po, (IApplet**)ppObj,
                (AEEHANDLER)Tiles_HandleEvent, (PFNFREEAPPDATA)Tiles_Cleanup) == TRUE)
        {
            return(AEE_SUCCESS);
        }
    }

    return(EFAILED);
}
```

There are no surprises here. We simply create our applet with the passed-in class ID, as defined in the tiles.bid file. Of course, we pass the message handler, **Tiles_HandleEvent**, and cleanup function, **Tiles_Cleanup**, as callbacks. Our message handler is very simple:

Listing 16-3: The message handler

```
static boolean Tiles_HandleEvent(IApplet * pi, AEEEvent eCode, uint16 wParam,
    uint32 dwParam)
{
    myapp_t* pApp = (myapp_t*)pi;
    AEEApplet * pMe = &pApp->a;

    switch (eCode)
    {
        case EVT_APP_START:

            //Clear screen (default color is white)
            IDISPLAY_ClearScreen(pMe->m_pIDisplay);

            //Get system info
            ISHELL_GetDeviceInfo(pMe->m_pIShell, &pApp->di);

            //Clear out our data arrays
            MEMSET(pApp->pAllocFlags, 0, sizeof(char) * MAX_IMAGES);
            MEMSET(pApp->pBMPImagePtrs, 0, sizeof(int) * MAX_IMAGES);
            MEMSET(pApp->pRawImagePtrs, 0, sizeof(int) * MAX_IMAGES);

            pApp->nTileX = 0;
            pApp->nTileY = 0;

            pApp->nNumTilesX = pApp->di.cxScreen / TILE_WIDTH;
            pApp->nNumTilesY = pApp->di.cyScreen / TILE_HEIGHT;

            pApp->playerSpr.nX = TILE_WIDTH;
            pApp->playerSpr.nY = TILE_HEIGHT;

            //Load bitmaps
            Tiles_LoadResources(pApp);
```

```
            //Load and parse map file
            Tiles_ReadMap(pApp, "mazemap.map");

            //Draw and refresh screen
            Tiles_DrawScreen(pApp);
            IDISPLAY_Update(pMe->m_pIDisplay);

            return(TRUE);
            break;

    case EVT_KEY_PRESS:

            Tiles_HandleInput(pApp, wParam);
            Tiles_DrawScreen(pApp);
            IDISPLAY_Update(pMe->m_pIDisplay);
            return(TRUE);
            break;
    }

    return(FALSE);
}
```

The handler for **EVT_APP_START** is where we do all of our applet setup. First, we clear the screen with a call to **IDISPLAY_ClearScreen**. Then, we get the device information with a call to **ISHELL_Get-DeviceInfo**. We will use the attributes recorded in our **AEEDeviceInfo** structure later. Next, we have three calls to **MEMSET** to clear out our predefined arrays. We cannot rely on BREW to clear out any declared arrays; therefore, they could be filled with random values at the start. Next we set the **nTileX** and **nTileY** values to 0. This positions our screen window at 0,0 in the tile array (world coordinates). We then calculate how many tiles we can display across the screen by taking the resolution in each dimension and dividing it by the tile's width or height. This will tell us how many tiles can fit on a single screen. We also position our player sprite one tile down and to the right. We know the map has a border of wall tiles around it, so we must make sure the player will start in an empty square by placing it down one tile in each dimension.

Now we make a call to **Tile_LoadResources**:

Listing 16-4: Loading resources

```
void Tiles_LoadResources(myapp_t* pApp)
{
    Tiles_LoadImage(pApp, BMP_FACE, IMAGE_PLAYER, &pApp->playerSpr);
    Tiles_LoadImage(pApp, BMP_TILES, IMAGE_TILES, NULL);
}
```

This is a simple function that calls **Tiles_LoadImage** to read in the BMP files. This is very similar to what we do in the Flarb example when we load and convert BMPs to device-dependent bitmaps. As you can see with the first call, we pass in the address of our sprite structure to

properly fill it in with attributes gleaned from the bitmap file. The
Tiles_LoadImage function should look familiar:

Listing 16-5: Loading the BMP file

```
void* Tiles_LoadImage(myapp_t* pApp, int nResID, int nIndex, sprite_t* pSprite)
{
    AEEBmp pbmSource;
    AEEImageInfo imageInfo;
    byte* pDataBytes;
    boolean bVal = FALSE;

    //Pull the BMP from the resource file
    pApp->pBMPImagePtrs[nIndex] = pbmSource =
        ISHELL_LoadResData(pApp->a.m_pIShell, RES_FILE, (short)nResID, RESTYPE_IMAGE);


    if (pApp->pBMPImagePtrs[nIndex] == NULL)
        return(NULL);

    pDataBytes = (byte *)pbmSource + *((byte *)pbmSource);
    pApp->pRawImagePtrs[nIndex] = CONVERTBMP(pDataBytes, &imageInfo, &bVal);

    //If we have allocated memory in the conversion, signify
    if (bVal)
    {
        pApp->pAllocFlags[nIndex] = 1;
    }
    else
        pApp->pAllocFlags[nIndex] = 0;

    //If we want to retain the sprite info, fill it in
    if (pSprite)
    {
        pSprite->nWidth = (unsigned char)imageInfo.cx;
        pSprite->nHeight = (unsigned char)imageInfo.cy;
        pSprite->nImage = nIndex;
    }

    return(pApp->pRawImagePtrs[nIndex]);
}
```

This is basically the same code as in the Flarb example. We load up the
BMP, convert it to a device-dependent bitmap, record the allocation flag,
and then optionally fill in the sprite structure.

Getting back to the **EVT_APP_START** handler, we see that after we
load the resources, we must load the map with a call to Tiles_ReadMap:

Listing 16-6: Reading the map file

```
boolean Tiles_ReadMap(myapp_t* pApp, char* pszFile)
{
    int i;
    int nNumTiles;
    IFile * pIFile;
    IFileMgr * pIFileMgr;
```

```
unsigned char * pBuf;

//load up the map file
ISHELL_CreateInstance(pApp->a.m_pIShell, AEECLSID_FILEMGR,
    (void **)&pIFileMgr);

if (!pIFileMgr)
    return(FALSE);

pIFile = IFILEMGR_OpenFile(pIFileMgr, pszFile, _OFM_READ);

if (!pIFile)
    return(FALSE);

//The first two words are the width and height
IFILE_Read(pIFile, &pApp->nMapWidth, sizeof(int));
IFILE_Read(pIFile, &pApp->nMapHeight, sizeof(int));

//Now read in the byte entries
nNumTiles = pApp->nMapWidth * pApp->nMapHeight;

//Allocate and read in tile map from file
pBuf = MALLOC(nNumTiles);
IFILE_Read(pIFile, pBuf, nNumTiles);

//Allocate tile array and fill it with values
pApp->pMapArray = MALLOC(nNumTiles * sizeof(tile_t));
for (i = 0; i < nNumTiles; i++)
{
    pApp->pMapArray[i].nIndex = pBuf[i];
}

//Free up resources
FREE(pBuf);
IFILE_Release(pIFile);
IFILEMGR_Release(pIFileMgr);

return(TRUE);
}
```

This function first reads the height and width of the map from the first two words of the binary map file. It then allocates a chunk of memory to hold the tile array and read it in. After allocating enough tile structures to hold the entire map, it sets each index to be the associated entry in the raw map array. We will look at this in depth.

First, we create the **IFileMgr** interface and read in our file as per the filename passed in on the **pszFile** parameter. We then read in the first two integers to our **nMapWidth** and **nMapHeight** members. This records the width and height of the map in tiles and advances the file pointer to the beginning of the raw tile array.

If you recall, the array was defined in the INI file for Mappy as being a simple array of single-byte entries. First we multiply the tile width and height and figure out the grand total of tiles in this map. Then we use this value stored in **nNumTiles** to allocate a correctly sized array of bites. We then read the map directly into this chunk of memory.

Next we use **nNumTiles** to allocate enough memory to hold that number of tile structures. Then, for each entry in our array of tile structures, we set the **nIndex** member to be the associated value in the raw array of tile indices. Finally, we release all interfaces and files that we were using and deallocate the memory used to read in the raw tile array.

Next, we draw the screen by calling **Tiles_DrawScreen**:

Listing 16-7: Drawing the screen

```
void Tiles_DrawScreen(myapp_t* pApp)
{
    //Draw tile map
    Tiles_DrawMap(pApp, pApp->nTileX, pApp->nTileY);

    //Draw player sprite
    IDISPLAY_BitBlt(pApp->a.m_pIDisplay,
        pApp->playerSpr.nX - (pApp->nTileX * TILE_WIDTH) + OFFSET_X,
        pApp->playerSpr.nY - (pApp->nTileY * TILE_HEIGHT) + OFFSET_Y,
        pApp->playerSpr.nHeight, pApp->playerSpr.nWidth,
        pApp->pRawImagePtrs[pApp->playerSpr.nImage],
        0, 0, AEE_RO_TRANSPARENT);
}
```

This function does two things. First, it calls **Tiles_DrawMap** to draw the background tiles. Then it directly blts the player sprite to the screen at its appropriate location. First, look at **Tiles_DrawMap**:

Listing 16-8: Drawing the map

```
void Tiles_DrawMap(myapp_t* pApp, int nX, int nY)
{
    int i;
    int j;
    int nIndex;
    int nYModifier;

    //Find out how much we need to add to get to our
    //Y level in the map
    nYModifier = nY * pApp->nMapWidth;

    //Go through each visible tile and draw it
    for (i = 0; i < pApp->nNumTilesX; i++)
    {
        for (j = 0; j < pApp->nNumTilesY; j++)
        {
            nIndex = pApp->pMapArray[(j * pApp->nMapWidth) + nYModifier + i
                + nX].nIndex;
            IDISPLAY_BitBlt(pApp->a.m_pIDisplay,
                (i * TILE_WIDTH) + OFFSET_X,
                (j * TILE_HEIGHT) + OFFSET_Y, TILE_WIDTH, TILE_HEIGHT,
                pApp->pRawImagePtrs[IMAGE_TILES],
                nIndex * TILE_WIDTH, 0, AEE_RO_COPY);
```

```
              }
          }
   }
```

This is where we figure out how to get a two-dimensional array of tiles out of a one-dimensional array. For example, we will say that the map is eight tiles wide. In a one-dimensional array, every eight tiles constitute a row, meaning array entry 0 is the first tile in the first row of the map and tile 8 is the first tile in the second row. Therefore, if you want to get tile 2,4 in the map, you have to multiply 4 by the width of a row and then add 2 to it. So tile 2,4 is really array entry 34 (that is, 4 * 8 plus 2 tiles in the x direction).

With this information in mind, let's look at how we draw the tiles. As you can see, we have a nested loop. The first loop cycles through all the columns, stopping at **nNumTilesX**, which signifies the maximum amount of tiles that can fit on the screen horizontally. We refer to this as the x dimension's index. The nested loop cycles through each column, stopping at **nNumTilesY**, which is the number of tiles that can fit on the screen vertically. This is known as the y dimension's index.

We use these two loop indices to index into our map array and draw each tile at the appropriate location on the screen. First, we must get the proper tile index out of the tile array. To do this, we must multiply the row index by the width in tiles of the map, as we detailed previously. Then we must add on the column number to this to arrive at the correct array index for the desired tile. Therefore, we retrieve the tile like this:

```
nIndex = pApp->pMapArray[(j * pApp->nMapWidth) + nYModifier + i + nX].nIndex;
```

Now that we have the appropriate tile index, we have to draw the desired bitmap in the correct position. You may recall that the **IDISPLAY_BitBlt** function declaration looks like this:

```
IDISPLAY_BitBlt(IDisplay * pIDisplay, int xDest, int yDest, int cxDest, int cyDest,
    const void * pbmSource, int xSrc, irt ySrc, AEE_RasterOp dwRopCode)
```

Looking at the parameters, we see that after the interface, we specify the upper left-hand corner of the destination rectangle. Following this is the width and height of said rectangle. We previously defined the tile width and height at 16, so that is no problem. But how do we determine where this tile should be drawn?

We simply multiply the x dimension tile index by the tile width and the y dimension tile index by the tile height. This will give us the coordinate of the upper left-hand corner of the tile rectangle. We also have a predefined offset value that we add to each coordinate to center the gameplay window. Therefore, we compute the x coordinate like this:

```
(i * TILE_WIDTH) + OFFSET_X
```

Then we calculate the y coordinate like this:

```
(j * TILE_HEIGHT) + OFFSET_Y
```

After the width and height parameters, we must pass a pointer to the source bitmap that we are using and the x and y coordinate of the upper left-hand corner of the source rectangle. The source rectangle is where the image will be copied from. Before we create this rectangle, pass a pointer to our device-dependent bitmap that contains the tile images. It is accessed like this:

```
pApp->pRawImagePtrs[IMAGE_TILES]
```

As you know, the tiles are stored in one long bitmap. This means we have to use **IDISPLAY_BitBlt**'s ability to copy smaller rectangles from the source bitmap to the frame buffer. This is where the tile index comes into play. Much like we use the loop indices to calculate our destination rectangle position, we use the tile index to compute the source rectangle position. Because our source bitmap has only a single row of tiles, we simply multiply the tile index by the width of the tile for the x coordinate and use 0 for the y coordinate. Finally, we pass the raster operation that we wish to use when copying this tile bitmap over to the frame buffer. We only need a simple copy operation, hence the **AEE_RO_COPY** value is sent. The end result of this nested loop is a screen full of tiles, as stored in our map file.

After drawing the map, the **Tiles_DrawScreen** function blts the smiley face image to the location specified by the player's sprite structure. This is a simple bitblt operation that we are all familiar with by now. Note that the player character is the same width and height as a tile. This makes collision detection much easier and will also come in handy when we optimize for handsets in Chapter 19.

Now that we have gone over the **EVT_APP_START** event handler completely, we will look at how we handle input. Every time a key is pressed, move the player accordingly, run some collision detection, update the state of the map view, and redraw the screen. Looking at the message handler, the case for **EVT_KEY_PRESS**, we see:

```
case EVT_KEY_PRESS:

    Tiles_HandleInput(pApp, wParam);
    Tiles_DrawScreen(pApp);
    IDISPLAY_Update(pMe->m_pIDisplay);
    return(TRUE);
    break;
```

Now, first make a call to **Tiles_HandleInput**. This is the function that is responsible for moving the player, depending on which key was pressed. The function looks like this:

Listing 16-9: Handling input

```
void Tiles_HandleInput(myapp_t* pApp, uint32 nKeyCode)
{
    //Send the appropriate move for each movement key
    switch (nKeyCode)
    {
        case AVK_UP:
            Tiles_MovePlayer(pApp, 0, -TILE_WIDTH);
            break;

        case AVK_DOWN:
            Tiles_MovePlayer(pApp, 0, TILE_WIDTH);
            break;

        case AVK_LEFT:
            Tiles_MovePlayer(pApp, -TILE_WIDTH, 0);
            break;

        case AVK_RIGHT:
            Tiles_MovePlayer(pApp, TILE_WIDTH, 0);
            break;
    }
}
```

This simple function calls the **Tiles_MovePlayer** function with the appropriate arguments if any of the direction keys are pressed. **Tiles_MovePlayer** looks like this:

Listing 16-10: Moving the player

```
void Tiles_MovePlayer(myapp_t* pApp, int nX, int nY)
{
    int nNewX = pApp->playerSpr.nX + nX;
    int nNewY = pApp->playerSpr.nY + nY;

    //Check to see if we've hit the edge of the map
    if (nNewX < 0)
        return;

    if (nNewY < 0)
        return;

    if (nNewX >= (pApp->nMapWidth * TILE_WIDTH))
        return;

    if (nNewY >= (pApp->nMapHeight * TILE_HEIGHT))
        return;

    //Now check the tile to see if it is a wall (tile index is 1)
    if (pApp->pMapArray[(nNewX / TILE_WIDTH) + (pApp->nMapWidth * (nNewY /
        TILE_WIDTH))].nIndex)
        return;

    //Move the window over one 'page' if we've gone off the edge
    if ((pApp->nTileX > 0) && ((nNewX / TILE_WIDTH) < pApp->nTileX))
        pApp->nTileX -= pApp->nNumTilesX;
    else if ((pApp->nTileX < pApp->nMapWidth) &&
        ((nNewX / TILE_WIDTH) >= (pApp->nTileX + pApp->nNumTilesX)) )
```

```
        pApp->nTileX += pApp->nNumTilesX;

    if ((pApp->nTileY > 0) && (nNewY / TILE_WIDTH) < pApp->nTileY)
        pApp->nTileY -= pApp->nNumTilesY;
    else if ((pApp->nTileY < pApp->nMapHeight) &&
            ((nNewY / TILE_WIDTH) >= (pApp->nTileY + pApp->nNumTilesY)) )
        pApp->nTileY += pApp->nNumTilesY;

        pApp->playerSpr.nX = nNewX;
        pApp->playerSpr.nY = nNewY;
}
```

Succeeding the obligatory applet pointer, this function takes the two arguments nX and nY, which specify the amount of pixels in the x and y direction, to add to the current position of the player. This function does several things before adding this displacement to the sprite's position. First, it checks to see if the player has hit the edge of the map. It does this by checking if the new x and y coordinates are either less than zero or greater than the width or height of the map in pixels. If we have hit any of the edges, return from the function without updating the player's position. This effectively leaves the player in place, not moving it at all.

After checking for edge collision, it checks to see if the new position is inside a wall tile. To do this, first get the index of the tile that the player character's new position is directly over. Use this index to see if the tile is a wall tile. If it is a wall, bail from the function.

Next, check to see if the player has moved off the screen. If so, update the tile coordinate of the viewport's origin to focus on the next section of map that the player has entered. Do this by checking if the tile that the sprite resides on with its new position is out of range of the tiles displayed by the viewport. If so, move the appropriate tile coordinate of the viewport back or forward a screen width or height. This will then make the viewport focus on the new section of map to which the player has moved.

Finally, assign the sprite's x and y coordinates to the new positions calculated by this function. The result is that we have a sprite that is moved to a new tile position and a viewport that has followed the player if it has moved off the screen.

Getting back to the EVT_KEY_PRESS event handler, we see that we draw the screen and update the display. This will refresh the screen with our player potentially in a location with perhaps the viewport moved over a screen.

The only other function left to document is our cleanup callback. This is fairly standard stuff:

Listing 16-11: Applet cleanup

```
void Tiles_Cleanup(myapp_t* pApp)
{
    int i;

    for (i = 0; i < MAX_IMAGES; i++)
    {
        //Release all bitmap resources
        if (pApp->pBMPImagePtrs[i])
        {
            ISHELL_FreeResData(pApp->a.m_pIShell, pApp->pBMPImagePtrs[i]);
            pApp->pBMPImagePtrs[i] = NULL;
        }

        //If we have allocated memory in the conversion, free that as well
        if (pApp->pAllocFlags[i])
        {
            if (pApp->pRawImagePtrs[i])
            {
                SYSFREE(pApp->pRawImagePtrs[i]);

                pApp->pRawImagePtrs[i] = NULL;
                pApp->pAllocFlags[i] = 0;
            }
        }
    }

//Free allocated memory and release interfaces
FREE(pApp->pMapArray);
}
```

As you can see, we release all the BMP images that we have loaded and free up the device-dependent bitmaps if the allocation flags have been set. We then free the memory that we allocated for our map array and thus have successfully freed all resources and memory taken up by this example applet.

Conclusion

In this chapter you learned how to use tiles to display a background and store information about the game world. The example applet shows a simple case where moving the character past the screen boundary results in the display being moved over one screen width or height. In more advanced games, you may want the viewport to smoothly follow your character. This means updating the viewport position in units of pixels, not tiles. However, smooth scrolling tile engines are hard to optimize when compiling code for real hardware. You will see how this "flick-screen" tile solution can be greatly optimized when we discuss speed issues in a later chapter.

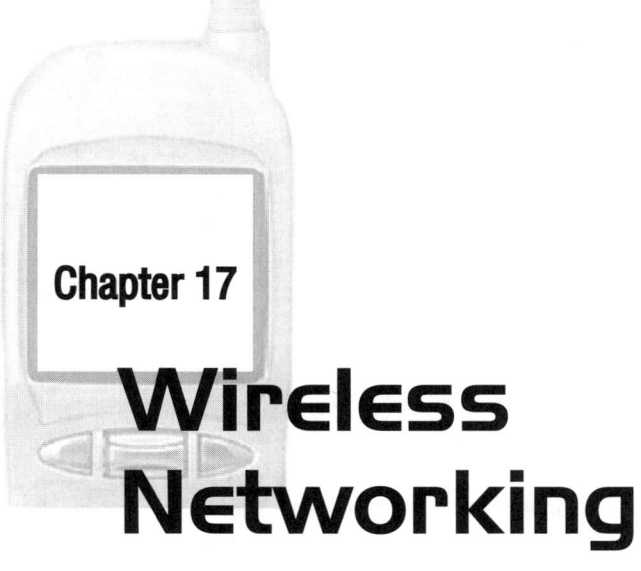

Chapter 17

Wireless Networking

Introduction

Mobile gaming is often referred to as "wireless" gaming. What is usually meant by the term "wireless" is the ability to communicate through networks without wires. This sometimes means communicating on a standard LAN via wireless protocols, such as 802.11b using radio waves. It could also mean communicating to other terminals via infrared beams. In the case of mobile phones, we are talking about digital wireless phone communication protocols, such as CDMA or GSM. This chapter will show how BREW can use the standard TCP/IP protocol to communicate over wireless phone networks for multiplayer gaming and data services. Most of the hard technical facts in this chapter come from Andy Dornan's absolutely excellent book, *The Essential Guide to Wireless Communications Applications*, Prentice Hall (ISBN 0-13-031716-0). I give this book my absolute highest recommendation if you are really interested in the nuts and bolts of wireless networks and technology.

The Wireless Network

When reading about wireless phone networks, you quickly become awash with strange acronyms and unfamiliar jargon. Words like GSM, CDMA, 1XRTT, and other scary-looking terms leap from the page. Yes, there is a wide variety of different standards and technologies that mobile phones use for communication. The good thing is you really do not need to know much about them. BREW encapsulates its wireless communication features in a familiar socket system that is similar to the tried-and-true socket APIs, such as the Berkeley Socket Interface network programmers are used to from regular desktop operating systems. However, it does not hurt to be familiar with the underlying technologies behind the scenes.

Bandwidth

"Bandwidth" is a term frequently used to determine the speed of an Internet connection—or perhaps more specifically, the amount of data your connection can accommodate. The bandwidth refers to how large the Internet connection is, or how "fat" the "pipe" is, that carries your data. For instance, if you have a cable modem, your bandwidth is much larger than someone who uses an ordinary dial-up modem connection. More data can reach your system per second via cable than dial-up.

There are typically two different bandwidth figures; one measures your transfer capacity for sending information (uplink) and another measures the capacity for receiving information (downlink). In the case of the average cable modem, downlink bandwidth usually ranges somewhere between 1 and 3 megabits per second. The uplink bandwidth usually ranges somewhere from 500 kilobits to 2.5 megabits per second.

To give you a little perspective, a megabit is 2^{20}, or 1048576 bits. A byte is 8 bits. There are 1024 bytes in a kilobyte. Therefore, one megabit is equal to 128 kilobytes. If you compare these figures to a typical local area network that uses 10- or 100-megabit hardware, you can see that these figures are significant but have a way to go until they reach the responsiveness that we are used to from using our own LANs.

Frequencies

In the case of digital wireless phone networks, the bandwidth usually refers to the range of radio frequencies on which the signals travel. This range of frequencies is often referred to as a *spectrum*. The wider the frequency range, the more data the signals can carry (well, sort of).

The range of frequencies available to wireless phone carriers depends on the country in which they are operating. Governmental organizations such as the Federal Communications Commission (FCC) in the United States or the International Telecommunications Union (ITU) in Europe determine which frequency ranges are available to private industry. They then determine how to auction, sell, or otherwise give away chunks of this spectrum to different private interests. This is one of the reasons why phones bought in the United States usually do not work on European networks. The frequency range used for many mobile phones in Europe is reserved in the United States for military use. Therefore, even if an American mobile phone network uses the same transmission technology as European carriers, the phones that the American carrier uses will most likely remain incompatible with European systems as it communicates on an entirely different frequency range. There are global phones that are able to operate on multiple frequencies as well as satellite systems to address this problem.

Multiplexing

Now that a carrier has a chunk of spectrum to play with, which technologies it uses to deploy mobile phone service determines how many customers it can support and what kinds of data transfer speeds users can expect. This is where all the acronyms come in.

The first principle is that most mobile phone networks are indeed cellular. Although the term "cellular" is typically used in reference to old analog networks, even new digital networks use cell architecture to provide service. This means that a wireless carrier has a series of base stations, or cells, that receive transmissions from customers inside a certain radius. There are a number of different types of cells that vary in the area of coverage that they provide. The way these cells communicate with the customers and the network as a whole vary depending on the type of technology used. Yet, the basic concept remains the same.

The number of customers each cell can serve is largely dependent on the type of multiplexing technology that the carrier uses. *Multiplexing* slices up the wireless spectrum that a carrier owns into "sub-bands" to address multiple customers. Each one of these sub-bands can then be accessed by multiple users. The two major competing multiplexing techniques are Time Division Multiple Access (TDMA) and Code Division Multiple Access (CDMA).

TDMA multiplexes a sub-band so that up to three users share the same frequency. It does this by allowing each user a small amount of time to use the sub-band. Once the user's time is up, control is released and another user can get a hold of the same sub-band. These time slice interruptions are so small that they are imperceptible to the user. Voice conversations continue seamlessly, despite all of these tiny interruptions.

CDMA is what is called a spread-spectrum system. That is, a CDMA phone sends the same signal on multiple sub-bands. However, each signal is encoded differently. The individual cell can pick out which signal is encoded for it to understand. CDMA was invented by Qualcomm, and hence all of the current phones using Qualcomm's BREW technology are also based on CDMA technology. Yet, we may see BREW implemented with alternative communication technologies. In fact, at a recent trade show in France, Qualcomm demonstrated BREW running on a GSM device.

Current Digital Wireless Phone Services

Digital wireless phone services are typically referred to as Personal Communication Services, or PCS. These PCS services are based upon TDMA or CDMA technology. The big three these days are Global System for Mobile Communications (GSM), General Packet Radio Service (GPRS), and cdmaOne.

GSM has been in use throughout Europe for quite some time. This is because the governments in Europe made it a standard early on. However, the United States took a free-market approach and allowed the market to find its own standard, with the decision ultimately being made by customer's wallets. GSM is based on TDMA technology and allows up to 14.4 kilobits per second transfer rates for data traffic in modern implementations.

The cdmaOne standard is, as you can probably guess, based on Qualcomm's CDMA technology. Transmission speeds on cdmaOne can reach 14.4 kilobits per second. This is comparable to GSM and most GPRS systems in action today.

GPRS is an upgrade to existing GSM networks that can yield data rates from 14.4 kilobits per second up to 115.2 kilobits per second on the upstream or downstream connection. The actual bit rates depend on the type of GPRS network implemented, and the higher speeds are rather uncommon. Generally, the upstream connection is the same or higher in speed than the downstream link.

The major change in GPRS is the fact that it is a packet-switched network. Both GSM and cdmaOne are circuit-switched. The term "circuit-switched" means that each user maintains an open communications stream, using resources even when there is no data transferring. Packet-switched networks send voice and data in little chunks, called *packets*. This is how the Internet works via the TCP/IP protocol. This means that if there is an awkward silence in a conversation or no data is being sent through a socket, there are no packets being transmitted. This allows a lot more users to use the same sub-bands, as they can take advantage of gaps in the packet flow. Because of the fact that both voice and data are just packets, more advanced GPRS implementations promise the ability to make voice calls while simultaneously maintaining a data connection. This means you can surf the web while talking.

Packet-switching also means that carriers can charge data access by the packet instead of the minute. With existing circuit-switched networks, you are charged by time. If you stare at the same WAP page for an hour, you get charged for an hour's worth of access. With a packet-switched system, you are only charged for the kilobytes downloaded in the transmission of that WAP page. This also has far-reaching implications for game developers, as the amount of data you are transferring in multiplayer games affects the consumer's billing.

The Future of Wireless: 3G

All of the digital wireless systems mentioned in the previous section are what is called second generation, or 2G systems. This is in contrast to the first generation, or 1G, which is composed of the stone-age analog systems of the '80s and early '90s. As for current 2G systems, GPRS is actually commonly referred to as 2.5G. Because of its ability to potentially deliver higher speed access, it is sort of in limbo between the slower second generation networks of current day and the third generation, or 3G, networks of the future.

3G networks are supposed to give the average user 144 kilobits per second transfer rates at minimum levels of service. They can provide up to 2 megabits per second under optimal conditions. Naturally, there are a number of different competing 3G standards. Of course, Qualcomm has their own standard, CDMA2000, which is a clear upgrade path for their current cdmaOne carriers.

CDMA2000 has three phases of deployment. In the first phase, you get the baseline 144 kilobits per second transfer rate, as required by the basic 3G definition. By the time you get to the third phase of CDMA-2000 implementation, you have up to 2 megabits per second and simultaneous voice and data transfers. The three phases of CDMA2000 installation provide existing cdmaOne carriers with a simple and inexpensive upgrade path to 3G.

Other competing standards include Enhanced Data for Global Evolution (EDGE). This is an upgrade path for existing GSM carriers to provide 3G services with up to 384 kilobits per second transfer rates. EDGE is a TDMA-based system that is noted for its conservative use of bandwidth.

Wideband Code Division Multiple Access (W-CDMA) is another 3G CDMA scheme that can provide up to 4 megabit per second speeds using a wider frequency band. Qualcomm has developed a fourth phase of CDMA2000 that can accommodate these higher speeds using the same small amount of bandwidth that it already consumes.

Japan has seen one of the earliest rollouts of W-CDMA technology—the Japanese 3G standard from the creator of the amazingly successful iMode service, NTT DoCoMo. Freedom Of Mobile multimedia Access (FOMA) from NTT DoCoMo is currently providing users with up to 384 kilobit per second transfer speeds with up to 2 megabits promised in the future. Customers have had many complaints of spotty service, defective handsets, and expensive fees, not to mention the handset recalls and other issues that have done much to tarnish NTT DoCoMo's sterling reputation in Japan. It is impressive that NTT DoCoMo has managed to install a small yet growing 3G network way ahead of the competition. Naturally, there are a number of other 3G standards, but a description of all of them is beyond the scope of this book.

Although 3G networks appear to be the wave of the future, it seems the current trend in the United States is to go with 2.5G solutions— even if many of them are dubbed "3G." This includes rollouts of CDMA-2000 phase 1 or enhanced GPRS networks. Spectrum is scarce as well as investment capital these days, so many carriers are being far more conservative with their technology plans, as opposed to the glory days of the dot com boom. Considering the diminutive size of most game applets and the meager capabilities of current handsets, 384 kilobits per second is more than enough to download a simple 50 kilobyte-sized game applet. Until full 3G, we may not get Dick Tracy-style video conferencing and other perks, but the mobile gaming industry can survive with a middle-of-the-road solution for now.

Note: Many European wireless phone carriers have been saddled with massive amounts of debt because of the huge fees paid to buy 3G spectrum. Revenue generated from games and other data services are being looked to as a way to make their money back.

So Who Cares?

You may be trying to wrap your brain around that whirlwind tour of buzzwords and raw statistics. Chances are your BREW handset currently uses cdmaOne or one of the phases of CDMA2000. The good thing is, even in the event that Qualcomm releases BREW chipsets using something like EDGE or W-CDMA, it does not matter to you as a developer.

The only thing you have to be concerned with is how to communicate with the network using the BREW API. BREW has two basic ways to communicate wirelessly: a TCP/IP socket API and a Short Message Service, or SMS, capability. This chapter focuses on sockets, but you can look in Appendix B for information on SMS.

Internet Game Programming

Programming multiplayer games that use the Internet is a rich subject worthy of its own book. The basic concept is that there are two basic elements: the client and the server. The client is the game itself. This is the program that runs on your handset (or PC, console, or whatever) that takes your input and acts upon it like any other game. Except in this case, the client sends the results of your input through the network to a server. The server takes your input, along with that of other players connected through their own clients, and sends back its own data. This data could be the new positions of all the other players in the game world or perhaps information telling the client that you have been defeated by an opponent.

Another network architecture for online games is peer-to-peer communication. Peer to peer (P2P) gained popularity as a buzzword in the glory days of Napster. Napster was a P2P network in that there was no central server handing out MP3 files to everyone. Instead, you connected directly to another user's computer and pulled files off it. That other user is considered a peer.

The concept of P2P is perhaps as old as network gaming itself. Some of the earliest multiplayer games on PCs required you to dial up your opponent directly with a modem for a head-to-head challenge. This directly contrasts with a client-server game, such as Quake, where every player is connected to a single server that takes the player's moves and spits out the state of the game world to all attached clients.

Regardless of the method you choose for your game, you need to use the TCP/IP protocol to communicate over the Internet and thus reach other BREW handset users. And to do this, you need sockets.

Sockets

For TCP/IP communication, BREW uses what we call *sockets*. Sockets have been around since the land before time, also known as the early '80s. Sockets were added to the UNIX operating system in the form of the Berkeley Sockets Interface as a way to communicate over networks with various protocols.

What Is a Socket?

A socket can be thought of as a standard telephone jack, or socket. Through this socket you can communicate with another terminal (phone) in a bidirectional manner. That is, you can send data (talk) and receive data (listen). When you create a socket using BREW's API, you are essentially creating one of these connections between your applet (the client) and a process running on another machine (the server). This process can be another mobile phone or an actual server machine listening for your phone's connection attempt.

Different Kinds of Sockets

There are two different kinds of sockets: streams and datagrams. Each type uses a different protocol to communicate. A stream socket uses TCP, which is a connection-oriented protocol. This means that you have an open connection that is maintained by the program to send and receive data. This connection is closed when the server or client explicitly closes the socket. Stream sockets have very much the same behavior as circuit-switched wireless network protocols, as discussed earlier in this chapter.

Stream communication means that you get the data sequentially in the order that you sent it. Because a stream is continuous, you can count on your data arriving in the order that you sent it. This is also known as a reliable protocol. This may seem like a fairly basic operation, but the fact is the Internet is a very complicated web of nodes and networks. Packets sent out through a socket often arrive at their destination out of order. There is a lot of work going on behind the scenes to make sure that the packets are arranged in the order you sent them. This overhead, and the fact that you must wait for missing packets to arrive in your stream before accessing them, means that TCP is not the fastest way to send data over the Internet.

The second kind of a socket is a datagram socket. Datagram sockets use the UDP protocol that is a record-oriented system instead of connection-oriented like streams. *Datagrams* are individual chunks of data, not continuous streams. This means that the packets sent will not necessarily be received by the server in the order that you sent them. In fact, some packets may be duplicated and others may simply not arrive at all. This is otherwise known as an unreliable protocol.

"Why should I use something considered unreliable in my program?" you may ask. The issue here is speed. UDP packets require much less overhead and do not have to worry about being ordered sequentially upon arrival. Therefore, they can be sent and received much quicker than data through a stream. The problem is that it is up to your application to deal with the fact that the data is potentially arriving out of sequence or not at all. This may involve sending acknowledgment messages, embedding a time stamp inside your data to order it yourself, and using various client prediction schemes to make educated guesses about what information the user may be sending if there is a gap in the packet transmissions.

BREW's Networking and Socket Interface

There are two interfaces that we use with socket communication: INetMgr and ISocket. INetMgr contains the functionality to create sockets, as well as gather data about the phone's network information and status. The ISocket interface is similar to the Berkeley Sockets Interface in that it allows the reading and writing of data from both stream and datagram sockets.

If you are already familiar with socket programming, you may find BREW's implementation of sockets somewhat confusing. This is because BREW does not allow for blocking function calls; it does not have any socket functions that hold up the execution of the program until they finish. When you are dealing with network communications, there are often periods of time where you must wait for a server to respond to a message or congested network traffic causes latency in the transmission or reception of packets. These delays cause traditional socket APIs to pause, or block, until the execution is complete.

Instead of blocking, BREW uses a callback system. You are already familiar with the concept of callbacks from using BREW timers in an earlier chapter. BREW allows for different callback functions to be registered for various socket operations. If a socket function cannot complete its operation, the function will return immediately with a status code telling you that the operation has not completed executing. Once the operation completes or otherwise encounters an error, the desired callback function will be called with the status of the operation and any other associated user data passed to it. The status can be a success code, an error code, or even another status code telling the user that the function is still waiting to complete its operation.

If you are unfamiliar with the use of callbacks in socket programming, Qualcomm's BREW web site provides an excellent document on how to convert your existing blocking socket code to the BREW callback system. You can find this at the developer section on www.qualcomm.com/brew. Also, in the code example section later in this chapter you will be able to see the contrast in methods when we detail the code for the Win32 Winsock-based server and BREW client.

The Net Manager

The first interface you need to know about is INetMgr. As explained previously, INetMgr allows you to set and read settings for BREW's network subsystem, as well as create stream and datagram sockets for communication. Naturally, the first thing you must do to use INetMgr is instantiate the class with ISHELL_CreateInstance. After this, you have access to all of the functions of the interface.

Finding the Address

There are two kinds of addresses used to access resources on the Internet. You can use either the host name, such as www.wordware.com, or the Internet Protocol address, like 209.240.155.150. Both of these addresses refer to the same computer on the Internet. You can verify this by using the Windows console command NSLOOKUP, as shown in Figure 17-1.

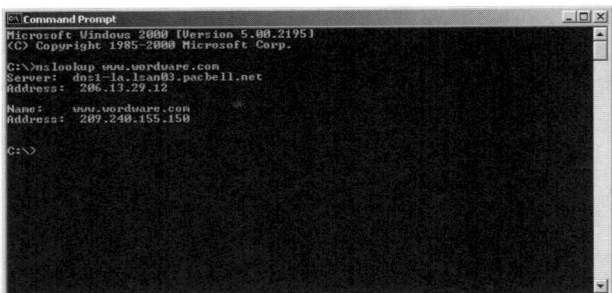

Figure 17-1: NSLOOKUP in action

Looking at the output from NSLOOKUP, you can see that the IP for the hostname www.wordware.com is indeed 209.240.155.150. You are familiar with the host name from using Uniform Resource Locators (URLs) on the World Wide Web. For instance, to connect to the Wordware web site, type the following URL in your web browser: http://www.wordware.com. However, you can also access the same site by typing http://209.240.155.150.

So, what exactly is an IP address? The IP address is how the Internet Protocol references computers on the network. It is a 32-bit number that is actually four 8-bit values, or octets, separated by periods. Because the full address is a 32-bit value, this means there is a grand total of 4,294,967,296 different addresses available. The reality is that there are actually fewer possible combinations, as certain addresses are reserved for special purposes and some companies buy large blocks of addresses for their organizations that they may or may not be actually using. With the rapid expansion of the Internet, we are quickly approaching the limit of addresses that we can fit in the 32-bit range. This is where the next generation IP protocol, IPv6, comes in. Among other enhancements, IPv6 contains an increased address space to solve this problem. IPv6 is being phased in slowly but surely, and BREW will most likely support this protocol when it becomes a real factor in Internet communications.

Note: It has been reported that some smaller countries just getting wired to the net are going straight to IPv6 technology. More information on the IPv6 can be found at www.ipv6.org.

If the Internet Protocol only uses IP addresses to refer to computers on the network, how does the host name also address the same server? This is through the magic of the Domain Name Service (DNS). On the Internet, there are many DNS servers whose entire job is to take a host name, such as www.wordware.com, and convert that mnemonic address to the 32-bit numeric IP address, like 209.240.155.150. To do this translation process, DNS servers contain huge tables that match up host names with IP addresses.

With the zillions of IP addresses on the Internet, it is impossible for any one DNS server to contain a table of every IP out there. Instead, each DNS server contains a portion of this list and can then query other DNS servers on the Internet if it cannot find the host name locally. You can avoid this lookup and translation process by typing the IP address directly into your browser, as we did earlier. In some cases this may speed up the loading of the web page as the DNS server is cut out of the loop.

This brings us to the first function in this chapter, INETMGR_GetHostByName. This function takes a host name and retrieves the IP addresses associated with it. Yes, it is possible for a host name to connect to multiple IP addresses. The function definition looks like this:

```
void INETMGR_GetHostByName(INetMgr * pNetMGr, AEEDNSResult * pres, const char * psz,
    AEECallback * pcb)
```

After the interface pointer, the next argument, **pres**, is a pointer to an **AEEDNSResult** structure that will be filled in with values upon the return of this function call. As defined in AEENet.h:

```
typedef struct
{
    int      nResult;
    INAddr   addrs[AEEDNSMAXADDRS];
} AEEDNSResult;
```

The first member in this structure, nResult, contains either the number of IP addresses retrieved or the error code of the INETMGR_

GetHostByName function call. Some of these errors are detailed in Table 17-1.

Table 17-1. GetHostByName errors

Error	Description
AEE_NET_BADDOMAIN	The host name is malformed.
AEE_NET_UNKDOMAIN	The host name could not be found.
AEE_NET_ETIMEDOUT	The DNS connection timed out.
EUNSUPPORTED	No DNS servers are configured for this device.
ENOMEMORY	Memory allocation failed in the execution of this functionality.

The final member is an array of **INAddr** types named **addrs**. The **INAddr** type is actually just an unsigned integer, as defined in AEENet.h:

```
typedef uint32 INAddr;
```

This 32-bit value contains the IP address. This array is predefined but contains only as many address entries as specified by the count in **nResult**. It is up to you to pick which one you want to use for your socket connection.

The next argument in **INETMGR_GetHostByName** is a pointer to a character string, **psz**. This is the host name, such as "www.word-ware.com," but must be a NULL-terminated string or end with a colon. Finally, we have a void pointer, **pcb**, which points to user data that will be passed as an argument to the callback function called in response to **INETMGR_GetHostByName**'s execution.

The callback used by **INETMGR_GetHostByName** is set via the function **CALLBACK_Init**:

```
void CALLBACK_Init(AEECallback * pcb, PFNNOTIFY pfn, void * pd)
```

This function is a general way to assign a callback to various SDK functions. The first argument is a pointer to the **AEECallback** structure. As defined in AEEShell.h, the structure looks like this:

```
struct _AEECallback
{
    AEECallback *  pNext;          // RESERVED
    void *         pmc;            // RESERVED
    PFNCBCANCEL    pfnCancel;      // Filled by callback handler
    void *         pCancelData;    // Filled by callback handler
    PFNNOTIFY      pfnNotify;      // Filled by caller
    void *         pNotifyData;    // Filled by caller
    void *         pReserved;      // RESERVED - Used by handler
};
```

This structure describes the callback function and is how you identify the function when you wish to cancel a pending callback that you registered with this call. There are three members that you need to be concerned with. The first is the function pointer, pfnCancel. This is a pointer to a function that gets called when you cancel this callback. The second is the pointer to the callback function, pfnNotify. Finally, there is the void pointer to user data passed to your callback as data, pNotifyData.

You can cancel this callback by calling **CALLBACK_Cancel**:

```
void CALLBACK_Cancel(AEECallback * pcb)
```

The **AEECallback** structure argument, pcb, points to a structure registered with a previous call to **CALLBACK_Init**. This function cancels the pending execution of the desired callback function and calls the function pointed to by **pfnCancel** inside the **AEECallback** structure.

Once **INETMGR_GetHostByName** completes its execution, the callback function registered with **CALLBACK_Init** will be called. From this callback, you can use it to take your applet to the next stage of establishing an Internet connection. Once a callback is executed, it is removed from the list of callbacks to call.

Getting back to **INETMGR_GetHostByName**, the second argument, pres, is a pointer to the callback function that you want to be called. This may seem redundant, given the fact that you specify this in your **AEECallback** structure as well. The final argument, pcb, is a void pointer to user data that will be passed as an argument to this callback.

Your handset also has a unique IP address. You can find out the address of your handset by calling **INETMGR_GetMyIPAddr**:

```
INAddr INETMGR_GetMyIPAddr(INetMgr * pINetMgr)
```

This function takes the interface as an argument and returns the IP address in the form of an **INAddr** type. When you use this on the emulator, it returns 127.0.0.1, which is a special reserved IP address that refers to your own machine. Therefore, it is not a good way to uniquely identify the handset if you are testing code in the emulator for this purpose.

Dealing with Errors

The result codes that any of these **INetMgr** functions can return may sometimes tell you that an error has occurred. The thing is, they may not tell you what kind of error. You can use the function **INETMGR_ GetLastError** to retrieve a more specific code for the last error that occurred with the use of the **INetMgr** interface:

```
int INETMGR_GetLastError(INetMgr* pINetMgr)
```

This function takes the INetMgr interface as a pointer and returns the last error code encountered as an integer. You can look at these error codes in AEENet.h to see what the returned value means.

Getting Information About the Connection

BREW keeps track of a number of different attributes about the handset's current data network connection. You can retrieve this information via a call to INETMGR_NetStatus:

```
NetState INETMGR_NetStatus(INetMgr * pINetMgr, AEENetStats * pNetStats)
```

After the interface pointer, pass another argument, pNetStats, which is a pointer to an AEENetStats structure. This structure will be filled with information about the network connection. The structure is defined in AEENet.h like so:

```
typedef struct
{
    uint32    dwOpenTime;
    uint32    dwActiveTime;
    uint32    dwBytes;
    uint32    dwRate;

    uint32    dwTotalOpenTime;
    uint32    dwTotalActiveTime;
    uint32    dwTotalBytes;
    uint32    dwTotalRate;
} AEENetStats;
```

As you can see, there are a variety of different members that provide information about the connection's time, data rate, and other statistics. This information may be valuable if you wish to compute billing information, such as letting the player know how many bytes have been set if she is being charged by the packet instead of the minute.

The function returns a NetState enmumerated type that is defined in AEENet.h like this:

```
typedef enum
{
    NET_INVALID_STATE,
    NET_PPP_OPENING,
    NET_PPP_OPEN,
    NET_PPP_CLOSING,
    NET_PPP_CLOSED
} NetState;
```

This gives you a general status of the network connection. It will tell you if it is in the process of opening or closing, or is open and ready to handle data.

Opening a Socket

Of course, getting information about a net connection is fine. But how do you actually make the connection? First, you must use INetMgr to create a socket. This is done via the function call INETMGR_ OpenSocket:

```
ISocket * INETMGR_OpenSocket(INetMgr * pINetMgr, NetSocket type)
```

The second argument after the interface, type, is an enumerated type, NetSocket. This is defined in AEENet.h like this:

```
typedef enum
{
    AEE_SOCK_STREAM=0,
    AEE_SOCK_DGRAM
} NetSocket;
```

As you can see, you can specify either a stream socket (TCP) or a datagram socket (UDP). The function returns a pointer to the ISocket interface of the socket that you create. The ISocket interface is used to communicate with the network.

The Socket

Now that you have created a socket, you have essentially installed a phone jack, plugged a phone into it, and picked up the receiver. You now have to make a call (or rather, make a connection to a server).

Opening a Connection

Opening a connection is done with the function ISOCKET_Connect:

```
int ISOCKET_Connect(ISocket * pISocket, INAddr a, INPort wPort, PFNCONNECTCB pfn,
    void * pUser)
```

The first argument is a pointer to the socket interface that you are trying to connect with. Next is the network address to which you are connecting. This can either be gained from INETMGR_GetHostBy-Name, as explained previously, or you can explicitly provide your own IP address. Next is the port number.

In addition to the IP address, you must also tell the socket which port on that address you wish to connect to. Because a server can run many different applications that listen for network traffic simulta-neously, each application must have its own place to look for its associated network traffic. These different locations are known as *ports*.

For instance, web servers traditionally listen for HTTP protocol connections on port 80. FTP operations commonly use port 21 to com-municate between clients and servers. This does not mean you cannot reconfigure the server to route web or FTP traffic to arbitrary port numbers. However, you can be sure that various well-known ports are used for popular services, such as FTP and web traffic. When you write a game server or any other server application, you must pick a port not used by any other applications on the machine. Since you never can be sure which ports are in use, often server programs will allow the port to be configured by the server's administrator.

The port parameter, wPort, is of type INPort. If you look in the AEENet.h header file, you will see it defined like so:

```
typedef uint16 INPort;
```

It is simply an alias for a 16-bit integer, or short. So, what if you want to communicate with a server application that is listening for traffic on port 50000? You cannot simply pass the value 50000 as the port parameter. Instead, you have to convert this value to be in network byte order.

Network byte order is the order in which your data must be organized in order to be properly handled by the TCP/IP protocol. There are two types of byte ordering commonly used on all computer systems today. They are called *big endian* and *little endian*.

These two systems order how the bytes are lined up left to right. In the case of little endian, the rightmost byte is most significant. For big endian systems, the leftmost byte is most significant. That means that to read a value on a little endian system, you read left to right. On a big endian system, you read right to left.

CPUs used in PC machines are little endian. This includes processors such as Intel's Pentium and AMD's Athlon. Processors such as Motorola's Dragonball series used in the Palm PDA systems are traditionally big endian. Some CPUs, such as Motorola's PowerPC processors, can read both formats.

It just so happens that TCP/IP uses a big endian system to order its bytes. Because the processor on your PC running the BREW emulator and the ARM CPU inside your BREW handset use the little endian representation of data, you must convert port numbers and IP addresses to the network byte order, which is big endian.

Luckily, BREW has a few handy macros for doing this. If you want to pass the port number 50000 in the wPort parameter, use the AEE_htons macro for this purpose:

```
AEE_htons(50000)
```

The resultant value is the number 50000 with its bytes reversed for network byte order. There are a series of macros used for converting various data types to network byte order; they are defined in the AEENet.h header file. You also have to convert the INAddr argument to network byte order if it is not already. The addresses generated by INETMGR_GetHostByName are automatically in network byte order.

Moving along to the next argument in ISOCKET_Connect, we see pfn, which is a pointer to a callback of type **PFNCONNECTCB**. This is the callback function that will be called upon completion of the ISOCKET_Connect operation. The final argument, pUser, is a void pointer that will be passed to this callback as a parameter. Naturally, it is common to pass a pointer to the applet cast as a void pointer for this argument.

If the socket connection happens immediately and there is no error, the callback will not be called. If for some reason the connection cannot be made, the callback will be called and the appropriate error code will be passed to the function along with the user data specified with pUser. The connection callback function definition looks like this:

```
typedef void (*PFNCONNECTCB)(void * pUser, int nError);
```

As you can see, the first argument is the user data that you specified in the ISOCKET_Connect function. The second argument is the error code generated by the same function call.

The ISOCKET_Connect call returns an integer that is an error code as well. If the call went through without a hitch, it will return AEE_NET_SUCCESS. If there is an error, it returns AEE_NET_ERROR. The callback will be called with more information. The error code in the nError argument to the callback can be one of several values defined in AEENet.h that you can use to further diagnose your network problem.

It is possible to open multiple sockets to different addresses. However, keep in mind that there are limits to how many sockets you can open on a handset. In the case of the Kyocera 3035, you can only have one stream and one datagram socket open at once. Therefore, you must note the device's capabilities when designing your network application.

Now that you are connected to another process on another machine, you are free to send and receive data. Depending on whether you have a stream or datagram socket, this is done in different ways. First, let's examine the stream socket.

Sending Stream Data

To send data on a stream socket, use the function ISOCKET_Write:

```
int32 ISOCKET_Write(ISocket * pISocket, byte * pBuffer, uint16 wBytes)
```

The first argument is the socket to which we are writing. Next is a pointer to an array of characters, pBuffer, which is the data we are going to send through the socket. The final parameter, wBytes, is the size of this array.

This function returns a 32-bit wide integer. If the call is successful, it returns the number of bytes written to the socket. If you are trying to send large amounts of data, the phone may not be able to transfer all of the array at once. Therefore, you must call ISOCKET_Write again while moving the pointer in your data buffer forward by how many bytes were written out in the previous call.

If the ISOCKET_Write call fails for some reason, you will get one of two error codes: AEE_NET_ERROR or AEE_NET_WOULDBLOCK. The first code signals that a general error has occurred in the process of writing to the socket. You can get more information on this error by calling ISOCKET_GetLastError:

```
int ISOCKET_GetLastError(ISocket * pISocket)
```

As you can see, this is similar to INetMgr's own error function. The value returned in response to an ISOCKET_Connect error can be any number of values defined in AEENet.h. Using these error codes, you can further diagnose what went wrong with your socket connection and handle the situation accordingly.

The other error code returned from ISOCKET_Connect, AEE_NET_WOULDBLOCK, relates to the non-blocking nature of BREW's socket interface. This means that the call could not go through for some reason. Because BREW does not block until the socket becomes free, you must call ISOCKET_Write later when the data is able to be written. BREW will call a callback function once the network resource becomes available and it is time to attempt a write again as an alternative to holding up the execution of the applet.

How do you assign a callback to be used with ISOCKET_Write? This is done through the function call ISOCKET_Writeable:

```
void ISOCKET_Writeable(ISocket * pISocket, PFNNOTIFY pfn, void * pUser)
```

This call registers the function specified in the function pointer pfn to be called when the socket pointed to by pISocket determines it can attempt another write operation. This goes for all write calls, including the datagram write functions that we will detail later in this chapter. The final argument, pUser, is a void pointer to user data, which will be passed to the callback upon its calling. As with other callbacks in BREW, a pointer to the applet cast to void is traditionally used.

You can cancel these callbacks by using the function ISOCKET_Cancel:

```
void ISOCKET_Cancel(ISocket * pISocket, PFNNOTIFY pfn, void * pUser)
```

Simply pass a pointer to the socket that the callback is registered for with pISocket, a pointer to the callback function you wish to cancel with pfn, and a pointer to the user data you are passing in as an argument to the callback. This is used to cancel all types of callbacks used with the ISocket interface.

Inside the callback function, you must attempt another ISOCKET_Write. You must be prepared to handle another AEE_NET_WOULD-BLOCK return code. There is a chance that the socket still cannot be written to for some reason. In this case, you must register the callback and try again next time.

Receiving Stream Data

Now that you know how to send information on a stream socket, how do you receive information? This is done through the ISOCKET_Read function:

```
int32 ISOCKET_Read(ISocket * pISocket, byte * pbDest, uint16 wSize)
```

The first parameter is a pointer to the socket from which we are reading. The second is a pointer to the buffer where we want the data to be placed. The final argument is the total number of bytes we wish to read from this stream.

The function returns an integer that will be either the number of bytes read from the stream or an error code similar to those of ISOCKET_Write. If you were not able to read all of the data you need in one go, you have to subtract the amount of bytes read from your byte count, move up the pointer in your destination buffer, and try again. If the number of bytes read is 0, the connection has been shut down and you may need to reconnect.

The error codes again are AEE_NET_ERROR and AEE_NET_WOULDBLOCK. You can get more information on an error by calling ISOCKET_GetLastError, as in the ISOCKET_Write example. To set up a callback to be used in the case of an AEE_NET_WOULDBLOCK result from this and any other socket writing functions, you must call ISOCKET_Readable. It works very much the same way as ISOCKET_Writeable:

```
void ISOCKET_Readable(ISocket * pISocket, PFNNOTIFY pfn, void * pUser)
```

The first argument is the socket for which we are registering a callback. The second is a pointer to our callback function. The final argument is a pointer to user data that will be passed in as an argument to our callback. This callback is called when BREW thinks the socket may be ready for another read attempt. As with the write example, you must be prepared to handle another stall inside this callback with an AEE_NET_WOULDBLOCK return.

Sending Datagrams

Because sending and receiving datagrams is fundamentally different from transferring data on streams, there are separate read and write functions for UDP sockets. Although you can use the same ISOCKET_Write function for sending datagrams as for sending streams, there is a separate function for writing to datagram sockets as well. To send data only through a datagram socket, use the function ISOCKET_SendTo:

```
int32 ISOCKET_SendTo(ISocket * pISocket, byte * pBuff, uint16 wBytes, uint16 wflags,
    IPAddr a, INPort wPort)
```

As usual, the first argument is a pointer to the socket through which we are sending a datagram. The second parameter, pBuff, is a pointer to the buffer containing the information we are sending. The third argument, wBytes, is the size of the buffer. The wflags parameter is ignored for now, perhaps for use with future versions of BREW. Finally, pass the IP address and port we are sending the datagram to with the final two parameters, a and wPort. Remember that the address and port must be in network byte order.

Why must you specify the address and port when you already did that in ISOCKET_Connect? Well, datagram sockets are never really connected. With a stream, ISOCKET_Connect actually opens up a stream connection with a peer. If the socket is of a datagram type, the address and port are stored for future use, but no constant connection is maintained with a datagram-based socket. When you use the standard ISOCKET_Write call to send data through a datagram socket, it uses the address and port that you specified with ISOCKET_Connect to send the datagram. By using ISOCKET_SendTo, you skip this extra step and simply specify the IP address and port yourself.

Receiving Datagrams

The same goes for receiving datagrams. It is possible to use ISOCKET_Read on a UDP socket with it using the address and port from ISOCKET_Connect implicitly. Or, you can use ISOCKET_RecvFrom:

```
int32 ISOCKET_RecvFrom(ISocket * pISocket, byte * pBuff, uint16 wbytes, uint16 wflags,
    INAddr * pa, INPort * pwPort)
```

This works much like ISOCKET_SendTo. The first parameter, pISocket, is the socket we are listening to. The second, pBuff, is a pointer to a buffer where we will put the received data. The third, wbytes, is the amount of data we wish to read in. The third, wflags, is an unused flags argument. Finally, pass the IP address and port in network byte order.

Both the reading and writing of datagrams can return an AEE_NET_ WOULDBLOCK value, which means you have to use the same callback scheme as seen in stream operations. This should be old hat to you by now, and it works in exactly the same manner.

Socket Wrap-up

Now you know how to connect to other machines with sockets, as well as how to send and receive data. As previously discussed, socket resources are extremely limited on some handsets, so you must be careful with how you use your sockets. Also, keep in mind that the user can suspend the applet at any time with an incoming call, SMS message, or other operation. This does depend on the network's ability to provide simultaneous voice and data, of course. Therefore, you must be prepared to terminate any socket connections upon suspension and resume them properly later.

Seasoned socket programmers may notice a certain function missing from the ISocket interface. The missing function is listen from the Berkeley Sockets Interface. The ability to listen to a socket allows a program to watch a port and act when a socket connection is made. This is useful for server programming, as we will see in our example. Because of this limitation, writing a server as a BREW applet is problematic.

Also, note that direct socket connections to other BREW handsets are considered rather unstable. It is advisable to connect to a proxy server, which sends data to all involved clients instead of making direct socket connections to another player's phone. This makes P2P-style network games virtually impossible on BREW handsets at the moment.

Testing socket programs can become exceedingly complex. This is why the True BREW testing fees for network-aware applets are more expensive than regular single-user applets. Synchronization issues with servers, dealing with unreliable networks, and handling latency issues is worthy of a book unto itself. In fact, many have been written on the subject.

Now that you have a brief overview of the major functions involved in socket communication, it is time to show an example program. By the time you are done going through the code, you should have a good idea about how to implement your own multiplayer game.

A Simple Example

The socket example in this chapter is actually two different programs: a BREW client and a Windows server. The end result is a simple Tic-Tac-Toe game that gets the computer's move from a remote server on the Internet. Although the server itself generates its own move, it theoretically could be receiving move information from another player and passing it to the BREW client. Although this is not a book about Windows network programming, we will delve into the Windows server briefly later so you know what the BREW client is dealing with. The code for both the client and server projects is found in the Source\Chapter 17 folder in the companion files.

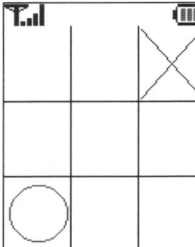

Figure 17-2: The Tic-Tac-Toe client running in the emulator

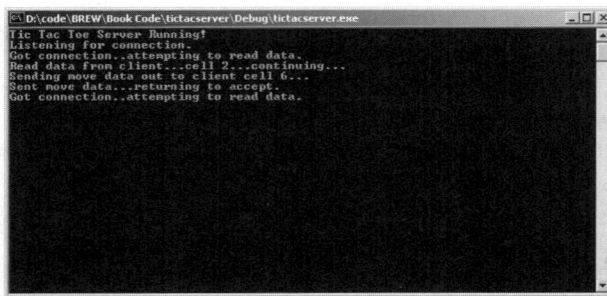

Figure 17-3: The Tic-Tac-Toe Win32 client running in the console window

The Server

The Windows socket API is known as Winsock. This is very similar to the Berkeley Sockets Interface with a few extras unique to Windows. We will not deal with any of the Windows minutiae at the moment. That is a topic for a whole other book. If you want a great explanation of Winsock, check out James A. Frost's excellent paper at world.std.com/~jimf/papers/sockets/winsock.html. That page contains skeleton code for both a client and server. The latter is used as a basis for the Tic-Tac-Toe Winsock server application.

The only source file of the server is tictacserver.cpp. At the start of the file, there are a few global variables and definitions. Yeah, I know, but I said this was quick and dirty:

Listing 17-I: Definitions

```
#define CONDITION_WIN     255
#define CONDITION_LOST    254
#define CONDITION_DRAW    253

#define PORTNUM 50000
```

```
unsigned char g_pnGameBoard[9];
int g_nWinner;
```

The first three defines are special codes that will be sent to the server to signify a win, loss, or draw. Next we have the port definition. Port 50000 is just an arbitrary number. Chances are, nobody is using it. Feel free to change it if you like.

Finally, we have the global variables. The array of nine characters, **g_pnGameBoard**, is used to hold the contents of the Tic-Tac-Toe game board. The **g_nWinner** variable is used to signal if someone wins. If it is set to 0, nobody has won yet. If it is set to 1, the client has won. If it is set to 2, the server has won.

Here is something we have not seen in a while: a main loop. Here is the server's main:

Listing 17-2: Main server loop

```
int main(int argc, char* argv[])
{
    SOCKET sockHandle;
    bool bBreakOut = FALSE;

    printf("Tic Tac Toe Server Running!\n");

    //Start up winsock
    WSADATA info;
    if (WSAStartup(MAKEWORD(2,0), &info) != 0)
    {
        printf("Cannot startup winsock!\n");
        WSACleanup();
        exit(1);
    }

    ResetGame();

    //Open socket on our port
    if ((sockHandle = CreateStreamSocket(PORTNUM)) == INVALID_SOCKET)
    {
        printf("Cannot create socket!\n");
        WSACleanup();
        exit(1);
    }

    //Infinite read/write loop
    while (1)
    {
        bBreakOut = false;
        printf( "Listening for connection.\n");

        //Wait for client to connect
        SOCKET newSocket = accept(sockHandle, NULL, NULL);

        if (sockHandle == INVALID_SOCKET)
    {
            printf( "Connection error!\n");
            WSACleanup();
```

```
            exit(1);
    }

        //While the game is in progress, run turns
        while (!bBreakOut)
        {
            bBreakOut = DoMove(newSocket);
        }

        closesocket(newSocket);
    }

  WSACleanup();
    return (0);
}
```

The first thing to do is declare a few variables. First, a **SOCKET**, sockHandle, is declared. Then our global **bool**, bBreakout, is declared right below. The socket is obviously used for communication, while the **bool** is used to tell the server to break out of its infinite loop of waiting for player communication when the game has ended.

Before we do any socket operations, we have to initialize the Winsock subsystem. Do this with a call to **WSAStartup**. This call tells Winsock which version we want to use. In this case, we specify Winsock 2.0. Next, initialize the global game variables with **ResetGame**:

Listing 17-3: Resetting the game

```
void ResetGame()
{
    //no winner
    g_nWinner = 0;

    //seed randomizer
    srand(GetTickCount());

    //clear the game board
    memset(g_pnGameBoard, 0, sizeof(char) * 9);
}
```

This is all very self-explanatory. Then, create a stream socket using our custom function, **CreateStreamSocket**:

Listing 17-4: Creating the server-side socket

```
SOCKET CreateStreamSocket(unsigned short nPortnum)
{
    char    pszHostName[256];
    SOCKET sockHandle;
    sockaddr_in sa;
    hostent *hp;

    gethostname(pszHostName, sizeof(pszHostName));
    hp = gethostbyname(pszHostName);

    if (hp == NULL)
```

```
            return(INVALID_SOCKET);

        memset(&sa, 0, sizeof(sockaddr_in));
        sa.sin_family = hp->h_addrtype;
        sa.sin_port = htons(nPortnum);

        sockHandle = socket(AF_INET, SOCK_STREAM, 0);

        if (sockHandle == INVALID_SOCKET)
            return INVALID_SOCKET;

        if (bind(sockHandle, (sockaddr *)&sa, sizeof(sockaddr_in)) ==
            SOCKET_ERROR)
        {
                closesocket(sockHandle);
                return(INVALID_SOCKET);
        }

    listen(sockHandle, 3);
        return(sockHandle);
}
```

This is all basic Winsock stuff. You probably recognize some of the calls from the ones in the BREW API. We get our host name by calling **gethostname**. Then we take our **socketaddr_in** structure and fill it with the appropriate settings, including our port number, which is passed in the short integer argument **nPortnum**. Of course, we convert that port number to network byte order using **htons**. Then, using the **socket** function, we create a stream socket. Because we are going to use **listen**, we also have to use the **bind** function. This associates the socket with the port and other attributes that we set in the **sockaddr_in** structure. BREW also has a **Bind** function, but it is not as useful, as there is no listening capacity in BREW. Finally, set the socket to listen on the socket for any incoming connection. Pass 3 as the argument to tell Winsock that we want a maximum of three waiting connections in queue.

Now, back to the main loop. After creating the stream socket, enter our infinite loop of waiting for connections and sending/receiving stream data:

```
while (1)
{
    bBreakOut = false;
    printf( "Listening for connection.\n");

    //Wait for client to connect
    SOCKET newSocket = accept(sockHandle, NULL, NULL);

    if (sockHandle == INVALID_SOCKET)
    {
        printf( "Connection error!\n");
        WSACleanup();
        exit(1);
    }
```

```
//While the game is in progress, run turns
while (!bBreakOut)
{
    bBreakOut = DoMove(newSocket);
}

closesocket(newSocket);
}
```

The first thing we do is call **accept**. This blocks until we receive a socket connection. Once the connection is made, we then call **DoMove** in an infinite loop until the **bool**, **bBreakOut**, is flagged as True. Once **bBreakOut** is flagged, close the socket and return to the top of the loop where we wait for another connection. **DoMove** is where all the magic happens:

Listing 17-5: Performing a move on the server

```
bool DoMove(SOCKET sock)
{
    //Read the socket to get the move
    unsigned char nCell;
    int nBytesRead = 0;
    int nTotalBytes = 0;
    int nBytesSent = 0;
    int nTotalBytesSent = 0;
    bool bFinished = false;

    char buf[32];

    printf("Got connection..attempting to read data.\n");

    //Read move data in from client
    while (nTotalBytes < sizeof(char))
    {
        nBytesRead = recv(sock, buf, sizeof(buf), 0);

        if (nBytesRead < 0)
        {
            int nErr = WSAGetLastError();
            ResetGame();
            return TRUE;
        }
        else
            nTotalBytes += nBytesRead;
    }

    //Copy the data into our integer
    memcpy(&nCell, buf, sizeof(char));

    printf( "Read data from client...cell %d...continuing...\n", nCell);

    //If there is a win condition, the game is over
    if (nCell == CONDITION_WIN)
    {
        ResetGame();
```

```
          printf("Server lost! Resetting server!\n");
          return TRUE;
    }
    else if (nCell == CONDITION_LOST)
    {
          ResetGame();
          printf("Server won!  Resetting server!\n");
          return TRUE;
    }
    else if (nCell == CONDITION_DRAW)
    {
          ResetGame();
          printf("Draw!  Resetting server!\n");
          return TRUE;
    }

    //Fill in the cell if it is unoccupied
    if (!g_pnGameBoard[nCell])
    {
          g_pnGameBoard[nCell] = 1;
    }
    else //Otherwise, something bad is going on
    {
          printf("Someone is cheating! Exiting!\n");
          WSACleanup();
          exit(1);
    }

    //Now do alleged AI (random placement)
    while (!bFinished)
    {
          //Keep generating cell positions until we get a good one
          nCell = rand()%9;

          //If the cell is not occupied, fill it and send it
          if (!g_pnGameBoard[nCell])
          {
                g_pnGameBoard[nCell] = 2;

                printf("Sending move data out to client cell %d...\n", nCell);

                //Send this cell ID out
                while (nTotalBytesSent < sizeof(char))
                {
                      nBytesSent = send(sock, (const char*)&nCell, sizeof(char), 0);

                      if (nBytesSent <= 0)
                      {
                            int nErr = WSAGetLastError();
                            ResetGame();
                            return TRUE;
                      }
                      else
                            nTotalBytesSent += nBytesSent;
                }

                bFinished = true;
          }
```

```
        printf("Sent move data...returning to accept.\n");
    }

    return FALSE;
}
```

This may seem like a large function, but it is actually very simple. Overall, this gets the turn information from the client in the form of a single byte. This byte tells the server where the client placed its O. If the client sends a message signifying a win, loss, or draw, terminate the game. Otherwise, pick a random cell to put the server's X and send that out to the client.

First, we declare and initialize a bunch of variables. The first, nCell, is used to hold the cell on the game board that we are receiving a move for or placing the server's X in. Although we are only going to send and receive a single byte for each turn, use the variable nBytesRead to determine if we have read or sent enough bytes with our socket calls. The nTotalBytesSent variable is used to determine how many bytes we have sent, while nTotalBytes is used to determine how many bytes we have received. As mentioned before, these should both equal 1 if we have successfully sent or received a move. Finally, buf is a general-purpose buffer that we use to place our data that we are sending or receiving.

After the variable declaration, wait in a loop calling recv to take in any socket communication from the client. Notice that we always check to see if the socket call has returned an error by returning a number of bytes read equal or less than zero. If so, we just bail. Sure, that is not very robust, but this is not a commercial-grade application anyway. Use this behavior for all socket errors in the server.

Next, check the win conditions. If the game is over, reset the game and return False from the function. This tells the main loop to break out and listen for a new client socket connection.

Then fill in the cell in the game board sent in from the client. If there is already something in that cell, then something is amiss. In this case, bail once again. Perhaps someone is cheating with a hacked client. This is a definite possibility in most PC networked games, although in a controlled environment such as BREW, it is unlikely. Perhaps I am just being a little paranoid. There is nothing wrong with a little defensive programming, however.

Now it is time for our astounding artificial intelligence. Well, not really. We stay in a loop, generating random cell positions until we find one that is empty. Once we get a valid cell, send this move out to the client as a single byte by using the send function. Loop until we have sent out all the bytes, and of course, bail upon any socket error.

That is about it. The server is exceedingly simple. You can use this code as a framework for a more complicated game server. In most cases, game servers are attached to player and world databases, sometimes using SQL and other database systems. Also, they may match up players with others who would like to play and broker games between them. Obviously this simple game of Tic-Tac-Toe is a far cry from the advanced commercial game servers of most real wireless multiplayer games. But it serves our purpose of getting to know the BREW networking APIs.

The Client

Now on to the game client. This is a traditional game applet, involving graphics, user input, and now socket communication. It is a simple game of Tic-Tac-Toe, but instead of playing against computer AI embedded into the applet, the moves are coming from the computer player residing on the server. For the sake of testing, the server and the client will reside on the same machine. That is, we will use the special IP address 127.0.0.1 to connect to the server that actually always refers to the host machine.

One thing to note with this applet is the MIF file. If you open the MIF in the MIF Editor and check the permissions, notice that the Network box is checked:

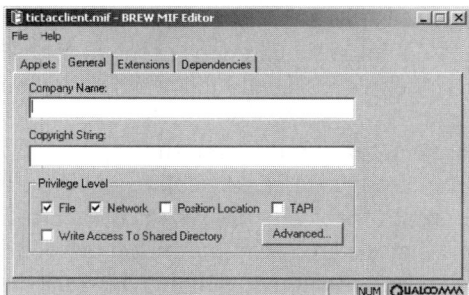

Figure 17-4: The MIF file for the client applet

If this box is not set, the applet will not have permission to make any network communication calls. Therefore, all of your socket operations will fail.

The header file contains the standard array of defines, function definitions, and structures. At the start, we have a few macros that are needed for an address conversion function that I got from the BREW SDK socket example. We also have the win condition defines that you should be familiar with from the server side. We also have some enumerated types and structures:

Listing 17-6: Client header file

```
enum
{
    NETSTATE_WRITE = 0,
    NETSTATE_READ,
    NETSTATE_CONNECT,
    NETSTATE_OK,
};

typedef struct myapp_s
{
    AEEApplet        a;
    AEEDeviceInfo    di;

    unsigned char nMove;

    int nNetstate;
    int nWinner;
    INAddr nodeINAddr;

    unsigned char pnGameBoard[9];

    ISocket *    pISocket;
    INetMgr *    pINetMgr;
    IGraphics * pIGraphics;

} myapp_t;
```

The enumerations are used to tell the applet which socket process we are in the midst of. Because we are merely connecting to a server application on the same machine, I do not anticipate any latency problems. But in the interest of learning techniques useful for real development, I have used them as a safeguard.

Next, we have the applet structure. Aside from the obligatory **AEEApplet** structure and our customary **AEEDeviceInfo**, we have a bunch of other members, the first of which is **nMove**, a character in which we will store the move that we are reading from the server. It can fit in a character because we are only reading a single byte from the client. However, we will treat it as an integer.

Next we have a series of integers starting with **nNetstate**. This will always be set to one of the previous enumerations to tell the applet which network communication operation we are in the midst of. The **nWinner** integer is used to determine if there is a winner. This behaves in the same way that the corresponding **nWinner** variable does in the server code.

After the integers, we have an IP address, **nodeINAddr**. This will be used to store the network byte-ordered IP address of the server to which we are connecting. Next we have an array of 9 characters, one for each cell in the Tic-Tac-Toe board. When one of the cells is set to 0, there is nothing in it. If it is set to 1, the player has put an X there. If it is set to 2, the server has placed an O in it.

The three interfaces at the end of the structure begin with a pointer to an **ISocket** interface, **pISocket**. This is the socket that we will use to communicate with the client. Next we have a pointer to the **INetMgr** interface, **pINetMgr**, to instantiate our socket. Finally, we have a pointer to the **IGraphics** interface, **pIGraphics**, to draw graphics, such as the lines of the Tic-Tac-Toe board.

Naturally, the code begins with the applet creation function that we all know and love:

Listing 17-7: Creating the applet instance

```
int AEEClsCreateInstance(AEECLSID ClsId, IShell * pIShell, IModule * po, void ** ppObj)
{
    *ppObj = NULL;

    if(ClsId == AEECLSID_TICTACCLIENT_BID)
    {
        if(AEEApplet_New(sizeof(myapp_t), ClsId, pIShell, po, (IApplet**)ppObj,
            (AEEHANDLER)TicTacClient_HandleEvent,
            (PFNFREEAPPDATA)TicTacClient_Cleanup) == TRUE)
        {
            return(AEE_SUCCESS);
        }
    }

    return(EFAILED);
}
```

There is nothing out of the ordinary here. We simply pass the **AEEApplet_New** function a pointer to our creation and cleanup functions as usual. Let's look at that message handler:

Listing 17-8: Event handler

```
static boolean TicTacClient_HandleEvent(IApplet * pi, AEEEvent eCode, uint16 wParam,
    uint32 dwParam)
{
    myapp_t* pApp = (myapp_t*)pi;
    AEEApplet * pMe = &pApp->a;

    switch (eCode)
    {
        case EVT_APP_START:

        //Initialize game variables
        pApp->nWinner = 0;
        pApp->nMove = 0;

        //Get device capabilities
        ISHELL_GetDeviceInfo(pApp->a.m_pIShell, &pApp->di);
```

```
                    //Clear out game board
                    MEMSET(pApp->pnGameBoard, 0, sizeof(char) * 9);

                    //Create interfaces we are using
                    ISHELL_CreateInstance(pApp->a.m_pIShell, AEECLSID_NET,
                        (void **)&pApp->pINetMgr);
                    ISHELL_CreateInstance(pApp->a.m_pIShell, AEECLSID_GRAPHICS,
                        (void **)&pApp->pIGraphics);

                    //Now we have to connect to the server socket
                    TicTacClient_OpenConnection(pApp);

                    //Clear and draw screen
                    TicTacClient_DrawScreen(pApp);
                    IDISPLAY_Update(pMe->m_pIDisplay);

                    //We've successfully handled this message
                    return(TRUE);
                    break;

                case EVT_KEY: //Key input (where to put your X)

                        if (pApp->nNetstate == NETSTATE_OK)
                        {
                                TicTacClient_HandleInput(pApp, wParam);
                        }

                        //Refresh screen
                        TicTacClient_DrawScreen(pApp);
                        IDISPLAY_Update(pMe->m_pIDisplay);

                        return(TRUE);
                        break;
        }

    return(FALSE);
}
```

Begin by looking at the **EVT_APP_START** handler. Initialize a few variables here, get the device information, clear out the game board, and create the interfaces we will use throughout the applet: **INetMgr** and **IGraphics**. Next, call the **TicTacClient_OpenConnection** function to get our socket connection going. Let's look at that now:

Listing 17-9: Opening connection to server

```
void TicTacClient_OpenConnection(myapp_t* pApp)
{
    int nRet;

    pApp->nNetstate = NETSTATE_CONNECT;

    pApp->pISocket = INETMGR_OpenSocket(pApp->pINetMgr, AEE_SOCK_STREAM);

    if (!pApp->pISocket)
    {
```

```
         int nErr = ISOCKET_GetLastError(pApp->pISocket);
         ISHELL_CloseApplet(pApp->a.m_pIShell, FALSE);
}

//Connect to the server
pApp->nodeINAddr = TicTacClient_ConvertToINAddr(HOST_ADDR);
nRet = ISOCKET_Connect(pApp->pISocket, pApp->nodeINAddr,
    AEE_htons(HOST_PORT), TicTacClient_ConnectCallBack, (void*)pApp);

//If we have an error on connection, quit
if (nRet && (nRet != AEE_NET_WOULDBLOCK))
{
         int nErr = ISOCKET_GetLastError(pApp->pISocket);
         ISHELL_CloseApplet(pApp->a.m_pIShell, FALSE);
}
else if (nRet != AEE_NET_WOULDBLOCK)
{
         pApp->nNetstate = NETSTATE_OK;
}
}
```

This creates a socket, gets an IP address, and attempts to connect the socket to the server application. First, put the applet in the connection netstate by setting nNetstate to NETSTATE_CONNECT. Then, open a socket with a call to INETMGR_OpenSocket and assign the resultant pointer to pISocket in the applet structure. If the socket cannot be opened, bail from the applet. Much like the server, this is not a very robust solution. For now, it will do.

We then have to create an IP address to which we connect the socket. We do not use INETMGR_GetHostByName because we already know the IP we want to use. Because the server will be running on the same machine as the emulator for the sake of testing, we will use the special IP address 127.0.0.1, which is used to refer to the host PC. We have defined the IP as a string known as HOST_ADDR. The thing is, this string has to be converted into a 32-bit integer. Do this with a function taken out of the BREW socket example called TicTacClient_ConvertToINAddr:

Listing 17-10: Address conversion utility function

```
INAddr TicTacClient_ConvertToINAddr(char *psz)
{
   INAddr ul = 0;
   int nByte = 0;
   char c;

   if(!psz)
      return 0;

   while (ISDIGIT(*psz)) {
      int n = 0;
      while (ISDIGIT(c=*psz)) {
```

```
        n = n*10 + (c - '0');
        ++psz;
      }
      ((char*)&ul)[nByte++] = n;

   if (nByte == 4 || *psz != '.')
        break;

   ++psz;
   }

 if (nByte < 4 || ISALNUM(*psz))
      ul = 0xFFFFFFFF;     // Invalid address

   return ul;
 }
```

This function uses the macros defined in the header to convert an IP address into a 32-bit integer representing the four octets of an IP address. Keep this function in mind; it will come in handy for your own use. There are gems like this all over the BREW example source.

Now that we have converted the IP address string into an integer and stored it in the nodeINAddr member, call ISOCKET_Connect. Of course, in the parameter list we convert the port number, 50000, to network byte order with AEE_htons. If we have an error, quit out of the applet. Otherwise, put the applet into NETSTATE_OK so that the applet knows we are not in the midst of any network communication. Also pass a pointer to a callback function that is called at the completion of INET_Socket:

Listing 17-11: Connection callback

```
void TicTacClient_ConnectCallBack(void * pUser, int nError)
{
    myapp_t* pApp = (myapp_t*)pUser;

    if (nError)
    {
        if (nError < AEE_NET_ENETISCONN)
        {
            int nErr = ISOCKET_GetLastError(pApp->pISocket);
            ISHELL_CloseApplet(pApp->a.m_pIShell, FALSE);
            return;
        }
    }
    else
        pApp->nNetstate = NETSTATE_OK;
}
```

If there is an error, quit out. Otherwise, the cost is clear and we continue on. This is a little redundant, since we already handle the error case in the calling function, but for the sake of completion, we will leave this in.

Continuing with the **EVT_APP_START** handler, draw the screen and update the display. One interesting thing about the client's graphics is that they are resolution independent. Instead of using bitmaps for the X's, O's, and game board, draw them with geometric primitives.

Lines are perfect for the Tic-Tac-Toe board. Draw two intersecting lines for an X, and use the **IGRAPHICS_DrawCircle** function for the Os. The great thing about this is the function takes the resolution of the screen into account and shrinks or expands the graphics accordingly. That way, this will work on the smallest or largest of screens just the same. The code may look a little messy because of this, but it is a great way to avoid having to do multiple versions of your game for different screen resolutions if you do not need bitmaps:

Listing 17-12: Drawing screen

```
void TicTacClient_DrawScreen(myapp_t* pApp)
{
    AEECircle circle;
    AEELine line;
    int i, j, nContents;
    int nSkipX, nSkipY;

    IGRAPHICS_SetColor(pApp->pIGraphics, 0, 0, 0, 255);
    IGRAPHICS_SetFillColor(pApp->pIGraphics, 255, 255, 255, 255);

    IDISPLAY_ClearScreen(pApp->a.m_pIDisplay);

    //Draw lines of tic tac toe board
    nSkipX = pApp->di.cxScreen / 3;
    nSkipY = pApp->di.cyScreen / 3;

    //Draw game board lines
    for (i = 1; i < 3; i++)
    {
        line.sx = i * nSkipX;
        line.sy = 0;

        line.ex = i * nSkipX;
        line.ey = pApp->di.cyScreen;

        IGRAPHICS_DrawLine(pApp->pIGraphics, &line);

        for (j = 1; j < 3; j++)
        {
            line.sx = 0;
            line.sy = j * nSkipY;

            line.ex = pApp->di.cxScreen;
            line.ey = j * nSkipY;

            IGRAPHICS_DrawLine(pApp->pIGraphics, &line);
        }
    }
```

```
//Draw Xs and Os
    for (i = 0; i < 3; i++)
{
    for (j = 0; j < 3; j++)
    {
        nContents = pApp->pnGameBoard[(j * 3) + i];

        if (nContents == 1)
        {
            //Draw blue X
            IGRAPHICS_SetColor(pApp->pIGraphics, 0, 0, 255, 255);

            line.sx = (i * nSkipX) + 2;
            line.sy = (j * nSkipY) + 2;

            line.ex = ((i + 1) * nSkipX) - 2;
            line.ey = ((j + 1) * nSkipY) - 2;

            IGRAPHICS_DrawLine(pApp->pIGraphics, &line);

            line.sx = ((i + 1) * nSkipX) - 2;
            line.sy = (j * nSkipY) + 2;

            line.ex = (i * nSkipX) + 2;
            line.ey = ((j + 1) * nSkipY) - 2;

            IGRAPHICS_DrawLine(pApp->pIGraphics, &line);
        }
        else if (nContents == 2)
        {
            //Draw red O
            IGRAPHICS_SetColor(pApp->pIGraphics, 255, 0, 0, 255);

            circle.cx = (i * nSkipX) + (nSkipX / 2) + 1;
            circle.cy = (j * nSkipY) + (nSkipY / 2);

            if (nSkipX < nSkipY)
                circle.r = nSkipX / 2;
            else
                circle.r = nSkipY / 2;

            circle.r -= 3;

            IGRAPHICS_DrawCircle(pApp->pIGraphics, &circle);
        }
    }
}

//If we are waiting for something, notify user
if (pApp->nNetstate != NETSTATE_OK)
{
    AECHAR szBuf[] = {'W','a','i','t','i','n','g','.','.','.','\0'};

    IGRAPHICS_SetColor(pApp->pIGraphics, 0, 0, 0, 255);

    IDISPLAY_DrawText(pApp->a.m_pIDisplay, AEE_FONT_BOLD, szBuf, -1, 0, 0, NULL,
        IDF_ALIGN_MIDDLE | IDF_ALIGN_CENTER);
}
    else if (pApp->nWinner)
```

```
{
    if (pApp->nWinner == 1)
    {
        AECHAR szBuf[] = {'Y','o','u',' ','W','i','n','!','\0'};

        IDISPLAY_DrawText(pApp->a.m_pIDisplay, AEE_FONT_BOLD,
            szBuf, -1, 0, 0, NULL, IDF_ALIGN_MIDDLE | IDF_ALIGN_CENTER);
    }
    else if (pApp->nWinner == 2)
    {
        AECHAR szBuf[] = {'Y','o','u',' ','L','o','s','e','!','\0'};

        IDISPLAY_DrawText(pApp->a.m_pIDisplay, AEE_FONT_BOLD, szBuf,
            -1, 0, 0, NULL, IDF_ALIGN_MIDDLE | IDF_ALIGN_CENTER);
    }
    else
    {
        AECHAR szBuf[] = {'D','r','a','w','!','\0'};

        IDISPLAY_DrawText(pApp->a.m_pIDisplay, AEE_FONT_BOLD, szBuf,
            -1, 0, 0, NULL, IDF_ALIGN_MIDDLE | IDF_ALIGN_CENTER);
    }
}
}
```

After setting the colors that we are going to use for drawing, calculate how big each cell is going to be. This is done by taking the resolution in the x and y directions and dividing it by 3—one for each cell in each dimension. The result of these computations is put in nSkipX and nSkipY.

Next, draw the game board by drawing two horizontal and two vertical lines. Very much like tile drawing, multiply the loop indices by the width and height of the cells to get the lines to appear in the right positions.

Now, loop through the game board array in much the same manner we do the tile array in the tile graphics example. The one-dimensional array of nine cells is used to draw the contents of a 3x3 grid. Using the size of the cell as a basis for the length of the X's lines and the radius of the O's circle, step through the array and draw the contents of each occupied cell. If there is a 0 in the cell, we draw nothing. If there is a 1, we draw a blue X. If there is a 2 in the cell, we draw a red O. Shave off a few pixels to make sure they do not get too close to the edge of a cell.

At the conclusion of the function, draw the status text. If there is a win condition, draw the appropriate game over text. If we are in the midst of a network operation, such as waiting for a server packet to arrive, print "Waiting..." on the screen.

Now that the **EVT_APP_START** handler is through, let's get into the meat of the applet by looking at the input handler:

```
case EVT_KEY: //Key input (where to put your X)

        if (pApp->nNetstate == NETSTATE_OK)
        {
            TicTacClient_HandleInput(pApp, wParam);
        }

        //Refresh screen
        TicTacClient_DrawScreen(pApp);
        IDISPLAY_Update(pMe->m_pIDisplay);

        return(TRUE);
        break;
```

Looking at this fragment, we see that if there is no network activity, we take the key input and pass it into TicTacClient_HandleInput, after which we redraw the screen. Looking at this handler we see:

Listing 17-13: Input handling

```
void TicTacClient_HandleInput(myapp_t* pApp, uint32 nKeyCode)
{
    int nCell = -1;

    //If we are in the win state, then reset game
    //and re-establish connection with server
    if (pApp->nWinner)
    {
        pApp->nWinner = 0;
        pApp->nMove = 0;
        MEMSET(pApp->pnGameBoard, 0, sizeof(char) * 9);
        TicTacClient_CloseConnection(pApp);
        TicTacClient_OpenConnection(pApp);
        return;
    }

    //Each key corresponds with a cell on the board
    switch (nKeyCode)
    {
        case AVK_1:
            nCell = 0;
            break;

        case AVK_2:
            nCell = 1;
            break;

        case AVK_3:
            nCell = 2;
            break;

        case AVK_4:
            nCell = 3;
            break;

        case AVK_5:
            nCell = 4;
            break;
```

```
        case AVK_6:
            nCell = 5;
            break;

        case AVK_7:
            nCell = 6;
            break;

        case AVK_8:
            nCell = 7;
            break;

        case AVK_9:
            nCell = 8;
            break;
    }

    if (nCell < 0)
        return;

    //If the cell is clear, fill it in
    if (!pApp->pnGameBoard[nCell])
    {
            pApp->pnGameBoard[nCell] = 1;

            pApp->nWinner = TicTacClient_CheckWin(pApp);

            if (!pApp->nWinner)
            {
                //Send move to server, receive server's move
                if (TicTacClient_WriteMove(pApp, nCell))
                {
                    TicTacClient_ReadMove(pApp);
                }
            }
            else    //Send win condition to server
            {
                if (pApp->nWinner == 1)
                    TicTacClient_WriteMove(pApp, CONDITION_WIN);
            else if (pApp->nWinner == 2)
                TicTacClient_WriteMove(pApp, CONDITION_LOST);
            else
                TicTacClient_WriteMove(pApp, CONDITION_DRAW);
            }
    }
}
```

This function is the heart of the game. First, if we are in a win state, reset the game. As you can see, we clear all of the variables, close the connection, and reopen it. The TicTacClient_CloseConnection function is very simple:

Listing 17-14: Closing connection to server

```
void TicTacClient_CloseConnection(myapp_t* pApp)
{
    if (pApp->pISocket)
    {
```

```
        ISOCKET_Release(pApp->pISocket);
        pApp->pISocket = NULL;
    }

    pApp->nNetstate = NETSTATE_OK;
}
```

All that is needed to close down a socket is to release it like we do any
other interface. Then set nNetstate to NETSTATE_OK, signifying that
we are not in the midst of any network operations.

Next, pick the appropriate cell for the button pressed. If you have
not noticed by now, Tic-Tac-Toe is a perfect match for the mobile phone
interface. You have nine cells on the game board and nine corresponding
number buttons in the exact same layout. Therefore, each number 1
through 9 represents their appropriate cell. If you are looking at your
keyboard's number pad and thinking you have it upside down, remember the number pad on a phone is inverted vertically.

If the cell chosen is not currently occupied, then fill its associated
cell with the value 1 and check for a win condition. This is done through
the function TicTacClient_CheckWin:

Listing 17-15: Checking for win condition

```
int TicTacClient_CheckWin(myapp_t* pApp)
{
    //check to see if we've won
    if (TicTacClient_ScanBoard(pApp, 1))
        return 1;

    //Check to see if server won
    if (TicTacClient_ScanBoard(pApp, 2))
        return 2;

    //Check for a tie
    if (TicTacClient_ScanBoard(pApp, 3))
        return 3;

    return 0;
}
```

This function checks for all three possible endings: win, lose, or draw.
Tic-Tac-Toe is won if you have occupied three adjacent spaces horizontally, vertically, or diagonally. The game ends in a draw if there are no
spaces left with no victors. The return value is a Boolean that is True
for a win, False if not. The function it calls to check for a win condition is
TicTacClient_ScanBoard:

Listing 17-16: Scanning board for positions

```
boolean TicTacClient_ScanBoard(myapp_t* pApp, int nPlayer)
{
    int i;
```

```
//Scan for a draw
if (nPlayer == 3)
{
    for (i = 0; i < 9; i++)
    {
        if (!pApp->pnGameBoard[i])
            return FALSE;
    }

    return TRUE;
}

//Check horizontals and verticals
for (i = 0; i < 3; i++)
{
    if ( (pApp->pnGameBoard[(i * 3)] == nPlayer) &&
        (pApp->pnGameBoard[(i * 3) + 1] == nPlayer) &&
        (pApp->pnGameBoard[(i * 3) + 2] == nPlayer))
    {
        return TRUE;
    }

    if ( (pApp->pnGameBoard[i] == nPlayer) &&
        (pApp->pnGameBoard[(1 * 3) + i] == nPlayer) &&
        (pApp->pnGameBoard[(2 * 3) + i] == nPlayer))
    {
        return TRUE;
    }
}

//Check diagonals
if ( (pApp->pnGameBoard[0] == nPlayer) &&
    (pApp->pnGameBoard[1 + 3] == nPlayer) &&
    (pApp->pnGameBoard[2 + 6] == nPlayer))
{
    return TRUE;
}

 if ( (pApp->pnGameBoard[2] == nPlayer) &&
    (pApp->pnGameBoard[1 + 3] == nPlayer) &&
    (pApp->pnGameBoard[6] == nPlayer))
{
    return TRUE;
}

return FALSE;
}
```

This function's main argument is an integer, **nPlayer**. If we pass a 1 to this function, it checks to see if the player has won. If we pass it a 2, it checks to see if the server has won. If we pass it a 3, it checks for a draw.

First, see if the user wants to check for a draw. If so, see if every cell on the board is occupied. If any one cell is empty, return False. Otherwise, we return True—there is a draw.

Next we check to see if the desired player has a victory with a horizontal or vertical row. Do this by looping three times. On each loop, check a row and a column for three consecutive player pieces (a 1 or 2, depending on the nPlayer argument). Because the array is a one-dimensional nine-element array representing a 3x3 grid, we have to do a little math trickery, as we did in tile drawing. For each column, add 3 to the index to get the next cell underneath. For every row, multiply the index by 3 and check each of the succeeding three cells. After this, we have a few hard-coded checks for the two diagonal win conditions. If there is a win, return True. Otherwise, return False.

Getting back to TicTacClient_HandleInput, if there is no winner, write out the cell that we placed our X in to the server via TicTacClient_WriteMove:

Listing 17-17: Writing move to server

```
boolean TicTacClient_WriteMove(myapp_t* pApp, int nMove)
{
    int nRet;

    pApp->nNetstate = NETSTATE_WRITE;
    pApp->nMove = nMove;

    if (!pApp->pISocket)
        return TRUE;

    //Cancel callback and write data to server
    ISOCKET_Writeable(pApp->pISocket, NULL, NULL);
    nRet = ISOCKET_Write(pApp->pISocket, (char*)&nMove, sizeof(char));

    if (nRet == AEE_NET_ERROR)
    {
        int nErr = ISOCKET_GetLastError(pApp->pISocket);
        ISHELL_CloseApplet(pApp->a.m_pIShell, FALSE);
        return FALSE;
    }

    //We are not done sending bytes
    if (nRet == AEE_NET_WOULDBLOCK)
    {
        //If this call blocks, assign callback
        ISOCKET_Writeable(pApp->pISocket, TicTacClient_WriteCallBack, (void*)pApp);
    }
    else
    {
        pApp->nNetstate = NETSTATE_OK;

        //If the game is over, kill the socket
        if (pApp->nMove >= CONDITION_LOST)
            TicTacClient_CloseConnection(pApp);

        return TRUE;
    }
```

```
        return FALSE;
}
```

This function attempts to write the byte of information passed to it by the **nMove** variable as a byte through the socket connection to the server. If the write went through, return True; if not, return False. First, clear any pending callbacks for the write operation with **ISOCKET_Writeable** using NULL arguments. Next, call **ISOCKET_Write**, sending the byte passed in through **nMove** to the server. If the return value is an error, call **ISOCKET_GetLastError** for some more information and quit out. Again, this is not robust but fine for demonstration purposes.

If the return value is **AEE_NET_WOULDBLOCK**, assign the callback and write again. If the call does not go through, the callback will handle it. Finally, if there are no errors or wait conditions, set **netstate** to **NETSTATE_OK** and kill the network connection if we have lost the game.

The callback function, **TicTacClient_WriteCallBack**, looks like this:

Listing 17-18: Network write callback

```
void TicTacClient_WriteCallBack(void *pUser)
{
    myapp_t* pApp = (myapp_t*)pUser;

    //If we have successfully written a packet
    //then update the screen and read
    if (TicTacClient_WriteMove(pApp, pApp->nMove))
    {
        TicTacClient_DrawScreen(pApp);
        IDISPLAY_Update(pApp->a.m_pIDisplay);

        TicTacClient_ReadMove(pApp);
    }
}
```

This function simply attempts to write the move again. If it works, we redraw the screen and try to read the incoming response from the server with a call to **TicTacClient_ReadMove**. Getting back to the **TicTacClient_InputHandler** function, you will see that we also call **TicTacClient_ReadMove** in response to a successful write:

```
if (TicTacClient_WriteMove(pApp, nCell))
{
    TicTacClient_ReadMove(pApp);
}
```

TicTacClient_ReadMove is a similar function:

Listing 17-19: Reading move from server

```
boolean TicTacClient_ReadMove(myapp_t* pApp)
{
    int nRet;

    pApp->nNetstate = NETSTATE_READ;

    //Cancel callback and read data from server
    ISOCKET_Readable(pApp->pISocket, NULL, NULL);
    nRet = ISOCKET_Read(pApp->pISocket, &pApp->nMove, sizeof(char));

    if (nRet == AEE_NET_ERROR)
    {
        int nErr = ISOCKET_GetLastError(pApp->pISocket);
        ISHELL_CloseApplet(pApp->a.m_pIShell, FALSE);
        return FALSE;
    }

    if (nRet == AEE_NET_WOULDBLOCK)
    {
        //If this call blocks, assign callback
        ISOCKET_Readable(pApp->pISocket, TicTacClient_ReadCallBack, (void*)pApp);
    }
    else
    {
        pApp->nNetstate = NETSTATE_OK;
        pApp->pnGameBoard[pApp->nMove] = 2;

        pApp->nWinner = TicTacClient_CheckWin(pApp);

        if (pApp->nWinner)
        {
            if (pApp->nWinner == 1)
                TicTacClient_WriteMove(pApp, CONDITION_WIN);
            else if (pApp->nWinner == 2)
                TicTacClient_WriteMove(pApp, CONDITION_LOST);
            else
                TicTacClient_WriteMove(pApp, CONDITION_DRAW);

            TicTacClient_DrawScreen(pApp);
            IDISPLAY_Update(pApp->a.m_pIDisplay);

            return FALSE;
        }
        else
            return TRUE;
    }

    return FALSE;
}
```

This function gets the cell that the server chose in the form of a byte read from the socket connection. An X is placed in the cell that the server chose, and if there is a win, loss, or draw, we tell the server with another call to TicTacClient_WriteMove and return False. Otherwise, if there is no end to the game and the network connection went through, we return True.

One of the first things the function does is set nNetstate to
NETSTATE_READ, notifying the applet that we are in the midst of a
socket read. Then we cancel any pending read callbacks by sending
NULL parameters to ISOCKET_Readable. After this, we attempt to
read data from the socket with a call to ISOCKET_Read.

If the value returned by ISOCKET_Read notifies us of an error, we
get more information on the error via a call to ISOCKET_GetLastError
and then quit out. Otherwise, if we get AEE_NET_WOULDBLOCK,
meaning that the network is jammed or perhaps no data has come yet,
we register the callback with ISOCKET_Writeable and try again.

If the read went fine, deposit the server's move into our game board
array and see if we have a winner. If so, notify the server by sending it
one of the win codes and update the screen. If we have a win condition
or the network communication did not go through, return False. Otherwise, return True.

The callback registered with ISOCKET_Writeable for stalled reads
is similar to the one we use for writes:

Listing 17-20: Read callback

```
void TicTacClient_ReadCallBack(void *pUser)
{
    myapp_t* pApp = (myapp_t*)pUser;

    //If we have successfully read data
    //from the server, we are done with this turn
    if (TicTacClient_ReadMove(pApp))
    {
            TicTacClient_DrawScreen(pApp);
            IDISPLAY_Update(pApp->a.m_pIDisplay);
    }
}
```

We simply attempt to read a move again. If it works, draw the screen
and update the display accordingly. The last function we need to document is the cleanup callback:

Listing 17-21: Applet cleanup

```
void TicTacClient_Cleanup(myapp_t* pApp)
{
    //Tear down net connection and release resources
    if (pApp->pISocket)
    {
        ISOCKET_Release(pApp->pISocket);
        pApp->pISocket = NULL;
    }

    if (pApp->pINetMgr)
    {
        INETMGR_Release(pApp->pINetMgr);
        pApp->pINetMgr = NULL;
```

```
    }

    if (pApp->pIGraphics)
    {
        IGRAPHICS_Release(pApp->pIGraphics);
        pApp->pIGraphics = NULL;
    }
}
```

This simply releases all of the interfaces that we used in the applet. No
other dynamic memory or resource allocation has gone on, so this is all
we need to do.

You now have a complete Internet-playable Tic-Tac-Toe game. I fall
short of calling it multiplayer because you really are playing against a
computer AI, as in a single-player game. The only difference is that the
AI resides on a server program outside of the game applet. I am sure
you can easily imagine this applet being extended to allow another
player to connect to the server and send his moves in as a response to
yours. You could even test this by running two instances of the emulator
connected to the same server application.

Making a Real Multiplayer Game

OK, so this example is not exactly a viable game. That is not the point.
It serves to illustrate the concepts of socket programming with BREW.
While I am on the subject of networking, I might as well discuss a few
points with developing real multiplayer games with BREW.

Although most BREW handsets suffer from graphics memory, there
is one advantage the average handset has. This is the fact that every
single one of them is connected to the Internet. As you may know from
reading this chapter, latency, speed, and bandwidth are all developing
issues with wireless connectivity; but even with these problems, it is
still possible to create a wide variety of multiplayer experiences even
with the most low-powered handset.

One thing that is necessary for most multiplayer games is a server.
You saw in this chapter an example of a basic Win32 gameplay server.
Most real games require something more robust. This includes tracking
a player's progress in the game, recording a player's account name and
password, and even maintaining the state of the world to serve as an
accurate game state to each connected client. Of course, you could write
this all yourself or use one of the many server products out there. Platforms, such as DemiVision (www.demivision.com), provide multi-player
server solutions that are appropriate for most any commercial game. Of
course, you are going to have to fork over dough for server licenses and
bandwidth costs. Some carriers (most notably in South Korea) provide
royalties on airtime or data transfers. Thus, the cost of setting and

maintaining your network infrastructure could be offset by the charges accumulated by a popular multiplayer game.

Note: In some cases, the revenue generated from network traffic is so profitable that the game clients are given away for free. In these cases, the publisher is banking on traffic royalties to more than pay for the cost of development.

This is why many wireless game publishers are hot on the trail of the next hit multiplayer game. In the United States, the billing situation for airtime and/or packet traffic is still not well established. It may prove to be too expensive for the average gamer to get involved in a multiplayer contest that accrues too many minutes or kilobytes transferred. Until carriers can get a sensible billing structure and also share revenue with developers and publishers, the multiplayer wireless gaming situation will lag behind those in such mature markets as Japan and South Korea.

Conclusion

You now know how to create games that communicate over the Internet for multiplayer experiences. BREW allows you to use the net for much more. For instance, you can stream in new BMP graphics or other content by reading data from a socket into a memory buffer or file for later use. The usage of network communication blows back the boundaries of what you can do with a game. With multiplayer gameplay and dynamic downloadable content, the possibilities are endless. However, this also gives the user an additional cost since she has to pay for network transmission fees. The development process also becomes more expensive for you as you need to develop a server architecture as well as deal with the complicated testing procedures of networked applications.

Chapter 18

Running on Hardware

Introduction

By this point we have thoroughly explored how to write and design games for BREW handsets. Now we have come to one of the final steps of BREW development: compiling your game for a BREW handset. You may think that once you have your game debugged and running smoothly in an emulator that you are almost there. The fact is getting your code to execute properly on real hardware can sometimes be an even bigger task than writing the game in the first place. This chapter explores this issue as well as provides some handy debugging tips when going through this rather painful process.

Inside BREW Hardware

Qualcomm has designed a series of chipsets on which BREW is used. Known as the MSM series, these chipsets range from the extremely primitive with 1-bit graphics and no sound, such as the MSM5000 platform, to the lavishly powerful that include 16-bit displays and digital sound, like the MSM5500. Future chipsets, such as the MSM6600, include extremely high-speed 3G communications support as well as 3D graphics and other advanced features. Over 55 percent of the MSM chipsets shipped in the first quarter of 2002 were 3G. Soon, the vast majority of new BREW handsets shipped will use high-speed digital networks ranging from 153Kbps to a whopping 2.4Mbps.

The one thing that all of these chipsets share is the use of ARM CPUs at the heart of their design. Advanced RISC Machines (ARM) Ltd.

is one of the original pioneers of Reduced Instruction Set Chip (RISC) design. They began in the mid-'80s as a group inside the late European computer manufacturer Acorn. Since then, they have been creating state-of-the-art CPUs using RISC technology for speed and efficiency increases over competing manufacturers.

Originally, ARM CPUs were found exclusively in Acorn's desktop computers. However, many modern ARM processors are not only fast, but they require very little power to operate. This makes the ARM line of processors ideal for such small battery-powered devices as PDAs, portable video game machines, and mobile phones. As a result, ARM now has a huge chunk of the embedded CPU market with their processors found in PocketPCs, Nintendo's Game Boy Advance, and all current BREW hardware, among other platforms.

Although Qualcomm provides several different sets of basic BREW chipset designs, the companies that actually manufacture the handsets (original equipment manufacturers, or OEMs for short) are free to implement their own custom extensions, as well as pull features from various additions to the basic BREW design offered by Qualcomm and its hardware partners. For instance, some manufacturers may choose to use the basic MSM5000 chipset but complement it with a MIDI sound chip or Global Positioning System (GPS) feature.

OEMs can also provide extensions to BREW that access unique features of their device. For instance, if one OEM chooses to include a digital camera in their phone before the BREW SDK supports such a peripheral, the OEM can write their own low-level extension to access this device. This extension will have an interface that can be used in BREW applications like any other BREW extension class.

In this race to create unique handset features to stand above the crowd of BREW OEMs, we are seeing a growing landscape of varying device features and performance. As a result, creating an application that runs on such a disparate array of devices can become quite a headache. OEMs seem to have no plans to agree on even basic standards, such as screen sizes and button layouts.

Despite fierce competition from OEMs in Japan, Japanese Java-based handset manufacturers have agreed on a few basic standards, such as the aforementioned screen and interface to ease the development of applications across different handsets. It remains to be seen if BREW OEMs will learn from the success of the Japanese market or if BREW will address these issues with an eye toward further hardware abstraction.

Until then, you will most likely find it necessary to create different builds for different handsets. Also, you will have to use the

ISHELL_GetDeviceInfo function to detect what the handsets' capabilities are and adjust your programs' behavior accordingly.

Getting a Handset

How do you actually get hold of a phone? This may seem like a silly question to ask, but it is a valid one. Before you do any compiling for handsets, you actually have to get one. The easiest way is, of course, to go down to your local BREW carrier's shop and buy a phone. In this case, Verizon Wireless is the only United States carrier supporting BREW at the time of this book's publication. It is possible to buy any number of different BREW handsets in their shops.

However, there is one minor hurdle. In many cases, you are designing games that do not require Internet access. Therefore, it seems like a waste to pay for wireless phone service when you merely need the handset to test your code on. In every Verizon store there are two prices listed for each handset. One is the price you pay when you buy a phone with service, and the other, higher price (usually in fine print) is what you pay without getting the phone activated. In my case, the Kyocera 3035 cost around $150 activated and $250 unactivated when I bought one in the early days of the device's release.

So, all you have to do is simply ask the clerk to sell you a phone at the higher price, right? Well, you would think so, but I have found this process to be very difficult. In many shops, the clerk will absolutely refuse to sell you a handset without service. "I cannot let you leave the building with an unactivated phone," said one adamant salesperson at my local Verizon shop. I had to go to two or three different locations (and in some cases visiting more than once) before I found someone willing to sell me a phone without service.

What about buying phones directly from the handset manufacturers themselves? I have found this even more difficult. Handset manufacturers do not want to undermine their customers' (the carriers) ability to sell hardware. Therefore, they are very reluctant to sell handsets directly to the consumers themselves. Unfortunately, most handset manufacturers make little distinction between a consumer and a developer. Perhaps as the mobile application industry expands, clues will begin to dawn on the management at these various companies.

It is possible to form a business relationship with a handset manufacturer or even Qualcomm itself to get prototype hardware on loan for a limited time. However, this involves a bit more wrangling than simply walking into a shop and picking up a unit. Your best bet is to break through the average Verizon clerk's stonewall and buy a device at retail.

You will also need a data cable to connect the phone to your PC for the uploading of your applet. I find that cables for certain handsets are not commonly available at retail locations. Therefore, you may have to buy the cable directly from the manufacturer's web site. In some cases, such as the Sharp Z-800, the handset comes with the data cable. Check the device's packaging to see if you need to purchase the cable separately or if you are getting it with the package.

Getting the Newest Installation of BREW

Now that you have a handset, there are some modifications that need to be made to it before you can start uploading your binaries to the device. In some cases, the version of BREW on the handset will not allow programs to be uploaded to it via the data cable. You need to have Qualcomm update the handset with a developer's version of BREW. Or you may want a more recent version of BREW if you want to test code on a device running a newer version of BREW not currently supported by your local carrier.

The easiest way to do this is mail your handset via trackable carrier, such as Federal Express or UPS, to Qualcomm's support center. The address you need to mail your handset to is:

QUALCOMM, Inc.
Attn: QIS Phone Center (Bldg R - 212P)
5775 Morehouse Dr.
San Diego, CA 92121

You must include a note with your return address and phone number, why you are sending the phone in, the phone number and carrier (if the phone is activated), whether the browser has been activated, and the ESN (electronic serial number). The ESN is a unique ID given to all handsets. This ESN is printed on a sticker on the back of the phone underneath the battery. In the case of the Kyocera 3035, the ESN is the number on the right side of the sticker underneath the first bar code, as seen in Figure 18-1. The turn-around time for this service is usually a few days. Check the developer FAQ at www.qualcomm.com/brew/developer/support/faq for more information.

Figure 18-1: The ESN on the Kyocera 3035

Compiling Native Code

Now you have a basic idea of what BREW hardware is like on the inside and how to get your hands on it. The next step is compiling your code into ARM instructions for execution on a real handset. Until now, you have been using Visual Studio's compiler to create a Windows DLL that will only run in the BREW emulator. The resultant DLL contains code native to the processor on your PC. Code that is designed to run on Intel CPUs will not run on ARM. Plus, Windows DLLs have a lot of information required by the Windows OS that is not supported in BREW, hence the bloated size of the Windows DLL compared to a real BREW application file. You have to use a completely different compiler to build an ARM native binary that will run on the phone itself.

The bad news is this compiler costs money—a lot of money. There are two choices available from ARM. First, there is the ARM Developer Suite (ADS), which clocks in at a cool $5,500, not including taxes and shipping. There is a 45-day free trial version of the ADS that you can get from ARM's web site at www.arm.com. Unfortunately, the currently available version (1.2) does not work with BREW. You will have to track down the old 1.1 version to try it for free.

The ADS is a complete tool suite for all sorts of ARM applications, including processors not used in BREW hardware. We are really concerned with ARM's C/C++ compiler and assembler suite for BREW. Therefore, the ADS is really overkill for our purposes.

The second option is a BREW-only solution. There is a subset of the ADS called the ARM BREW Builder available for a considerably lower price. This contains just the tools that we need to create ARM native binaries of our BREW code. This is still quite expensive, at $1,500 for a single-seat license. If you have multiple developers, you will have to either cough up the dough for multiple licenses or find a way to share the machine that ADS is installed on. This can be quite annoying when you have multiple developers finishing a product at the same time.

Of course, ARM is not the only provider of compilers for its chips. The free open-source ARM GCC compiler from GNU should be more than capable of creating BREW binaries. Unfortunately, Qualcomm has provided no information on how to use GCC with BREW. I am almost positive that this can be done, although I have not managed to find a way to do it. ARM maintains that their set of tools creates more efficient code than GNUs. However, GCC is used in many commercial applications, such as PlayStation 2 and Game Boy Advance development. Even if GCC is a little inefficient, it surely beats paying thousands of dollars in compiler licenses for developing mobile games. The high price of ARM's tools keeps BREW development out of the reach of many small shops and hobbyists. Considering the scope of the average mobile game, it is in everyone's interest to lower the barrier of entry by allowing the use of as many free tools as possible.

Installing the Compiler

Let's make the assumption that you have purchased the ARM BREW Builder. Installing it is fairly straightforward. However, if you blindly follow the on-screen instructions of ARM's installation wizard, you will find it difficult to get the compiler to actually work. The key is the directory in which you install the tool. You should make sure there are no spaces in the path name, as the tools have a hard time dealing with them. As recommended by Qualcomm's own documentation, I would suggest installing the compiler to "C:\apps\ads." Feel free to put it anywhere you like, but just make sure that there are no spaces in the path.

Unfortunately, BREW's default installation will place the BREW SDK in a path with spaces in the filename. Because the ARM compiler needs to use BREW's headers and source files to create a native binary, you should copy the inc and src folders underneath your "BREW v101" folder to another location with no spaces in it. In my case, I copied them to "C:\brewincs." You will use these paths frequently as we attempt to build our native binary. Alternatively, it is possible to use the tilde naming convention to avoid using spaces in file paths. For instance, it is possible to use "Progra~1" instead of "Program Files" to avoid the spaces.

Once you have installed the compiler, you need to generate a license file. The wizard should guide you through this process. If not, there is also a License Installation Wizard in the compiler's group on the Start menu. The wizard will generate a file that is based off of the MAC address of your network card or serial number of your hard drive. This is how the compiler will identify the machine that it is running on. E-mail the resultant file to ARM and they will send back a DAT file, which you then place in the licenses folder. The turn-around time on this process is usually a day or so.

Creating the Makefile

Now that you have installed the compiler and obtained a license, you must create what is called a makefile. A *makefile* is a script that tells the compiler how to build your binary. These files usually end with the .mak extension. Those of us who were around in the pre-GUI days of programming may be well-versed in the art of makefiles. In fact, many commercial compilers actually use makefiles behind the scenes. Most modern compilers do most of the dirty work for us, insulating the programmer from the headache-inducing task of keeping an up-to-date makefile for the project.

To execute our makefile script, use Microsoft's nmake utility. This is a command-line application that comes with Visual Studio. Although it shares the same name, it is no relation to AT&T/Lucent's own nmake program, even though they do similar things. Nmake will execute the commands in our makefile and create a native BREW binary with the file extension .mod. Think of the MOD file as the BREW equivalent of a DLL.

Although not present in BREW 1.0, later versions of the SDK come with a sample makefile. This makefile still requires significant modifications to work with even the simplest program. Also, the makefile scripts for compiling C are slightly different from compiling C++ code. This chapter will focus on creating a makefile to build our C version of Hello World. Read Appendix E to see an example of how to create a makefile for the C++ version.

The script used in makefiles is very complicated and rather arcane. Entire books have been written on the subject of creating makefiles. Describing make scripts in detail is well beyond the scope of this book. Therefore, I will simply illustrate the bits that you may have to deal with inside the makefiles included with the companion files. The first makefile we will deal with is for our original Hello World program. It is the makefile.mak file found in Source\Chapter 05\hello.

 Tip: Microsoft's nmake documentation is available at http://msdn.microsoft.com/library/default.asp?url=/library/en-us/vc ug98/html/_asug_overview.3a_.nmake_reference.asp.

Hello World Makefile

The first thing you will notice when you open the makefile is all of the text starting with # symbols. The # indicates a comment in make language, basically the equivalent of // in C++. The introductory comments provide some information about the makefile and who wrote it. Some of the comments are actually inaccurate, and you can basically ignore most of it.

Directly following these comments are the following lines:

Listing 18-1

```
BREW_HOME     = C:\brewincs
ARM_HOME      = C:\apps\ads
TARGET        = hello
OBJS          = hello.o
APP_INCLUDES  =
```

Basically, we are creating the equivalent of macros in C/C++. We are setting the macro **BREW_HOME** to be the path to which we copied the inc and src directories from BREW's default installation folder. Next, the **ARM_HOME** definition is the path where we installed the compiler itself. Finally, **TARGET** specifies the name of the resultant MOD file. **OBJS** is the name of the object file that will be generated by this script. Object files are sort of temporary files in an intermediate state of processing generated by the compiler. These can be deleted once we are done compiling. We do not set the **APP_INCLUDES** value to anything. You can use this to specify special directories from which you wish to pull header files.

If you look toward the middle of the script, you will see this line:

Listing 18-2

```
ZA  = -zo    # LDR may only access 32-bit aligned addresses
```

This is noteworthy because it is required to be changed from the default makefile provided by Qualcomm. Earlier versions of the SDK required the ZA parameter to be set to a different value. However, for 1.1 and beyond, we need to set ZA to -zo.

The major thing that the sample makefile omits is the commands to compile the common BREW source files included with all BREW applications (namely, the AEEAppGen and AEEModGen files). Therefore, we have to add some code to this script to build and link these two files.

The first step is setting another definition to indicate these required files. Near the end of the makefile is this script:

Listing 18-3

```
APP_OBJS =    AEEAppGen.o \
              AEEModGen.o \
              $(OBJS)
```

This includes the compiled AEEAppGen and AEEModGen with our hello object file, so they can be used in the linking process. If you look at the bottom of the makefile, you will see the following code:

Listing 18-4

```
# ----------------------------
# BUILD APPLET FILES

AEEAppGen.o : $(SUPPORT_DIR)\AEEAppGen.c
AEEAppGen.o : $(SUPPORT_INCDIR)\AEEAppGen.h

AEEModGen.o : $(SUPPORT_DIR)\AEEModGen.c
AEEModGen.o : $(SUPPORT_INCDIR)\AEEModGen.h
```

```
$(TARGET).o : $(TARGET).c
$(TARGET).o : $(SUPPORT_INCDIR)\AEEAppGen.h
$(TARGET).o : $(SUPPORT_INCDIR)\AEEModGen.h
```

This code creates object files out of the obligatory BREW source files that can then be used to link our BREW application with SDK function calls. Without these commands, you will get a link error because the compiler will not be able to find basic BREW calls, such as AEEApplet_New.

The Hello World applet is very simple and only has one source file. If you have more than one source file, you are going to have to generate an object for every source file that you wish to compile and link. You can add additional source files in a way similar to how we added the BREW files. Copy and paste the above changes and use filenames of your additional source files. Since you are executing nmake in the directory in which your code resides, you do not need the path prefix on the additional source file. Therefore, you can remove the "$(SUPPORT_INCDIR)\" and "$(SUPPORT_DIR)\" part of the above paths when adding files that are in the same directory as your target.

There is one other set of references to these obligatory BREW files. If you look at the part of the script that determines the "clean" functionality, you will see this script:

Listing 18-5

```
clean :
        @echo ----------------------------------------------------------
        @echo CLEAN
        -del /f AEEAppGen.o
        -del /f AEEModGen.o
        -del /f $(OBJS)
        -del /f $(TARGET).$(EXETYPE)
        -del /f $(TARGET).$(MODULE)
        @echo --
```

Modified from the original code, this adds the extra object files to be deleted when we perform a clean. By passing **clean** as an argument of nmake, we can delete the files listed in this part of the script to remove all of the temporary objects generated in the build process.

Now, if every file is where it should be, the makefile is right, the compiler is installed and licensed properly, and the planets are in the proper alignment, we can compile our MOD file. To do this, simply type this command:

```
nmake makefile.mak
```

You should see output in the DOS window like this:

Listing 18-6

```
Microsoft (R) Program Maintenance Utility   Version 6.00.8168.0
Copyright (C) Microsoft Corp 1988-1998. All rights reserved.

----------------------------------------------------------------
OBJECT AEEAppGen.o
        C:\apps\ads\bin\armcc -c -DDYNAMIC_APP -I. -ID:\brewincs\inc  -cpu ARM7T
DMI -apcs /ropi/interwork -littleend -zo -zas4 -fa -g -Ospace -O2 -I. -ID:\brewi
ncs\inc  -o AEEAppGen.o D:\brewincs\src\AEEAppGen.c
"<command line>", line 1: Warning: C2067I: option -zas will not be supported in
future releases of the compiler
D:\brewincs\src\AEEAppGen.c: 1 warning, 0 errors, 0 serious errors
----------------------------------------------------------------
----------------------------------------------------------------
OBJECT AEEModGen.o
        C:\apps\ads\bin\armcc -c -DDYNAMIC_APP -I. -ID:\brewincs\inc  -cpu ARM7T
DMI -apcs /ropi/interwork -littleend -zo -zas4 -fa -g -Ospace -O2 -I. -ID:\brewi
ncs\inc  -o AEEModGen.o D:\brewincs\src\AEEModGen.c
"<command line>", line 1: Warning: C2067I: option -zas will not be supported in
future releases of the compiler
D:\brewincs\src\AEEModGen.c: 1 warning, 0 errors, 0 serious errors
----------------------------------------------------------------
----------------------------------------------------------------
OBJECT hello.o
        C:\apps\ads\bin\armcc -c -DDYNAMIC_APP -I. -ID:\brewincs\inc  -cpu ARM7T
DMI -apcs /ropi/interwork -littleend -zo -zas4 -fa -g -Ospace -O2 -I. -ID:\brewi
ncs\inc  -o hello.o hello.c
"<command line>", line 1: Warning: C2067I: option -zas will not be supported in
future releases of the compiler
hello.c: 1 warning, 0 errors, 0 serious errors
----------------------------------------------------------------
----------------------------------------------------------------
TARGET hello.elf
        C:\apps\ads\bin\armlink -o hello.elf -ropi AEEAppGen.o  AEEModGen.o  hel
lo.o -first AEEMod_Load
Warning: L6305W: Image does not have an entry point. (Not specified or not set d
ue to multiple choices.)
Finished: 0 information, 1 warning and 0 error messages.
----------------------------------------------------------------
TARGET hello.mod
        C:\apps\ads\bin\fromelf  hello.elf -bin hello.mod
Warning: Q0115W: Deprecated command syntax will not be supported in future versi
ons. Use -output to specify the output file.
Finished: 0 information, 1 warning and 0 error messages.
```

There are plenty of warnings and other information here. You can ignore
them for now. The good thing is we made it out alive! If you look at the
contents of the directory, you will see that you now have a hello.mod
file. This is your native ARM binary. Notice how small it is compared to
the Windows DLL. The MOD is around 1 to 2K versus over 200K for
the DLL. You cannot use the size of the DLL to judge the size of your
actual MOD file at all.

ARM vs. Thumb

The ARM CPUs used in BREW hardware are what we call Thumb-aware. This means that they can use two different instruction sets: ARM and Thumb. ARM code is the regular 32-bit instructions used on all ARM CPUs, whereas Thumb instructions are smaller 16-bit instructions. Because there are not as many instructions available in 16-bit space, Thumb code is a subset of the most popular ARM instructions in 16-bit form.

On current MSM architectures, code executes out of RAM. When space is tight, it may be necessary to cut out some features or break up the executable into modules loaded in and out of memory at different times. However, 16-bit Thumb instructions are half the size of native 32-bit ARM opcodes. Therefore, it may be useful to compile your code into Thumb instructions to shrink the size of the compiled MOD file.

It is possible to use the ARM BREW Builder to create a Thumb binary. To compile Thumb code, use the tcc and tcpp compilers for C and C++ code, respectively. These tools are used to compile your source into 16-bit Thumb instructions instead of 32-bit ARM. The command-line arguments for this compiler are identical to armcc and armcpp, so the makefile is easy to modify. However, with all Thumb compilations, you must use the INTERNETWORK compiler option. Without this flag, the code may crash on the handset. Also, the **AEEApp_Load** function must be isolated in its own file and compiled in 32-bit mode with armcc. This is the only function that must use ARM instructions.

Note: ARM's official figures state that Thumb can reduce code density by up to 35 percent. This means it can crunch the size of your MOD file down by over a third.

Digital Signatures

Now that you have built a MOD file, you should be ready to get it on a phone and test it out, right? Not so fast. BREW requires each applet or extension class to be digitally signed. This is a way to verify that the provider of the application is indeed a recognized and registered BREW application publisher. In essence, you are providing your "signature" on a contract that states this applet is an authentic product from you or your company. Even when testing your application before commercially publishing it or sending it in for testing, you must generate test signature files in order for the code to run on the handset. This test signature is called a Test Enabled Signature, or TES.

VeriSign Authentic Document IDs

VeriSign is the partner Qualcomm has chosen to provide digital certificates for BREW developers and publishers. VeriSign is one of the pioneers of digital security on the Internet, providing security solutions for web transactions, e-commerce payment schemes, and other services. In this case, you need to subscribe to VeriSign's Authentic Document ID for BREW service. You can get the service information at www.verisign.com/products/brew.

Yes, this means that you have to pay for yet another tool to get your application running on a handset. I am still not totally sure why BREW requires test signatures to simply run code on your handset for testing, but it is somewhat of an annoyance and is yet another hurdle for hobbyist developers and small shops. The good news is for testing you only need to generate one signature file per phone. You can reuse it for every application on that handset.

Once you sign up for the VeriSign service, you gain access to Qualcomm's Developer Extranet web site. This site is a private area separate from the general developer web site that provides extensive information on current and upcoming handsets, as well as tools to generate not only BREW signature files, but legitimate BID values for your applet. We will get into the BID generation process a little later.

The Developer Extranet's test signature tool is a web-based application, where you fill in a few forms and it provides a signature file for you to download. As seen in the following figure, the first field is the name of the application. This is the name of the MOD file generated by the compiler. The second is the ESN in hexadecimal form (prefixed with "0x"). This number is usually on a sticker on the back of the unit, as described previously in this chapter. Simply click the Generate button and a button will appear allowing you to download the file. This file will have the same name as the application with the .sig file extension. The signature is good for 90 days; after that you will have to generate a new one. Also, you are only allowed to generate signatures for up to ten different handsets. If you need to support more devices, you must make special arrangements with Qualcomm to do so.

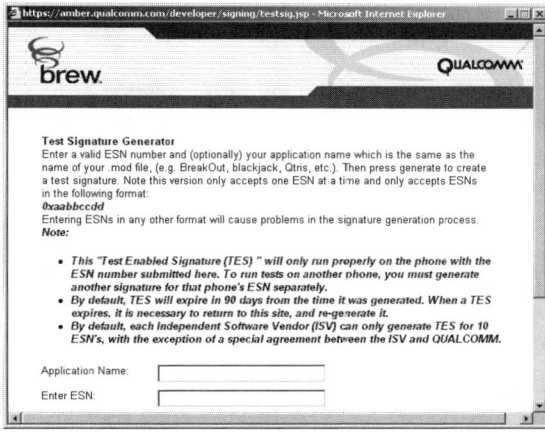

Figure 18-2: This is the form that you must fill out to generate a TES on the BREW Developer Extranet.

Creating Real BID Files

Generating a BID file containing a valid class ID for your applet is an even simpler process than creating a TES. There is another web-based tool for generating BID files. Simply type in the name of the applet or extension class that you want to generate a BID for and click the Submit button. Your browser will then prompt you to download the resultant BID file. Write over the temporary one that you made while developing the emulator code, and you are good to go. You do not necessarily need to do this while testing your game on hardware. Just as long as your BID file does not conflict with any BREW classes already installed on the handset, you can still use your test BID until you need to commercially release your applet.

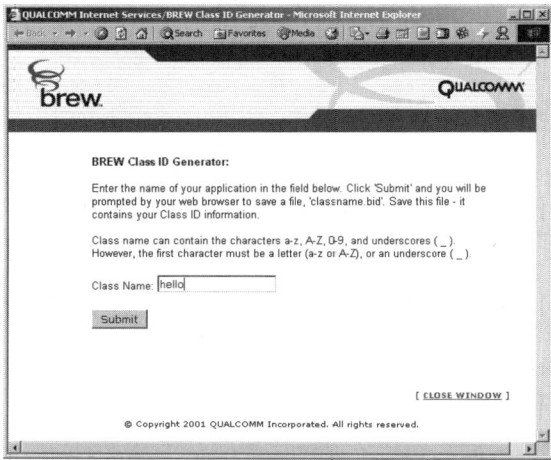

Figure 18-3: This is the web tool on the BREW Developer Extranet for generating a valid class ID.

Uploading to the Device

Now that you have your applet binary and signatures generated, it is time to upload your code to the device. One of the applications available on the Developer Extranet is the BREW AppLoader. This tool allows you to upload files and manage directories on a device connected to your host PC. In the case of the Kyocera 3035, use their commonly available data cable that is usually used to sync contacts with Outlook and such. Some devices may have other ways of connecting, such as USB. Some handset manufacturers actually provide their own custom tools for uploading and managing the BREW file system. You will have to contact the manufacturer directly to acquire these tools. Otherwise, Qualcomm's AppLoader works with pretty much every BREW handset, including prototypes that are not yet available to the public.

Note: You may want to keep older versions of AppLoader as a backup. In some cases, the newest releases of AppLoader do not work with older handsets.

When you start the AppLoader application, it brings up a dialog that displays all supported devices. You can also change the DLL used to connect to the device as well as the COM port. There is no need to change the DLL from the default QCOMOEM.dll. You may have to fiddle with the COM settings, depending on which port you plugged the phone into on your PC. Once you click the OK button, the AppLoader will attempt to connect to the device and return from this operation by displaying the contents of BREW's root directory.

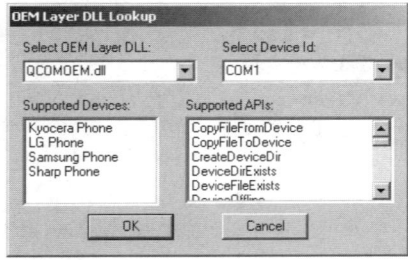

Figure 18-4: The first dialog visible when AppLoader is started

The BREW AppLoader has an interface similar to that of any standard file explorer utility in Windows. On the left-hand side of the application window is a tree control displaying the directory structure of the BREW file system. On the right are the contents of the currently selected directory.

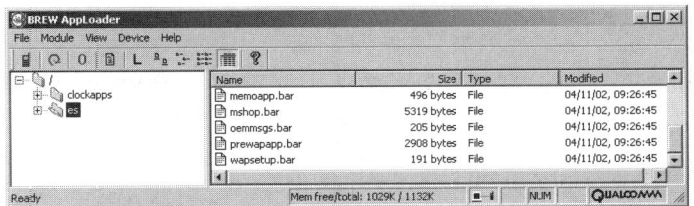

Figure 18-5:
An example of
browsing
folders with
AppLoader

In this root directory resides the MIF files for each application installed
on the device. Then, for each application is a folder with the same name
as the MIF and MOD files. This folder contains the MOD and any other
associated files needed for the applet's operation. This includes things
like BAR files and any BMPs the applet may load directly.

Assuming we have a MOD file for the Hello World example we
walked through with the makefile, upload this to the handset and see
what happens. First, find the MIF file named hello.mif in the example's
folder. Simply click on that file and drag it to the AppLoader window.
This file will be uploaded to the phone and should be visible in the
refreshed directory display. Next, we have to create a folder for the
applet. Simply click File|New|Directory, and you will be presented with
a dialog asking for the directory name.

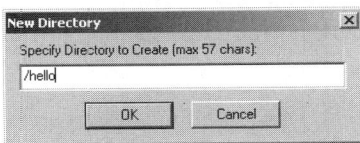

Figure 18-6: The directory creation dialog

Type in "/hello" as the directory. Always remember that the MIF, signature, and MOD must have the same name. Now that the directory is
created, copy hello.MOD and hello.SIG to this folder. Now, click
Device|Reset to reset the phone. Once the phone has been fully reset,
you are ready to test out the applet. In some cases, this means powering
on the handset. Depending on your handset, click on BREW Apps, and
find the icon for "hello." Click on this icon, and you should get a simple
display of "Hello World" centered on the screen in simple black and
white. Congratulations, you have completed the process of code to
completion!

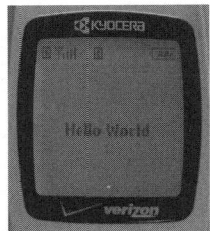

Figure 18-7: Hello World
running on my Kyocera 3035

Hardware Debugging

When running your game in the BREW emulator, you have plenty of debugging options. Microsoft's debugger in Visual Studio allows you to stop the execution of the program at any point and check the contents of variables and arbitrary memory locations. As you know, the emulator is hardly an accurate representation of how your code will perform on a real device. In many cases you will encounter crashes and other bugs not found while running the applet in the emulator. Thus, it becomes necessary to debug your code as it executes on real hardware. Considerably fewer debugging options are available on the handset. Say goodbye to most of the cushy tools provided by Microsoft.

Note: Not all native bugs are your fault. There are bugs in various OEM implementations of BREW. For instance, the Sharp Z800 does not properly use the −1 value in the width and height arguments with the IDISPLAY_BitBlt function. If you use −1 (which is supposed to automatically determine the size of the bitmap), it will simply refuse to draw the image. Even though these are the manufacturers' bugs, you still have to code around them as users are unlikely to upgrade their firmware.

Avoiding Crashes

Perhaps the simplest solution to hardware debugging is not to create bugs! Of course, you will run across bugs in your development, but by strictly following BREW's programming guidelines, you can avoid many bugs that plague novice BREW developers. These guidelines include not using floating-point values, keeping stack variables to an absolute minimum, and allocating as little memory as possible. Also, make sure that your MIF file has the permissions set correctly. For instance, if your applet does not have the file permission set and you try to open a file, the call will fail. If you do not handle this failing case properly, your code could blow up.

This also brings up an issue of code robustness. Because a BREW game is a commercial application to be used by potentially millions of customers on many different devices, you have to prepare for almost any bizarre usage case that may lead to bugs or errors. This means handling every possible error case thrown at you from the API. For instance, when allocating memory, make sure that you gracefully exit the program if the call to MALLOC fails for some reason. Even if you know you could not possibly be running out of memory, perhaps a poorly behaved application was run before yours and failed to clean up after itself properly. You have absolutely no idea under which conditions your applet will be run. You must expect the unexpected.

Handling these situations and following the BREW programming guidelines also means that you can avoid costly retesting by NSTL for True BREW certification. Every time you submit a new version, even for simply changing a bitmap or a portion of text, the applet must be retested. Testing fees can make mobile application development quite an expensive process.

What a Crash Looks Like

When encountering a crash, two things usually happen: Either BREW will throw up an error message or the phone will reset. Both cases are bad. In the former, you can sometimes find the meaning of this code on the BREW web site to give you a greater perspective into what is causing the problem. Otherwise, it is time for some old-fashioned detective work.

Common Error Codes

As mentioned previously, one way BREW is helpful in hardware debugging is in the display of various error messages. The problem is, most of these errors are caused by the same problems: stack overruns, writing to wild pointers, and such. This can make these problems sometimes hard to diagnose. Here is a brief list of error messages and what they mean:

- **Undef Inst Exception**—This is an "Undefined Instruction Exception." It means that an instruction that is not known has attempted to execute. Usually this means the program counter has jumped into a random area and perhaps the CPU has tried to read garbage data as an instruction. This also could mean that you have overrun the meager 500-byte stack or are attempting to run code compiled for a later version of BREW on an incompatible handset.

- **Re-entrant Data Abort**—This error is usually caused by a stack overrun. In some rare cases, it can be attributed to a BREW version incompatibility.

- **Pref Abort Exception**—This error usually means that you have somehow written over memory used by the operating system. This occurs if you have overrun an array, written data to an invalid pointer, or corrupted memory in some other way.

- **SWI Exception**—This is another memory error usually related to what Qualcomm's FAQ calls "heap memory mismanagement." This is described as mistakes such as freeing memory that has already been freed, writing past array bounds, and writing to invalid pointers. As usual, a stack overrun can also cause this error.

Recovering from an Error

In some cases the only way to recover from these errors is to yank the battery out and power the phone up again. In extreme situations you may actually corrupt the handset's firmware and have to mail it back to Qualcomm for a reflash. It may be a good idea to purchase at least one backup phone for these instances.

What a Reset Means

When the phone resets as the result of your program's execution, it could mean a number of different things. In many cases, accessing an invalid memory pointer or trouncing over memory with garbage data will cause the phone to reset. In other cases, it is due to the code being stuck in a tight loop.

BREW has what is called a watchdog. The watchdog kicks in when the program is caught in a tight loop and not giving any processing time to the OS. Perhaps you have an infinite loop in your code or a function you are calling is taking an enormous time to execute for some reason. When the watchdog detects this, it resets the phone to stop the runaway program's execution.

How to Debug

Now that the phone has started resetting or displaying obscure messages, how do you address the issue? You are not going to like the answer. BREW provides very little in the way of debugging support, but this is getting better with each new release of the SDK.

The "Old School" Way

The simplest way to debug a program is to insert some debugging code that lets you know where you are in the execution of the program before it crashes. For instance, one technique I do is insert code to vibrate or ring the phone at different points in my program. I might put a different sound or vibration length at the start of each suspected function to see which one gets called right before the crash. By using techniques like this, it is possible to localize the problem until you can narrow down the bug to a few lines of code.

On most devices, sound is asynchronous. So simply beeping or vibrating the phone may not give you the most accurate hint as to where the program is crashing. One other technique is to open up a simple text file on the device and write log messages as the program executes. You could provide a simple menu option to display this log as a static text control. It is possible to provide much more detailed information this way, including the contents of variables, the time of execution, and other potentially useful information.

Leaders of the New School

At the BREW 2002 conference, Sharp showed an interesting new debugging system for its Z-800 color handset. This solution involves a box manufactured by a German company called Lauterbach (www.lauterbach.com), which is connected to a Z-800 handset modified by Sharp for debugging purposes. Using this configuration with Lauterbach's TRACE32 debugging software, it is possible to step trace through native BREW code as it executes on the handset. This includes setting breakpoints and examining the contents of variables, and a set of features very similar to what is available in Microsoft's Visual Studio. The price of the package is quite high, clocking in at around $4,000. You must also send your Z-800 handset to Sharp for modification to work with this system. The turn-around time for this process is roughly a month. Regardless, this is an absolute godsend to developers, and I sincerely hope that other manufacturers make their devices work with this system.

The Logger

Perhaps a more middle-of-the-road solution is the Logger. Included in the tools downloadable from the BREW Developer Extranet, the BREW Logger is a massively useful new tool designed to aid in debugging code when running on a handset. The BREW Logger will read messages from the handset as code runs, outputting them to both the screen and a text file. The Logger reads all sorts of messages, but the most important is the output from the DBGPRINTF function. Now you can finally print messages from the device to your PC's screen to track down those hard-to-find bugs in native code. I cannot express enough how useful and comprehensive this tool is when making sure your code runs flawlessly on hardware.

Conclusion

This is perhaps the most difficult part of developing applications with BREW. There are many steps and hurdles to getting your code running on a handset. In this chapter we have used the simplest code possible. Getting actual real production code running on a handset is a much larger task. Regardless, you now know the basics of hardware compilation and debugging. There are few steps left on the road to getting your game out to the entertainment-starved public.

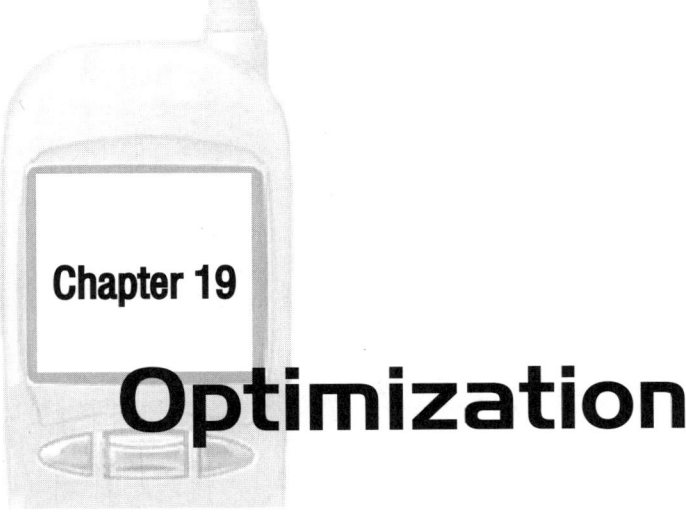

Chapter 19

Optimization

Introduction

Now that you know how to compile your game to run on BREW handsets, you have probably found out that the hardware is less than speedy. Although the CPUs in BREW handsets are fast enough to handle some sizable game calculations, the bottleneck occurs with graphics. This chapter will show you some basic optimization techniques, focusing on graphics operations.

Timing Your Code

There are two major resources that you must manage in a BREW application: CPU resources and memory. BREW handsets typically use ARM CPUs that are far slower than those multi-gigahertz monsters in the average PC these days. Therefore, you must be careful about making expensive API function calls and using other techniques that require a lot of CPU cycles to execute.

How do you determine which functions and sections of code are using up most of your CPU resources? In mature development environments, such as on the PC or major consoles, you have a tool known as a Profiler. The Profiler measures how long it takes to execute all the functions in your program over a certain period of time. It can then tell you how much time your application spends in each function, so you can determine which sections of your code are taking up the most execution time. These problem functions are fine candidates for optimization and reorganization.

Unfortunately, BREW does not yet have a profiler in its suite of tools. Therefore, you must time function execution yourself. You can do this by using BREW's **GET_TIMEMS** function:

```
uint32 GET_TIMEMS()
```

This function returns the current time in milliseconds, as detailed in previous chapters. By calling this function once before executing a function and once after executing a function, you can find out how many milliseconds it took to execute. The code looks like this:

```
nTime1 = GET_TIMEMS();
App_FunctionCall();
nTime2 = GET_TIMEMS();

nTotalTime = nTime2 - nTime1;
```

First, get the current time and store it in the integer variable **nTime1**. Then, we execute the function. After the function returns, put the current time in **nTime2**. Then calculate how long it took for this function to execute by subtracting **nTime1** from **nTime2**.

Of course, this timing is only valid when the code is running on a real device. Because you cannot step through native code in the debugger, how can you possibly see the value of **nTotalTime**? You could write the results into a text file stored on the device's local file system. Unfortunately, you cannot copy files from the device to the host PC via the BREW AppLoader. Instead, you might display the time on the screen with a static control. Or you could transmit this timing value via socket communication to a server running on your PC. Because of BREW's extremely limited tools, you need to get creative with debugging techniques like this.

Basic Optimization Techniques

Now that we know how to time our code, what are some basic methods that you can use to speed it up? What are some handy tips to reduce memory usage? There are many optimization techniques that you may read in books about general programming that are applicable to BREW development. But today's compilers are pretty good about optimizing code. So, some old tricks may be obsolete by today's modern compiler technologies. Here are a few descriptions of basic techniques that may come in handy.

Avoid Redundant Computations

If you use the same calculation frequently, such as **TILE_SIZE** * i as an array index in tile drawing, save the result of this computation in a local variable. Then use this variable to replace all the instances of this computation. This also goes for return variables from API function calls. Instead of calling **GET_TIMEMS** multiple times during the same function, just store the initial result in a variable and reuse that variable (that is, unless you are trying to time specific portions of the function).

Use Bit Flags

I have seen many programmers use dozens of chart-sized variables or Booleans to store simple on or off flags. This is a tremendous waste of space, as you only need a single bit to represent the true or false value of a flag. Instead, I like to use a lot of bit flags. Here is a hypothetical example:

We have a sprite structure for our game. This structure needs to keep track of whether the sprite is invisible and if the sprite can be collided against. If you were to have one member per flag, your structure might look like this:

```
typdef struct sprite_s
{
.
.
.
char nHidden
char nCollide
.
.
.
} sprite_t;
```

Sure, you are pretty crafty using bytes instead of whole integers to store this flag. But if you use bit flags, you can reduce this by half:

```
#define BITFLAG_HIDDEN 0x01
#define BITFLAG_COLLIDE 0x02

typdef struct sprite_s
{
.
.
.
unsigned char nFlags
.
.
.
} sprite_t;
```

Now we only need a single character, **nFlags**, to hold both flags. A byte is 8 bits. We use two of these to represent the hidden and collision states. For instance, if we want to set the hidden bit, simply OR in the bit flag defined up top:

```
nFlags |= BITFLAG_HIDDEN;
```

Now, the first bit is turned on. If we want to turn this bit on, we AND the flags with the inverse of the bit flag that we wish to remove:

```
nFlags &= ~BITFLAG_HIDDEN;
```

The first bit is now cleared. The great thing is that we can combine the flags like this:

```
nFlags |= (BITFLAG_HIDDEN | BITFLAG_COLLISION);
```

To determine if a bit flag is turned on, use the bitwise AND operation:

```
if (nFlags & BITFLAG_HIDDEN)
    return;
```

In this case, bail out of the function if the sprite is not visible. This code fragment would be something that you may see in a sprite drawing loop.

Of course, we have only defined two bit flags. A byte can hold 8. If you need more, you can use a short for 16 bits or an integer for 32 bits. Make sure you define the flags field as unsigned. Otherwise, you will have issues reading the values in the debugger when you turn on the sign bit.

Reduce Memory Allocations

The process of finding a free block of memory and giving it to the user is often a slow process. Therefore, we must reduce the amount of memory allocations in time-critical loops as much as possible. One way to do this is to use preallocated arrays instead of dynamically allocated memory. An example of this is how we use an array of sprites in our graphics examples to store our images instead of allocating each one every time we load a BMP.

Also, it is possible to make what is called a *memory pool*. You can allocate a reserved block of memory at the beginning of the applet's execution and then hand out chunks of this buffer to your pointers through your own memory allocation scheme. This also allows you to add debugging information to your memory blocks to track memory leaks and other hard-to-track bugs. The debugging time saved with custom memory allocation schemes can really be huge on complicated projects.

Avoid Memory Fragmentation

As we all know, memory is scarce on BREW handsets. Compounding this problem is the fact that the allocation of memory is often a slow process. Also, if you allocate and free memory in random patterns, you can get what is called *fragmentation*.

When you fragment memory, you leave a lot of small holes of memory vacant that BREW cannot fill with new allocations. You can reduce this by deallocating memory in first in, last out (FILO) order.

Examining this code fragment, we see a fragmentation situation:

```
char * pBuf1 = MALLOC(15);
char * pBuf2 = MALLOC(20);
char * pBuf3 = NULL;

FREE(pBuf1)

pBuf3 = MALLOC(30);
```

First, allocate **pBuf1** and then **pBuf2**. The block of memory pointed to by **pBuf2** is located further up in memory than **pBuf1**. When we deallocate the first buffer, we leave a hole in memory above **pBuf2**.

Now, let's say we have 50 bytes of memory in total. First we allocate 15 bytes and then 20 for a grand total of 35 bytes. We have 15 bytes left. We now want to allocate 30 bytes. Therefore, we must free up the 15 bytes held by **pBuf1**. After freeing this, we should have 30 bytes left. However, the **MALLOC** of 30 bytes at the end can fail. Why?

This is because we have one block of 15 bytes above the memory held by **pBuf2** and a block of 15 bytes underneath **pBuf2** remaining in the heap. BREW cannot find a single contiguous 30-byte block to give us. Therefore, even though we have 30 free bytes, **MALLOC** fails because it cannot find a hole big enough to give us a 30-byte allocation.

The solution to this is to make sure we deallocate memory in a FILO fashion, meaning we should deallocate **pBuf2** instead of **pBuf3**, which will keep the available block of memory contiguous. Sure, in some cases it is impossible to do this, but you should reorganize your code as much as possible for FILO memory allocation/deallocation if you are having memory usage problems.

Release Memory on Suspension

If you have many resources loaded into memory, such as bitmap graphics, it may make sense to free all of these resources upon applet suspension. You may also want to destroy any interfaces you may be using that consume a lot of memory. This way, any other applet that may be run while yours is suspended will have more memory available to it. When your applet receives a resume event, you must load up all of the resources freed and construct any interfaces released. It helps to do this in a FILO manner, as discussed previously.

Pack Structures

Your structures may be larger than you think they are. For instance,
let's say you have a structure that looks like this:

```
typedef struct whatever_s
{
    char nByte;
    int nInt;

} whatever_t;
```

An int is 4 bytes, so this structure should be 5 bytes in size, right?
Wrong. Many CPUs require pointers to start on certain memory boundaries; therefore the compiler may have to expand structures so that
they remain aligned for proper (and fast) memory access.

In this case, if you do a **sizeof** on Win32, it will report this structure
as being actually 8 bytes in size. You can tell the Visual Studio compiler
to pack structures on single-byte boundaries to force this to 5 bytes
using the **pack** pragma, but these compiler settings will not work with
the ARM BREW Builder. You have to use the ARM BREW Builder's
__pack compiler directive instead. However, in order to maintain com-
patibility with Visual Studio's Win32 compiler, you must make sure that
you use pragmas for the emulator but **__pack** for ARM. The following is
an example of this syntax:

```
#ifdef WIN32
#pragma pack(push, whatever, 1)
#define PACKED
#else
__packed
#define PACKED __packed
#endif
typedef struct whatever_s
{
    PACKED char nByte;
    PACKED int nInt;

} whatever_t;

#ifdef WIN32
#pragma pack(pop, whatever)
#endif
```

I only pack structures that I know I am reading from a file in an
unaligned format. Otherwise, packing structures may gain you a size
advantage, but the CPU may have to perform extra instructions to
access memory on unaligned memory locations. This can affect the
overall speed of your application, not to mention the hassle of defining
packed structures and dealing with their related restrictions on pointer
access and such.

For general cases, it is best to take the structure padding rules into account when designing your data structures. Start with the largest types (integers and pointers) and place the smaller types toward the end (shorts, bytes, chars). Otherwise, the compiler will have to pad out smaller variables in order to start the following larger ones on the correct memory boundaries.

There are plenty of other general optimization techniques that you can look up in many programming textbooks. Things like loop unrolling, shifting, reducing the amount of else clauses, and other familiar methods may allow you to get the most out of the limited ARM CPU of your average BREW chipset.

Graphics Optimizations

All of these memory and CPU cycle optimizations are great, but the true bottleneck performance is graphics. The fact is the typical ARM CPU is actually quite speedy and can handle a surprising amount of calculations. However, BREW's graphics hardware is less than stellar. Drawing images to the screen is one of the slowest operations in any BREW applet, not to mention the massive amounts of memory occupied by a graphics-rich applet.

In all of the examples in this book, we simply redraw the entire screen when needed. This is an incredibly expensive operation that will often make your game look like a slide show on an actual handset. The key to keeping your game running smoothly is to reduce the amount of pixels drawn on the screen. We will focus on a few of our example applets and how we can use a few tricks to keep the screen updates to a minimum and memory usage for graphics small.

Use 16-Color BMPs

Although the temptation to use 256-color graphics on 8-bit handsets is great, you may look into using less colorful 4-bit BMPs for your sprite images. They only have 16 colors, but their size is cut in half from an equivalent 8-bit image. Especially for large images, such as splash or title screens, the memory savings can be tremendous. Eight-bit color BREW handsets also support the 16-color palette. Therefore, you can use 4-bit BMPs without a hitch. If you really need those extra colors, use a mix of 4- and 8-bit graphics for various images. The more images you can reduce to the handset's 4-bit palette, the smaller your resources will be. Not only does this reduce the amount of memory you use to load in BMPs, but the size of the game download will be smaller as well.

Dirty Rectangles

The first major technique we will look at is dirty rectangles. The dirty rectangle technique is one where you keep track of which portions of the screen need to be redrawn. These rectangular portions are considered "dirty" because they have changed between frames. By only redrawing these sub-sections, or rectangles, of the screen, you can reduce the amount of pixels copied to the screen dramatically.

This is important because copying pixels to the screen is perhaps the single most expensive graphics operation in the BREW API. Early versions of the Kyocera 3035 monochrome handset took half a second to blt a screen-sized (89x99) monochrome bitmap. That is a pathetic two frames per second without executing any other code in between frames! Newer versions of firmware in these phones frequently speed things up, but pixel copying is still a major issue in even the latest BREW revisions.

Of course, if most of the screen changes frequently, you will end up having dirty rectangles that equal the same amount of pixels as a complete screen refresh. Therefore, you will end up with the same disappointing performance as the aforementioned full-screen refresh. This is why this technique is not that useful for games that have many moving objects.

Optimized Flarb

For this chapter, I have created a version of Attack of the Flarb using dirty rectangles. A few changes to the source code had to be made, including adding to existing structures and creating a few new functions. The source for this project can be found in the Source\Chapter 19\ flarbopt folder in the companion files. We will now go over the major differences between the old and dirty rectangles versions.

Dirty rectangles require us to keep track of the previous and current position of every moving object in the game. This demands a few more bytes per sprite object because we need two pairs of x and y coordinates. However, it is a small price to pay for the speed enhancements brought on by the technique. As you can see, we added a few members to the old sprite structure:

Listing 19-1: Optimized sprite structure

```
typedef struct sprite_s
{
    unsigned char nX;
    unsigned char nY;
```

```
unsigned char nOldX;
unsigned char nOldY;

unsigned char nWidth;
unsigned char nHeight;

unsigned char nFlags;

unsigned char nImage;

} sprite_t;
```

This is the same sprite structure, but we have added a new pair of coordinates, nOldX and nOldY, along with a bit flag member, nFlag. Before we change the x and y location of our sprite, copy the nX and nY variables into nOldX and nOldY. Then we are free to modify nX and nY with new positions. We also must set a bit flag that marks the sprite as "dirty" and thus ready to be redrawn. We do this by ORing in the SPRITEFLAG_BITMAP bit flag that is defined in the applet header like this:

```
#define SPRITEFLAG_DIRTY    0x01
```

Adding dirty rectangles to this applet requires us to change the screen drawing function. Here is the new one:

Listing 19-2: Screen draw routine

```
void FlarbOpt_DrawScreen(myapp_t* pApp, boolean bForce)
{
    AEERect rect;
    AECHAR szTempBuf[] = {'%','d','\0'};

    //clear background
    if (bForce)
            IDISPLAY_ClearScreen(pApp->a.m_pIDisplay);

    //clear screen area underneath score
    rect.x = 0;
    rect.y = 0;
    rect.dx = pApp->di.cxScreen;
    rect.dy = 10;
    IGRAPHICS_DrawRect(pApp->pIGraphics, &rect);

    //draw score and number of lives
    WSPRINTF(pApp->szBuf, (MAX_STRING_LENGTH * sizeof(AECHAR)),
        szTempBuf, pApp->nScore);
    IDISPLAY_DrawText(pApp->a.m_pIDisplay, AEE_FONT_NORMAL,
        pApp->szBuf, -1, 5, 0, NULL, IDF_TEXT_TRANSPARENT);

    FlarbOpt_DrawSprite(pApp, &pApp->sprPlayer, pApp->bForce);
    FlarbOpt_DrawSprite(pApp, &pApp->sprAlien, pApp->bForce);

    if (pApp->bBullet)
        FlarbOpt_DrawSprite(pApp, &pApp->sprBullet, pApp->bForce);
```

```
//draw game over text
if (pApp->bGameOver)
{
    ISHELL_LoadResString(pApp->a.m_pIShell, RES_FILE,
        STR_GAMEOVER, pApp->szBuf, sizeof(pApp->szBuf));
    IDISPLAY_DrawText(pApp->a.m_pIDisplay,
        AEE_FONT_NORMAL, pApp->szBuf, -1, 0, 0, NULL,
        IDF_TEXT_TRANSPARENT | IDF_ALIGN_CENTER | IDF_ALIGN_MIDDLE);
}

pApp->bForce = FALSE;

//refresh screen
IDISPLAY_Update(pApp->a.m_pIDisplay);
}
```

The first thing you may notice is that we have a new argument in this function, **bForce**. This is a Boolean used to force an entire screen refresh. When **bForce** is set to True, clear the display as we did in the original version of this function. This is for when we need a complete screen refresh, such as coming out of an **EVT_APP_RESUME** event or something. Also, notice that we now draw a rectangle underneath the score display. We need to clear the area underneath the score so that remnants of the previous score do not show through when the value changes. Note that the fill color is set to white in the **EVT_APP_START** handler so that all geometry drawing operations are drawing with the same color as the background.

The next major change is the usage of a new function, **FlarbOpt_DrawSprite**. Instead of drawing the sprites directly with **bitblt**, we need to handle some special processing for dirty rectangles. This special code is encapsulated in a drawing function for ease of use.

Listing 19-3: Optimized sprite draw routine

```
void FlarbOpt_DrawSprite(myapp_t* pApp, sprite_t* pSprite, boolean bForce)
{
    AEERect rect;

    if (!pSprite)
            return;

    //if it is not dirty, and we are not forcing a redraw, bail
    if ( (!(pSprite->nFlags & SPRITEFLAG_DIRTY)) && !bForce)
        return;

    //make clearing rectangle
    rect.x = pSprite->nOldX;
    rect.y = pSprite->nOldY;
    rect.dx = pSprite->nWidth;
    rect.dy = pSprite->nHeight;

    IGRAPHICS_DrawRect(pApp->pIGraphics, &rect);

    //now draw sprite in new location
```

```
IDISPLAY_BitBlt(pApp->a.m_pIDisplay,
    pSprite->nX, pSprite->nY, -1, -1,
    pApp->pRawImagePtrs[pSprite->nImage], 0, 0,
    AEE_RO_COPY);

    pSprite->nFlags &= ~SPRITEFLAG_DIRTY;
}
```

Now that we have the previous position of the sprites, we know what the dirty regions are. If a sprite's current coordinates and old coordinates are different, we know we have to redraw this sprite. Before we redraw, we have to clear a rectangle at the previous position using the sprite's width and height as the dimensions. In the case of our Flarb game, the background is simply white. Therefore, draw a white rectangle with its origin at nOldX and nOldY of nWidth and nHeight dimensions. If we are using a background image instead, simply bitblt this rectangle from the background bitmap to the screen. After dealing with the previous position, simply draw the sprite at its new location defined by nX and nY. If we were to change the background color to something other than white, perhaps this concept will be easier to see:

Figure 19-1: Attack of the Flarb using black as the fill color for dirty rectangles. Notice the trail of cleared rectangles behind the Flarb ship.

Looking at this new function, you will first notice it takes several arguments. Aside from the pointer to the applet itself with pApp and a pointer to the sprite we are drawing in pSprite, it takes a Boolean parameter, bForce. If bForce is set to True, the sprite will be redrawn whether it is dirty or not. We need this option because when we need to refresh the whole screen, such as after an applet resume, we need to draw every sprite regardless if it is dirty or not.

Therefore, we first check if the sprite is dirty via the bit flag or if we are forcing a redraw. Next, make a rectangle out of the previous x and y position and the sprite's dimensions. The color is set to white in the EVT_APP_START handler, so this ends up drawing a blank rectangle erasing the screen image of the sprite in its previous position. After drawing the rectangle, bitblt the sprite image as usual. Finally, clear out the SPRITEFLAG_DIRTY bit flag, leaving it to be set the next time that the sprite moves.

The SPRITEFLAG_DIRTY bit is set whenever we move an object. If you look in FlarbOpt_MoveBullet, FlarbOpt_MoveEnemy, and FlarbOpt_GameInput (where we move the player), you will see that we flick on SPRITEFLAG_DIRTY. As described previously, this will trigger the sprite drawing routine to redraw the sprite. Otherwise, if a sprite has not moved, there is no need to redraw it.

This applet posed one interesting problem. When we hide the bullet after a collision with an alien ship or if it goes off the screen, we turn off the bBullet Boolean, causing it not to draw. Since we do not clear the screen every frame, the old image of the bullet will just hang there on the screen. For this case, we have to add a function that is called to hide the bullet. In the process of hiding it, it draws a blank rectangle on the screen starting at its current position and stretching it to cover its previous one as well, just to be safe.

Listing 19-4: Optimized bullet removal routine

```
void FlarbOpt_HideBullet(myapp_t* pApp)
{
    AEERect rect;

    pApp->bBullet = FALSE;

    //clear the bullet's current position
    rect.x = pApp->sprBullet.nX;
    rect.y = pApp->sprBullet.nY;
    rect.dx = pApp->sprBullet.nWidth;
    rect.dy = pApp->sprBullet.nHeight + (pApp->sprBullet.nOldY - pApp->sprBullet.nY);

    IGRAPHICS_DrawRect(pApp->pIGraphics, &rect);
}
```

If you look at the code, you will see that aside from the applet initialization functions, everywhere we used to simply set bBullet to False, we call this instead.

Other than these changes and additions, the applet is pretty much the same as it was. The great thing is that now we have reduced the blting tremendously. If you run this game on an actual handset, you will see a dramatic increase in speed.

Dirty Tiles

Another variation of the dirty rectangles technique that I like to use is one called "dirty tiles." Well, at least that's what I call it. If you are using tiles instead of a solid color or large screen-sized image backdrop, the dirty rectangles technique does not work entirely. This is because you need to redraw the tiles that the character previously occupied instead of simply copying over a rectangle to the previous location. This is almost the same thing, except instead of erasing the background from moving characters, simply redraw only the dirty tiles instead of arbitrary dirty screen regions.

Figure 19-2: The optimized version of the tile demo only draws tiles that we have traveled over. This picture shows what it looks like if we do not do the initial full-screen draw.

Optimized Tile Map

To demonstrate this, we will create an optimized version of the tile map example. With minimal effort, we can convert this applet to use dirty tiles and sprites. To illustrate this, we will focus on only the major changes in the tile map code. You can see the entire source in Source\Chapter 19\tilesopt of the companion files.

If you recall, our tile structure had an unused field, nFlags:

Listing 19-5: Optimized tile structure

```
typedef struct tile_s
{
    unsigned char nIndex;     //Index into bitmap array
    unsigned char nFlags;     //Tile properties

} tile_t;
```

The nFlags member has a byte-sized flags field. In this case, we can create a flag that will mark a tile as "dirty." Then, in our tile map drawing routine, we only draw tiles marked as dirty. This flag is defined like so:

```
#define TILEFLAG_DIRTY     0x01
```

In the case of our tile example, our player character sprite was the size of a single tile. It also moved in complete tile-sized steps. Therefore, we simply need to divide the sprite's previous x and y world-coordinate positions by the tile height and tile width to determine which tile the sprite occupied previously. If we have sprites that are larger than a tile or move in pixel increments so that they could potentially cover many different tiles, we need to compute the tile location of each of the four corners of the sprite in a similar manner. Then mark all of the tiles in this range as dirty for the next loop through the map drawing routine. For the purposes of this example, we are only concerned with single tiles. Compute which tiles are dirty in a modified version of the player movement function:

Listing 19-6: Optimized player movement function

```
void TilesOpt_MovePlayer(myapp_t* pApp, int nX, int nY)
{
    tile_t * pOldTile;
    tile_t * pNewTile;
    int nNewX = pApp->playerSpr.nX + nX;
    int nNewY = pApp->playerSpr.nY + nY;

    //Check to see if we've hit the edge of the map
    if (nNewX < 0)
        return;

    if (nNewY < 0)
        return;

    if (nNewX >= (pApp->nMapWidth * TILE_WIDTH))
        return;

    if (nNewY >= (pApp->nMapHeight * TILE_HEIGHT))
        return;

    //Now check the tile to see if it is a wall (tile index is 1)
    pNewTile = &pApp->pMapArray[(nNewX / TILE_WIDTH)
        + (pApp->nMapWidth * (nNewY / TILE_WIDTH))];

    if (pNewTile->nIndex)
        return;

    //Get the old tile to make both dirty
    pOldTile = &pApp->pMapArray[(pApp->playerSpr.nX / TILE_WIDTH)
        + (pApp->nMapWidth * (pApp->playerSpr.nY / TILE_WIDTH))];

    //Move the window over one 'page' if we've gone off the edge
    if ((pApp->nTileX > 0) && ((nNewX / TILE_WIDTH) < pApp->nTileX))
    {
        pApp->nTileX -= pApp->nNumTilesOptX;
        pApp->bForce = TRUE;
    }
    else if ((pApp->nTileX < pApp->nMapWidth) &&
        ((nNewX / TILE_WIDTH) >= (pApp->nTileX + pApp->nNumTilesOptX)) )
    {
        pApp->nTileX += pApp->nNumTilesOptX;
        pApp->bForce = TRUE;
    }
    else
    {
        pOldTile->nFlags |= TILEFLAG_DIRTY;
        pNewTile->nFlags |= TILEFLAG_DIRTY;
        pApp->playerSpr.nFlags |= SPRITEFLAG_DIRTY;
    }

    if ((pApp->nTileY > 0) && (nNewY / TILE_WIDTH) < pApp->nTileY)
    {
        pApp->nTileY -= pApp->nNumTilesOptY;
        pApp->bForce = TRUE;
    }
    else if ((pApp->nTileY < pApp->nMapHeight) &&
        ((nNewY / TILE_WIDTH) >= (pApp->nTileY + pApp->nNumTilesOptY)) )
```

```
        {
            pApp->nTileY += pApp->nNumTilesOptY;
            pApp->bForce = TRUE;
        }
        else
        {
            pOldTile->nFlags |= TILEFLAG_DIRTY;
            pNewTile->nFlags |= TILEFLAG_DIRTY;
            pApp->playerSpr.nFlags |= SPRITEFLAG_DIRTY;
        }

        pApp->playerSpr.nX = nNewX;
        pApp->playerSpr.nY = nNewY;
}
```

There are a few changes to this function. At the start of the function, we define two pointers to tile structures, **pOldTile** and **pNewTile**. These will be used to reference the tile that our sprite resides on before being moved and then the tile it is over after we move it. This way, we can mark both as being dirty. It is also much easier than using an array reference every time that we want to access the contents of the desired tile structure.

So, instead of checking the tile's index value by simply going into the array, we assign the pointer to it and then check the **nIndex** member through the pointer:

```
pNewTile = &pApp->pMapArray[(nNewX / TILE_WIDTH) +
        (pApp->nMapWidth * (nNewY / TILE_WIDTH))];

if (pNewTile->nIndex)
    return;
```

Now that we have a pointer to where the tile is going to go and we have determined that it is not a wall, do the same computation but with the old coordinates to get the previous tile (or current one, depending on how you look at it):

```
pOldTile = &pApp->pMapArray[(pApp->playerSpr.nX / TILE_WIDTH)
        + (pApp->nMapWidth * (pApp->playerSpr.nY / TILE_WIDTH))];
```

Now we have to mark the tiles as dirty. But before we do this, check if we have moved over a page in the map either horizontally or vertically. If so, we need to redraw the entire screen. We can tell the applet to redraw by using the familiar **bForce** Boolean from the applet structure. We used this technique in the Flarb optimizations as well. If we have not moved over a screen, then simply mark the old, new, and player sprite as dirty:

```
pOldTile->nFlags |= TILEFLAG_DIRTY;
pNewTile->nFlags |= TILEFLAG_DIRTY;
pApp->playerSpr.nFlags |= SPRITEFLAG_DIRTY;
```

Looking at the drawing loop for the map, you see that we only draw a
tile if it is dirty or if we are being forced to:

Listing 19-7: Optimized map drawing function

```
void TilesOpt_DrawMap(myapp_t* pApp, int nX, int nY, boolean bForce)
{
    int i;
    int j;
    int nYModifier;
    tile_t* pTile;

    //Find out how much we need to add to get to our
    //Y level in the map
    nYModifier = nY * pApp->nMapWidth;

    //Go through each visible tile and draw it
    for (i = 0; i < pApp->nNumTilesOptX; i++)
    {
        for (j = 0; j < pApp->nNumTilesOptY; j++)
        {
            pTile = &pApp->pMapArray[(j * pApp->nMapWidth) + nYModifier + i + nX];

            if (bForce || (pTile->nFlags & TILEFLAG_DIRTY))
            {
                IDISPLAY_BitBlt(pApp->a.m_pIDisplay,
                    (i * TILE_WIDTH) + OFFSET_X,
                    (j * TILE_HEIGHT) + OFFSET_Y, TILE_WIDTH,
                    TILE_HEIGHT, pApp->pRawImagePtrs[IMAGE_TILESOPT],
                    pTile->nIndex * TILE_WIDTH, 0, AEE_RO_COPY);

                pTile->nFlags &= ~TILEFLAG_DIRTY;
            }
        }
    }
}
```

The dirty tile system is also why we do not smoothly scroll the screen
in our tile example. If we were to scroll the viewport across the tile
map, we would have to redraw the entire screen every frame. No dirty
rectangle or tile scheme would work because the entire background
shifts over with each movement. By restricting tile-based games to the
"flick-screen" type, we have developed a smart idea until newer phones
hit the market with faster drawing speeds. Even then, you will have to
accommodate the least common denominator of hardware for mass-mar-
ket acceptance.

Geometric Substitutes

Geometric substitutes is not necessarily worth its own chapter, but it
still deserves a mention. Some developers have noticed that filling the
screen area with a geometric primitive, such as a square or circle
instead of a similarly sized bitmap, is much faster. This means that if you

can replace any of your bitmaps with geometric graphics, you could get a big speed bonus. Unfortunately, you do not get pixel-level details with geometry, but if you have large areas of solid color in your backgrounds or sprite objects, you may consider replacing them with simple geometry objects.

Conclusion

In this brief chapter you have learned a few handy tricks for optimizing your code for memory usage, CPU cycle consumption, and graphics overhead. Optimizing code is an art in itself and many books have been written on the subject. Hopefully, you have learned enough to get yourself out of some common BREW performance bottlenecks. If not, do not be afraid to experiment with developing your own optimization schemes and tricks.

Chapter 20

Certification and Publishing

Introduction

Now that you have the game up and running on a real BREW handset, it is time to get it released to paying customers. Welcome to the crazy world of business development. This chapter focuses on the steps that you need to finalize the production of a commercial BREW application and get it released on carrier networks. Most of these issues are covered thoroughly in the documents Qualcomm provides on their Developer Extranet. I highly recommend reading over these documents before doing anything.

True BREW Certification

The first step to getting your completed game out to customers is to get it certified. Qualcomm requires each applet to be True BREW certified. This tells carriers that the applet is bug free and will not adversely affect their network. This process involves sending your applet to NSTL, an independent testing lab in Pennsylvania, which will make sure that the applet conforms to the basic standards outlined by Qualcomm for a commercial application. This includes making sure it does not crash or otherwise interrupt the normal functioning of the handset. This process costs between $750 and $2500 per test run, depending on the complexity of the application. Every time a bug is found, you will receive a bug report from NSTL. Then fix the specified bugs and send the applet in again for another test run. This process again costs between $750 and $2500. If time is of the essence, the testing process can be expedited, which raises fees to the $950 to $3200 range. Also, if you have higher developer status, it is possible to get 5 or 10 percent discounts on these fees. Therefore, it is a good idea to thoroughly test your applet internally before releasing it to NSTL for certification.

There are two phases of testing. This includes True BREW testing and True BREW functionality testing. True BREW testing is done mostly at NSTL's labs, with some interaction from Qualcomm. Here, the standard tests for stability and compatibility, as mentioned earlier, are performed. After you pass this phase, Qualcomm performs the functionality testing, which determines if your applet can be properly downloaded and installed on the handset. The testing process is actually quite detailed and complicated. Qualcomm has a series of documents available on their Developer Extranet that outlines the process and what steps you should take to ensure a trouble-free testing procedure.

Note: It is worth mentioning that sometimes when a new BREW handset comes out, NSTL will run a "First Time Free" special where you can submit your applet for that new handset absolutely free. If they find a bug and thus have to retest, you must pay. However, if you skate through the first time, you essentially have free True BREW certification. Contact NSTL or participating carriers for more information.

Self-Testing

Every time NSTL or Qualcomm finds a bug, you must resubmit the entire application for testing and are thus charged again for another test run. Naturally, it is in the developer's best interest to perform as many tests as possible before submitting it to make sure there are no problems. Thankfully, Qualcomm provides a document on their web site that outlines every single bug that is checked in the True BREW process. The True BREW test guide details every bug with the test performed and what the desired results are for a bug-free operation. If you meticulously test your application for every single one of these test cases, you should be well prepared for the True BREW certification process.

The NSTL Test Process

After performing your own internal self-tests, you must package your application up for delivery to NSTL. This involves an Application Specification Document describing your application, as well as all of the MOD, BAR, MIF, and signature files required for your application's operation. You must also provide a Windows emulator binary, as they perform some of the test in the emulator as well as on the actual device.

As with anything else in the test process, there is a desired format and a list of required items for submission to NSTL. Qualcomm also provides a guide, as well as some Microsoft Word templates for the submission documents to smooth over this process. If you do not properly package your applet when delivering it to NSTL, you will fail the first part of the testing process.

The next step at NSTL is the standard test. Here they use various tools including Qualcomm's Grinder, which randomly presses keys in the emulator trying to blow up your application. The Grinder is available to developers through the Extranet, so it is possible to see how your applet holds up before submitting it. They also test for things like abnormal delays with user input, icon sizes, and standard interface issues like the operation of the send and clear keys.

Next up is the functionality test. Here they check for various carrier requirements. These include seeing how the application holds up when timers and alarms execute. For instance, they check to see if your application properly resumes from being suspended by the handset's alarm clock function. They also check if your application works as described in the Application Specification Document submitted with your binary.

Qualcomm Testing

The last testing phase is the final verification test. This is the phase that Qualcomm handles. They check to see if your application properly suspends and resumes from different events, and they randomly check how it reacts to keypresses and such. Also, Qualcomm makes sure it properly downloads and installs over-the-air from the mobile shop and checks if your signature is valid. Once this phase is completed, your application is True BREW certified. As with any stage of testing, if you fail this part, it has to be resubmitted. Otherwise, it is then made available to BREW carriers for evaluation. Also, keep in mind that you must resubmit your product for testing whenever you create a new version in a different language or for a different BREW handset. Dozens of BREW handsets are slated for release in the near future; this can become quite expensive, even if you do not have any bugs.

Carrier Testing

Although we have walked through the True BREW certification process, there is one final test that may be performed. Each carrier that wants to make your application available on their network can perform carrier-specific tests to see if it meets their own requirements. Each carrier has their own guidelines, and you must contact their developer support department to find out what they look for. In some cases, bugs found in this phase can revoke your True BREW status previously declared by Qualcomm.

General Testing Tips

Qualcomm has information on common bugs and other things to look for in the interest of avoiding snags in the testing process. These include:

- Read the guides and perform QA on the test cases. For cases that you cannot perform internally, consult NSTL on the best course of action.
- Pay attention to suspend and resume events and how your application handles them.
- Focus on "file system full" or "max files" test cases (section 5.2 in the guide).
- Examine how the application handles "tight loop" situations. For instance, give continual and rapid keypad input for 90 seconds and see if your application hangs or resets.
- Look at how the application handles incoming SMS events, especially if it uses text controls.
- If the application includes a server component, make sure the server is available at the time of testing.
- Make sure the graphics display correctly. Look for things like screen overlap and other possible glitches.

Grinder

One of the tools used in True BREW certification is Qualcomm's Grinder. This is an application that works in conjunction with the emulator to test the robustness of your BREW application. It is designed to randomly press buttons and send events in an attempt to see how your application holds up under adverse conditions. When you sign on to the Developer Extranet, you are able to download this tool. Try it out and see how your code stands up to this rigorous test before submitting for certification. If you cannot break it with Grinder, chances are that NSTL will not.

Grinder is totally configurable. You can set the kinds of events sent to the device as well as use the built-in Shaker tool to set environment attributes, such as the amount of memory available, space left in the file system, and number of usable sockets. Log files can be generated from the usage of this tool to further diagnose how and why your application crashed. As with all of Qualcomm's BREW tools, extensive documentation of Grinder is available on Qualcomm's web site.

Carrier Distribution

Now that your application is True BREW certified, it is time to get it out to paying customers. This involves developing a relationship with one or many BREW carriers. The great thing about the BREW distribution system is regardless of how many carriers carry your product, you only have to collect money from Qualcomm instead of managing many different billing relationships. However, you still have to deal with each carrier individually when trying to get them to pick up your game for distribution.

Verizon Wireless

Verizon Wireless is currently the largest wireless phone service provider in the United States. Their own web site boasts 90 percent service coverage of the entire United States with over 29 million subscribers. They launched their BREW service, called "Get It Now," across the nation in the spring of 2002 as well as the first phase of their 3G CDMA 2000-based "Express Network" in select locations. At developers.verizonwireless.com you can register for their excellent developer's program. Here you can access information on Verizon's application standards, as well as extensive information on their unique distribution and marketing opportunities. Depending on the developer, it is possible to gain marketing support via web page listings, newsletter exposure, and even in-store displays. The developer's web site also provides a mechanism to submit your True BREW certified application for Verizon's evaluation process. This includes submitting marketing materials, including logos of various sizes and colors, as well as descriptions of your application and the test results from the certification process.

Other Carriers

Qualcomm has been inking deals with additional carriers both in the United States and around the globe. With Qualcomm's centralized billing system, BREW makes it easy to distribute your applications with carriers both domestically and internationally.

The United States

The next two carriers to implement BREW in the U.S. are Alltel and U.S. Cellular. Both of these carriers are smaller regional companies that will add around another 8 million to 10 million potential customers to the domestic BREW audience. Although it is possible for these carriers to request unique features of handset manufacturers, chances are they will have the same handsets as Verizon. Therefore, it should be easy to bring your already existing Verizon applets to these two new players. Check their respective web sites (www.alltel.com and www.uscellu-lar.com) for development information and handset specifications.

Overseas

Perhaps the most successful BREW carrier in the world is South Korea's KTF. KTF was actually the first carrier in the world to provide BREW service and has had tremendous success with it. Hundreds of local developers have been creating many different applications, with gaming being among the most popular. BREW has jumped to the number 2 mobile applications platform in South Korea, with J2ME a distant third and the Korean-specific GVM platform from SK Telecom in first place.

As for other carriers, Japan's KDDI is in the early phases of rolling out BREW service in addition to their already existing Java handsets. It will be interesting to see how the Japanese launch fares, because Java is well entrenched in the Japanese market. Qualcomm has also made a deal with China Unicom to bring BREW to another potential 50 million customers in China, not to mention the moves Qualcomm has been making in both Latin American and Europe. Europe is proving the toughest nut to crack, as Qualcomm's CDMA technology and the GSM/GPRS based systems in the EU are diametrically opposed. We shall see how successful Qualcomm is in introducing BREW to Europe.

In order to get your applications distributed on these networks, you will most likely have to localize them in their native languages. This is where keeping all of your strings in resource files comes in handy. Also, keep in mind that there are export restrictions when dealing with products that use certain kinds of encryption as well as foreign tax issues to consider.

Paid in Full

What about the money? The one step we have not covered is charging for your application and collecting the cash. Once you are a registered BREW developer and gain access to the Developer Extranet, you can access a variety of different pricing templates. These allow you to specify how and what you want to charge for your application. For instance, you can simply use a retail model by charging per download. Or, it is possible to "subscribe" to your game by charging the user a monthly fee for the use of your application. A variety of other methods exist, including the ability to charge for a number of uses. If you want to only allow the user a certain number of plays per payment, you may have to use the ILicense module provided in BREW 1.1 and beyond to manage the number of uses left.

Out of the applet cost you specify, Qualcomm takes 20 percent and gives you the rest. The carrier makes money not only on the data fees incurred by downloading the applet but also by marking up the cost. Depending on your relationship with the carrier, it is possible to have some control over the final cost of the application.

Once your application is out on carriers and available to paying customers, Qualcomm will deliver payments at regular intervals. At the lowest level of developer support, you get paid twice a year. If you pay Qualcomm's fees for higher levels of developer status, it is possible to get paid on a more frequent basis in addition to other perks. Be warned that reaching the highest echelons of developer status can be quite expensive. This can easily negate any advantage of receiving your payments at shorter intervals.

The Developer-Publisher Model

Most of this chapter has dealt with the process of publishing your own game. Arguably, this is the way to go, as the distribution costs are nil, especially when compared to the expense of publishing a traditional game on CD-ROM or cartridge. However, in many cases, having a publisher works in your favor. The benefits of having a publisher sometimes include royalty advances as well as testing and marketing services. This means that the publisher will generally handle the details of the certification process and all you worry about is fixing the bugs. Also, publishers often perform their own localization for foreign markets as well as advertise and promote your game. Once the market becomes saturated and games struggle for the spotlight among an onslaught of competing products, marketing and promotion muscle will be a huge advantage. Also, some carriers absolutely refuse to work with small developers and instead require you to be published by a major. Thus, in some cases you have no choice.

Be aware that many contracts count the costs of promotion and advertising against your royalty advance. As with any publishing agreement, most contracts in this industry are not in favor of the creator/developer. It helps if you have a good lawyer on your side when entering negotiations with any publisher, or even carriers for that matter. However, I find that many contracts done on these "small" games are often informal affairs with very loose guidelines. The costs associated with hiring a lawyer to work on contract issues can quickly deplete the small advance received on the average mobile contract.

Right now, most mobile games are released by publishers specializing in wireless gaming. You will find that many giants of traditional game publishing are wary of investing any amount of money in the market at this point. Outside of Japan and South Korea, the mobile gaming industry has not proven very profitable. A handful of large game publishers have dipped their toe into this market, including Sega, THQ, and Namco. However, you will find that the field is currently dominated by new companies specializing in mobile gaming, such as nGame, Cybiko, and GameLoft. These are just a few of the names that you might want to look up in your quest to find a publisher.

Conclusion

Although brief, this chapter has guided you through the testing and publishing process of your BREW game. Qualcomm's Developer Extranet provides extensive documentation on the test requirements, pricing models, and other issues faced when trying to get your product out on carrier networks. This chapter only serves as an introduction to the process. Read every document on both Qualcomm and the carriers' web sites thoroughly before beginning this process. Failure to properly submit and test your application can result in a rapidly expanding budget and unrecoverable development costs.

Appendix A

BREW and Java

Introduction

In the wireless application industry, there is a raging debate over the pros and cons of BREW and Java 2 Micro Edition (J2ME). It seems that both platforms are diametrically opposed, competing not only for mobile gaming platforms but all other interactive mobile data applications as well. The battle lines are drawn. In the United States, Verizon Wireless is using BREW, while SprintPCS, another CDMA carrier, has chosen J2ME as their mobile application standard. In South Korea, J2ME usage trails behind BREW service from KTF. In Europe, BREW has yet to make a splash, while J2ME is all the rage. Are they really exact opposites? Can they possibly coexist? We will explore the differences between the platforms as well as the clever way that BREW has managed to incorporate Java into its suite of applications, thus making this issue multiple shades of gray instead of black and white.

Introducing Java

If you have read this far, you know all about BREW. Now it is time to give readers a crash course in mobile Java technology. This is not meant to teach you how to create applications in Java (for that, buy my next book). Instead, we will discuss the general technology and concepts behind the Java platform.

What Is J2ME?

When we mention "mobile Java," we mean Java 2 Micro Edition, or J2ME. J2ME is a slimmed-down version of Sun Microsystems' popular Java programming language. This means they have removed many packages and existing features of standard Java (J2SE) to not only fit inside the limited memory and processor constraints of mobile devices but also to address the interface concerns of such miniature hardware.

Much like regular Java, J2ME is an interpreted language that uses a virtual machine. This means that the code is not compiled directly for the hardware. Instead, J2ME code is complied into bytecodes, which are then executed by the Java virtual machine. Theoretically, you can write Java code and it will run on any machine that has a suitable virtual machine written for it. Of course, any programmer with Java experience knows that the so-called "Write Once, Run Anywhere" Java motto is largely a myth. This is especially true for game programmers.

One of the disadvantages of Java is that this virtual machine adds an extra layer of abstraction between your code and the hardware. The VM has to go through an extra step of translating your code from a Java bytecode to native instructions. Theoretically, this means Java code is slower than native compiled C/C++ (such as when using BREW). However, I have been very impressed with the performance of many J2ME handsets.

The virtual machine that J2ME uses is called the KVM. The K stands for kilobyte, meaning that this virtual machine is much smaller than the one used with J2SE. Hence, KVM fits in the small memory allotted in most wireless handsets. The KVM also has some changes from the regular virtual machine, most notably the lack of floating-point support. As we have seen with BREW, many sacrifices have to be made to get code up and running on inexpensive mobile hardware.

When referring to "wireless Java," we also typically mean J2ME using the Connected Limited Device Configuration (CLDC) and Mobile Information Device Profile (MIDP). The CLDC and MIDP define a set of packages and services useful for wireless devices. Soon, different configurations and profiles will be made available to address different classes of devices and industry needs. In fact, an update to the existing MID Profile called MIDP-NG is being finalized as you read this. MIDP-NG will have many game-related functions in conjunction with other new features and improvements. Because of the usage of MIDP, wireless Java programs are typically referred to as MIDlets instead of applets.

Developing MIDlets

So, how do you create a MIDlet? You can get started immediately by downloading SUN's J2ME Wireless Toolkit from java.sun.com/j2me. The Wireless Toolkit allows you to write, compile, run, and debug J2ME code with a suite of very capable free tools. Many commercial Java IDEs, such as Metrowerks' CodeWarrior, IBM's WebSphere, Borland's

JBuilder, and Sun's own Forte have been updated to include support for J2ME development. Several IDEs have also been released with a heavy focus on J2ME such as CodeWarrior's Wireless Studio and Zucotto's WHITEboard SDK. The cost of these kits varies from hundreds to thousands of dollars. However, it is entirely possible to develop a complete product using only Sun's free tools included with the Wireless Toolkit.

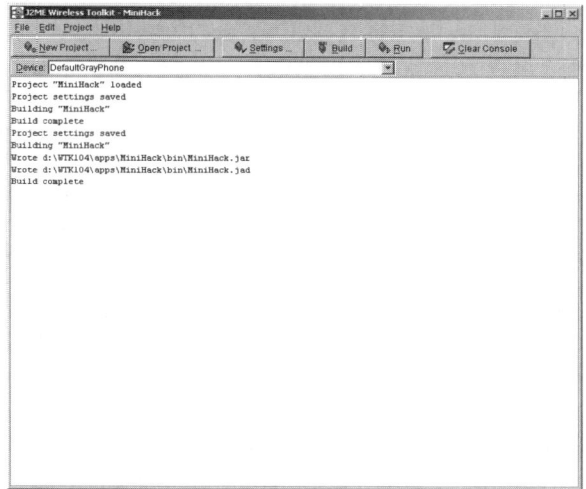

Figure A-1: Sun's KToolbar is a free and convenient J2ME development tool.

If you are already familiar with Java, programming with wireless Java should be easy to get into. You still have Java's object-oriented approach to programming, garbage collection, and sandbox security model. Although the set of packages used with MIDP has been heavily modified and in some cases completely removed, it should take a seasoned Java programmer no time to get used to these modified classes and new packages.

One example of these changes is the complete removal of the Abstract Windowing Toolkit for GUI design. Instead, Sun has replaced it with LCDUI, which is a new GUI package designed for providing interfaces suited for the small display of the average wireless phone. There are many useful GUI controls, such as menus, progress bars, and buttons that are suitable for most any application. Compared to BREW, the positioning and layout of controls is quite limited, however.

As for graphics, MIDP uses PNG as its basic bitmap image file format. One area of confusion, however, is pixel transparency. With BREW, simply use the obnoxious purple color to represent clear pixels in 8-bit images. How pixel transparency is dealt with in J2ME is actually an optional part of the standard. Therefore, many MIDP implementations choose to implement pixel transparency in their own way. Even worse, some do not use it at all. J2ME also has the standard array of geometric primitive and text drawing commands.

Once you have your code compiled, you must verify and package your MIDlet and then run it in the emulator or test it on an actual device. The verification and packaging process makes sure that your MIDlet does not use any illegal commands unsupported by MIDP and then places your classes and all the assorted resources into a compressed JAR file.

When using J2ME emulators, many of the same issues faced with BREW emulation are apparent. Each hardware vendor generally provides their own emulator, and thus each of these emulators vary in accuracy. In many cases, things like the screen resolution or pixel transparency support may vary wildly from the emulator to the actual hardware. Also, many VM bugs may be apparent in the emulator but not in the phone, and vice versa. Much like BREW's emulator, the performance is not device-accurate and instead depends on your host machine's CPU. Some newer emulators allow you to tweak the execution speed to better reflect the device's performance, however. Also, the standard set of emulators that comes with Sun's toolkit allows you to modify different properties to emulate different devices, much like BREW's Device Configurator. However, there is no substitute for testing your code on a real handset.

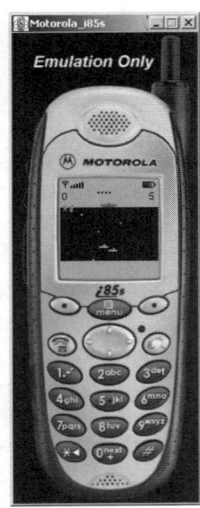

Figure A-2: A MIDlet game
running in Sun's emulator

To execute your MIDlet on hardware, the JAR must then be uploaded to your handset and executed. The tools to upload the JAR to the handset are provided by the handset manufacturer and not by Sun; therefore these loader tools are not provided by Sun's Wireless Toolkit. This process is fairly straightforward and does not require you to muck around with license files, MIFs, and class IDs. The process is infinitely more convenient than testing BREW code on real hardware.

Java for BREW?

So, are BREW and J2ME two warring factions battling for supremacy over the mobile application industry? It sure looks that way. Some carriers, such as SprintPCS in the United States, are supporting J2ME with no plans to incorporate BREW handsets into their lineup. However, others, such as KDDI in Japan, are supporting both BREW and J2ME. In South Korea, BREW is used exclusively by KTF, but both BREW and J2ME take a backseat to the overwhelmingly popular South Korean GVM standard used by SK Telecom.

In a true testament to BREW's versatility, a new development has made this battle even more complicated. At least two companies have developed Java virtual machines that run on top of BREW. Because BREW is compiled native code, it is possible to write a full implementation of J2ME that fits into the BREW system as just another application. I have personally seen IBM's J9 Virtual Machine developed by OTI (www.oti.com) in emulation and Insignia's (www.insignia.com) full MIDP implementation running on average BREW handsets with impressive results.

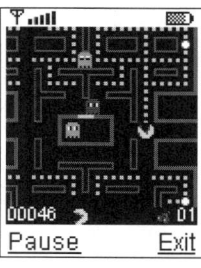

Figure A-3: Karl Hornell's MIDP Man is one of the applets that has been executed with excellent performance via Java VMs running on top of BREW.

The advantage to J2ME running on BREW is not just a wider variety of applications. At least with Insignia's solution, it is possible to use Qualcomm's provisioning system to distribute and bill for MIDlets internationally. Of course, there are still many issues; most BREW handsets do not adhere to the basic J2ME interface standards, and many low-end BREW handsets are simply not capable of running the Java VM. However, BREW's ability to run MIDlets should make BREW an even more attractive mobile application solution for many carriers and handset manufacturers.

Conclusion

What does the concept of J2ME running on top of BREW mean to you as a developer? Well, if you have gone through the trouble of reading this entire book, you have invested a lot of time in learning BREW development. Thus, I would not suggest giving up on BREW and going with Java just yet. However, it is interesting to know that if you have Java coding skills, you can still apply them to BREW devices. With BREW already rolled out by major carriers in the United States, South Korea, and Japan, the future of BREW-native applications is bright. You will still be able to get the best performance out of BREW handsets by writing native code. It is not guaranteed that J2ME implementations will be available on all BREW handsets. Therefore, I would not suggest using Java as an alternative language for writing applications on BREW hardware.

Appendix B

SMS

Introduction

Short Message Service, or SMS, may be familiar to you as a kind of instant messenger service for mobile phones in the same vein as ICQ or AIM. Although SMS is wildly popular in Europe and other countries, it is just beginning to catch on in the U.S. as a convenient way to communicate with others. BREW contains limited functionality to deal with SMS messages. This chapter will focus on what an SMS message is and how to read SMS messages from inside a BREW applet.

What Is an SMS Message?

Short Message Service began as part of the GSM specification. Using some technical properties of the GSM system, short text messages can be sent and received by a GSM handset, even while a voice call is in progress. These text messages are limited in length to 160 bytes, which equals 160 characters if you are using standard 8-bit ASCII codes.

Now most every digital PCS system has a form of SMS messaging. Qualcomm's cdmaOne system has its own SMS scheme that can accept messages of up to 256 bytes in length. Some networks can only receive SMS messages, while others such as cdmaOne and GSM can both send and receive messages from a handset.

In Europe, SMS has taken off. People are using the SMS service much like instant messenger applications such as ICQ or AIM. Regardless of the plain text and small character length restrictions, the GSM Association recently reported that more than 30 billion SMS messages were sent in December of 2001 alone. People have even developed their own language of word abbreviations and symbols to accommodate the short message length and restricted typing interface of the average mobile phone's number pad.

In order to send an SMS message, you must know the phone number to which you want to send the message. You then use the handset's interface to write out a short message and send it to that phone number. There is no guarantee as to when this recipient will get its message, and you will not get any confirmation of its arrival. Yet this seems good enough for most casual communications purposes. Each SMS message that you send costs a small fee. With billions of messages sent per month, some carriers are beginning to make back some of the ridiculous amounts of money they blew on bandwidth licenses with SMS fees.

SMS Gaming

Naturally, game developers have taken advantage of this booming phenomenon by creating games and entertainment services that use SMS as a way to interact with the user. For instance, many developers have created simple text games, like trivia contests where a server sends an SMS message out to you with a question and you send the answer as a response SMS message. It may seem ridiculous, but there are many different SMS games out there of varying degrees of popularity. SMS developers can also sometimes get a portion of the SMS fee as a royalty, making it one of the few ways to generate real revenue from a wireless game outside of Japan and South Korea.

Figure B-1: BBC's Walking with Beasts SMS game

SMS and BREW

How does this relate to you, the BREW developer? It is possible to send an SMS message to a BREW applet instead of the user of the handset. This is a convenient way to communicate between a server and a client if you do not want to use socket connections. Or perhaps you just want to generate revenue from SMS fee royalties.

What can a BREW applet do with an SMS message? Well, there is a lot of information you can pack into 160 bytes. In our Tic-Tac-Toe example, every possible move and condition in the game was represented with a single byte. You now have 160 times that space to play around with. For instance, SMS could be used to send out player turns. The thing is SMS messages have high latency issues. Do not rely on them for time-critical network messaging. Use sockets for that. For slow-paced, turn-based games (i.e., Tic-Tac-Toe), this may be a valid option.

The biggest problem is that the early versions of BREW only support SMS receive. Unlike voice calls on American carriers, users do not pay for SMS messages that they receive. Fortunately, BREW 2.0 does support SMS sending. This will open up new possibilities for peer-to-peer networking and revenue generation. Considering the fact that cdmaOne and CDMA2000 both support SMS sending and receiving via handsets, it seems odd that the ability to send SMS messages was not included in BREW's initial release. Regardless, we must wait to use SMS messaging to entirely replace socket communication in the appropriate head-to-head multiplayer games.

There are still things that you can do with just SMS receive. As for real-world examples of this, I have seen some games that allow new levels to be downloaded in the form of SMS messages. You can fit a screen-sized tile map in 160 bytes at current device resolutions. Or perhaps you might cram statistics for a new monster to fight against in an RPG. To send SMS messages to your players, you have to develop a server application that sends SMS messages out to handsets. This also involves getting access to a carrier's SMS gateway and developing a billing relationship. Good luck on that one.

How to Receive an SMS Message

So now you know about SMS messages and their limitations. How exactly do you receive one in your BREW applet? It is one of the AEEEvent messages sent to the event handler. The AEEEvent message is defined as AEE_EVT_MESSAGE in the AEE.h header file.

When the event handler receives an AEE_EVT_MESSAGE event, the dwParam is actually a pointer to the ASCII string of the SMS message's contents. Simply cast dwParam to a char pointer, and you can read the incoming message.

The great thing about this is that an SMS message can be sent to an applet not currently running. The message event can "wake up" the applet for whatever use you may have for it. If you want to start the applet upon receiving this message, call ISHELL_StartMessage inside the event handler and the applet will run. Otherwise, the SMS message is processed without the user knowing. Keep in mind that when the AEE_EVT_MESSAGE event is received in an inactive applet, none of your startup code has run. This can have major implications on how you program the message handler.

How does BREW know that the SMS message is designated for an applet and not the user? The SMS message has to begin with the following text:

```
//BREW:[Class ID]:[Message]
```

After the "//BREW:" prefix, place the hexadecimal value of the class ID specified in your applet's BID file (without the surrounding brackets). This number is followed by a semicolon, which is then followed by the contents of the message. BREW will look at the class ID, match it up with the applet associated with that Class ID, and send the EVT_APP_ MESSAGE event.

Unfortunately, the BREW 1.0 emulator cannot simulate the sending of an SMS message to an applet. This makes debugging SMS-based applets rather difficult, if not impossible. However, the emulator included with BREW 1.1 includes this and many other enhancements. See Appendix C, "BREW 1.1," for more information.

Beyond SMS: EMS and MMS

With its strict character length limitations, lack of graphics, and rather byzantine interface, SMS is a fossil of early digital wireless services, which is why a few new flavors of SMS-style services have been developed recently.

The first, Enhanced Message Service, behaves much like SMS, except you now have the ability to send tiny monochrome or color bitmap graphics, animations, sound effects, and ringtone music, along with your short text message. The text can also be formatted with special codes to change the font or position of the message's text. EMS messages can also be concatenated so that you can create larger messages, getting around the tight size restrictions. The great thing about EMS is that SMS phones can still read the text in an EMS message without the graphics, sound, and text formatting enhancements.

MMS, or Multimedia Message Service, is the true next generation of SMS. Much like EMS, MMS supports graphics, sound, and animation, along with formatted text. MMS takes it to the next level by supporting high-quality JPEG or GIF images, as well as MPEG4 full-motion video clips. With all of this rich media involved, MMS obviously benefits from 3G bandwidth. Regardless, new 2.5G handsets on the market have started to implement the MMS standard with an eye toward future expansion into 3G services.

Whether BREW will support the sending and receiving of EMS or MMS messages remains to be seen. The ability to pack more data as well as images and sound in EMS and MMS messages may provide a convenient delivery mechanism for additional game graphics or other content to BREW applets. Granted, with MMS's amazing multimedia capabilities, it is possible that MMS game applications may one day rival the popularity of downloadable applet technologies such as BREW.

Conclusion

You now know what SMS messages are and how they can be used in an applet, complementing BREW's robust socket communication interfaces. BREW's current SMS capabilities are limited, but they are bound to play a major role in future versions. If you would like more information on using SMS with BREW and how to create an SMS message to send to a BREW applet, see Qualcomm's knowledgebase entry for SMS: www.qualcomm.com/brew/developer/support/kb/52.html.

Appendix C

BREW 1.1

Introduction

Until this appendix, we have focused solely on the original version of BREW, known as BREW 1.0. Although there have been a few slight revisions to BREW 1.0, the next major upgrade is BREW 1.1. This new version of BREW is shipping on most new handsets in the U.S., including the Motorola T720 and Toshiba/Audiovox CDM 9500. BREW 1.1 adds many brand new interfaces, as well as additional functionality to the existing ones that we are now familiar with. Qualcomm has also improved the tools and included a few new ones that help smooth out the development process and provide support for the new features added to the API. This appendix discusses some of the major enhancements of BREW 1.1, as well as the new tools and features present.

The New 1.1 Interfaces

Earlier in this book, we did an anatomy of BREW that listed each interface, along with a brief description of its purpose. Here we will detail all the new interfaces in the same way.

- **IAddrBook**—This interface allows you to access the phone's address book. You can access the various fields in the address book (for instance, work phone number, mobile phone number, etc.), as well as create new records. This is a real boon for messaging applets that may need access to the phone numbers stored in the handset's directory.

- **IAddrRecord**—This is the interface to an individual address book record, as accessed by IAddrBook. Through this interface you can add, remove, and modify fields, as well as alter other attributes related to an individual address record.

347

- **ICipher**—The ICipher interface is a simple way to encrypt or decrypt data. Currently, ICipher supports the simple but robust ArcFour method of encryption. Qualcomm has hinted at the possibility of support for future encryption methods, such as DES. For more information on the ArcFour encryption method, see this site: home.earthlink.net/~neilbawd/arcfour.html.

- **IGetLine**—This mysterious little interface is used for simple data parsing of "lines" of text. You can do things like scan for EOF or carriage return codes with it.

- **IHash**—The IHash interface is a hash table collection class. You are able to feed data into the table for storage as well as generate hash table indices using the MD5 hashing algorithm.

- **IHtmlViewer**—Finally, you no longer have to use unformatted static controls to display large amounts of text. Instead, the IHtmlViewer interface allows HTML data to be displayed in a similar fashion. It parses a subset of HTML 3.2, which is good enough for most basic text displays. This includes forms, hyperlinks, text colors, and more.

- **ILicense**—The ILicense interface allows the applet to access and modify its user license information. For instance, if you want the user to only be able to run the applet a fixed number of times before she must purchase another usage license, you can use ILicense to decrement the number of runs left after the applet has started. You can also use dates to only allow the user to run the applet for a certain length of time before she has to purchase another license.

- **IPeek**—IPeek is used to peek at, or view, the contents of various data streams, including socket connections.

- **IPosDet**—The IPosDet interface allows the applet to get position information about the handset. For instance, depending on the handset's capabilities, user preferences, and carrier network options, it may be possible to identify in which base station the user is currently active. Depending on the density of base stations in the user's area, you may be able to pin down the user to a very small physical area. Position data has very interesting applications for games but also very controversial privacy implications that continue to be debated throughout the industry.

- **IQueryInterface**—IQueryInterface is a new way to make API extensions, otherwise known as extension classes. With IQueryInterface, you can get a pointer to an object's interface in a Microsoft COM-like manner.

- **IRingerMgr**—Some phones have the ability to download different ringtones. For instance, instead of the standard beep, you might want your favorite song to play. The IRingerMgr interface allows you to access the ringer system of the phone. This includes the ability to add and remove tones from the phone, as well as play tones from external files or the phone's own library of ringtones.

- **IRSA**—Much like ICipher, IRSA also allows you to encrypt or decrypt data. With the IRSA interface, you are able to use RSA with varying key lengths. For more information on RSA, go to the site www.rsasecurity.org.

- **ISource**—The ISource interface creates data sources that can be used for peek operations. Look at IGetLine and IPeek for more information on peek operations.

- **ISourceUtil**—ISourceUtil has a bunch of functions used to get access to sources for use with IGetLine and such.

- **IWeb**—Instead of having to roll your own HTTP protocol on top of BREW's TCP/IP socket support, you can now use the IWeb interface to directly interact with remote web applications.

- **IWebOpts**—The IWebOpts interface is used to set various options for use with the IWeb interface.

- **IWebUtils**—The IWebUtils interface contains a number of helper functions that do common web-related tasks. These include encoding strings, parsing URLs, and URL string construction.

And so ends our whirlwind tour of new BREW interfaces. As you can see, there are quite a few new functionalities in BREW 1.1. Some of it fulfills needs that developers have had since version 1.0; others satisfy needs that nobody knew they had yet! Many of these interfaces may not be directly game-related, but I'm sure game developers will make use of them.

API Additions

The existing interfaces have also undergone some improvement. There are a few new functions, helper macros, and data structures in BREW 1.1. With graphics, perhaps the biggest addition is the new Brew Compressed Image, or BCI, support. The IGraphics interface now supports this custom image format in conjunction with the existing BMP functionality. We will discuss BCIs later on when we get into the new tools.

Perhaps the most important graphics addition is the **StretchBlt** function. An operation familiar to those of you from the DirectX world, IGRAPHICS_StretchBlt allows you to blt a bitmap from a source to a destination rectangle of a difference size. BREW will scale the bitmap to fit inside this rectangle. For instance, if the destination rectangle is smaller than the source, the image will shrink, and vice versa. The ability to scale bitmaps is a big boon to developers who are dealing with the many non-standard screen resolutions of BREW.

There are a few GUI additions as well. For instance, now you can sort entries in a menu control. You can also modify the font and jump to specific lines of text in a static control. The static control now has a scroll bar and allows you to move the text window with the direction keys.

BREW 1.1 also includes a lot of new helper functions and macros. These range from string operations like **STRDUP** and **STRSTR** to converting floats to wide-character strings. BREW 1.1 also includes helper functions for **MEMMOVE** and **MEMCPY**, adding further to BREW's list of Standard C Library function replacements. You can even search for characters or strings in memory buffers.

The sound capabilities remain largely the same, except for the addition of support for QCP PureVoice digital sound files. These are highly compressed digital samples suitable for voice, sound effects, and other sounds. This provides a significant advance over MIDI and tone lists for game sound effects. PureVoice files are used by the SoundPlayer interface on devices that support digital sound.

There are a few other odds and ends addressed in the new API enhancements. Yet, it still remains largely the same. Read Qualcomm's BREW documentation for more information. Most of the big changes are being saved up for BREW 2.0. The future of BREW, including these future upgrades, is discussed in the next appendix.

Tool Enhancements

In addition to new interfaces and API enhancements, BREW 1.1 comes with a few new tools and improvements on the old ones. Most of the changes have been to the emulator that now sports enhanced debugging in the form of a text window to which your applet can output text. The emulator also includes enhanced sound emulation for DTMF tones and tone lists. SMS support is also built into this emulator, allowing you to send SMS messages to your applet. Also, you can now invoke your own suspension and resume events. Finally, there is a speed option that allows you to tweak the running speed of the emulator to better reflect the performance of an actual handset. It is a far cry from true hardware emulation, but at least we are heading in the right direction. As for the other existing tools, the Resource Editor now supports the new BCI image format. Both the Resource Editor and MIF Editor are able to run

on a command line. There are also a few new features in the Device
Configurator and MIF Editor, such as new permissions and emulator
options to support new features in both the API and emulator.

New Tools

BREW 1.1 includes several new tools that make development easier
and provide ways to support the new features of the API.

The AppWizard

The first new tool of note is the AppWizard. Now you no longer have to
copy over and edit old projects to create a new BREW program. Instead,
Qualcomm has created a handy Visual Studio AppWizard to simplify this
process. Here is how it works:

Select the BREW Application Wizard from File|New and enter the
directory and name of the applet.

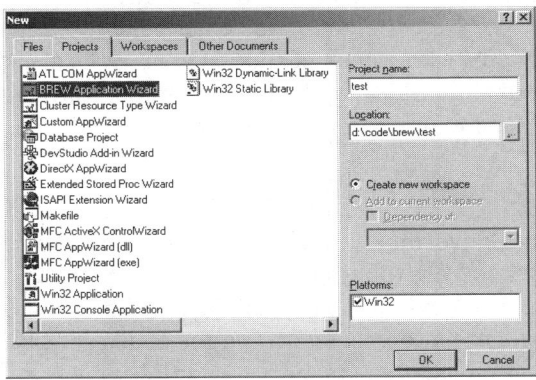

Figure C-1: Selecting the BREW
Application Wizard

Check the features that you want in this applet.

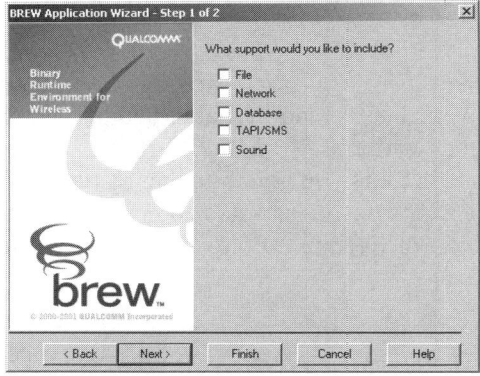

Figure C-2: Setting the applet features

Now launch the MIF Editor to create your MIF file.

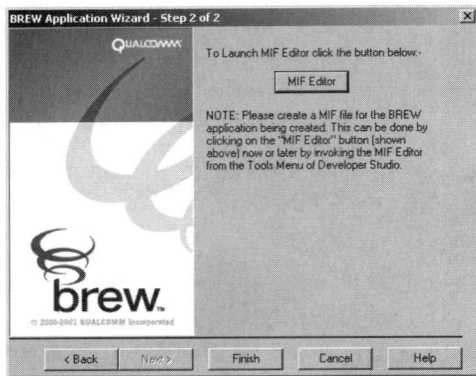

Figure C-3: Launch the MIF Editor from within the Application Wizard

After all of this is done, you have a skeleton program framework with a basic message handler and applet creation function. Because this code is so generic, you can also use the resultant files with BREW 1.0 applets.

As you can see by perusing the generated code, this Application Wizard does not really do a whole heck of a lot. Hopefully, Qualcomm will add things like automatic BID and MIF file generation through the wizard so you do not need to use external programs.

BREW PureVoice Converter

The BREW PureVoice Converter is a simple command-line DOS tool that converts WAV files to PureVoice QCP files. PureVoice is Qualcomm's own compression method for making small files out of voice samples. You can get more information on the codec at www.cdmatech.com/solutions/products/purevoice.html. There are some restrictions in the kinds of WAV files that can be converted, but other than that, there is not much else to this tool. As mentioned previously, the ISoundPlayer interface in 1.1 allows the playback of these files.

Visual C++ Add-In

BREW 1.1 also includes a Visual C++ add-in that expedites the process of compiling native ARM code with the ARM BREW Builder. By using the add-in, you get a toolbar that lets you generate and manage ARM makefiles with the touch of a button. You can also compile a native MOD file and launch the MIF Editor or Resource Editor from the toolbar.

BREW Compressed Image Authoring Tool

The final new tool is for dealing with the new BCI files, but just what is a BCI? BCIs are an alternative file format to BMP that Qualcomm has created for BREW. BCI files are compressed using ZLIB, which makes them ideal for the mobile phone environment where memory is at a premium. BCIs can be composed of multiple bitmaps for animation sequences as well.

With the BCI tool you can import BMP or PNG files and convert them to BCI. You can also edit various attributes of the resultant BCI before saving it to disk. The tool also has support for creating and editing animated BCIs if you wish to use the built-in animation functions of BREW.

Figure C-4: The BCI tool

To Use or Not to Use?

If BREW 1.1 gives you all of these goodies, why should you hesitate to use this new version? Chances are the version of BREW that ships with a handset will remain the version of BREW on that handset until the end of time. Although it is possible to flash the ROM with new OS revisions, this ability is only given to developers. Therefore, frequent BREW OS upgrades on at least the current generation of handsets are unlikely. Because of this, there are and will continue to be many handsets out there with BREW 1.0 installed.

Because BREW code written with the 1.1 SDK is not backward compatible with BREW 1.0, your applets will not work on the vast majority of 1.0 phones if you use 1.1. However, if you write code using BREW 1.0, it will run on 1.1 phones in addition to all of the 1.0 handsets already on the market. To reach a mass audience, it may be better to stick to 1.0; however, most new BREW handsets shipping today come installed with 1.1 and beyond.

Conclusion

As you can see, there are quite a few new additions to BREW 1.1 but nothing that makes your knowledge of BREW 1.0 obsolete. BREW 1.0 may be the most common BREW platform out now, but this is quickly changing.

Appendix D

BREW 2.0

Introduction

The future is here! Well, sort of. At the BREW 2002 Developers Conference in San Diego, Qualcomm announced the release of BREW 2.0. The SDK is freely available on Qualcomm's web site and will soon launch on handsets worldwide. Qualcomm also displayed some impressive prototype hardware that natively incorporates the features of BREW 2.0, not to mention the sneak preview of technologies to be incorporated in further revisions. This appendix discusses the major features of BREW 2.0.

The New 2.0 Interfaces

Much like BREW 1.1, BREW 2.0 features many new interfaces, as well as major enhancements to the old ones. Let's walk through the list of new BREW 2.0 interfaces.

- **IBitmap**—This new interface is a higher-level way to access bitmap data. Through IBitmap, it is now possible to get pixel-level access to bitmaps as well as manipulate the contents of the frame buffer.
- **IBTAG**—Allows usage of the headset as an audio device
- **IBTSDP**—Used to find and identify external peripherals attached to the handset
- **IBTSIOPORT**—This is the Bluetooth communication interface. Bluetooth is a wireless communications technology for short-range connections. It may be useful for things like local head-to-head play between two handsets.
- **IClipboard**—Used for cut and paste operations between documents

355

- **IDIB**—Derived from IBitmap, IDIB is a device-independent bitmap. Similar to DIBs in Windows, IDIB is used to modify bitmaps at a higher level.

- **IDNS**—This gives the programmer much more control over the usage of DNS servers when using TCP/IP.

- **IFont**—An interface used for drawing and otherwise manipulating the display of text. Each handset can have its own unique set of fonts.

- **IImageCtl**—A new GUI control that allows the scrolling display of bitmap images

- **IMedia**—This is the abstract base class of media objects, such as MIDI or MPEG streams. By using this class, you can play back MIDI, MPEG, and other media files with the same interface. This is a more general way to play media files than ISoundPlayer from the original SDK.

- **IMediaMIDI**—Derived from IMIDI, this allows for the playback of MIDI music, much like the old ISoundPlayer interface.

- **IMediaMIDIOutMsg**—Used to send MIDI messages to an external MIDI device. Using a BREW handset as a MIDI sequencer for recording studio equipment? Quite possibly.

- **IMediaMIDIOutQCP**—This sends QCP-formatted data to an external MIDI device.

- **IMediaMP3**—Obviously used to play MP3 files

- **IMediaQCP**—Handles the playback of QCP PureVoice files. As described in Appendix C, these are highly compressed digital sound files.

- **IMediaPMD**—Used for the playback of PMD WebAudio files

- **ISprite**—The ISprite interface is a new object used for drawing both sprites and tiles. Through ISprite you can control the drawing and transparency of the sprite, as well as transform it with rotation and scaling operations.

- **ISSL**—Used for SSL/TLS secure protocols with network communication

- **ITransform**—The ITransform interface is used to rotate, scale, and perform other transformations on bitmaps (including sprites).

- **IUnzipAStream**—This allows for the easy decompression of compressed streams.

- **IVocoder**—Allows the capture and playback of vocoder data. A vocoder is a device that encodes the user's voice into digital form when making a voice call. Therefore, you can sample and play back the user's voice for a variety of different applications.

- **IX509Chain**—Used to manage x.509 certificates for public-key encryption. This comes in handy for SSL communication.

Graphics Enhancements

As you can see, there are quite a few new interfaces in BREW 2.0. Perhaps the interface of the most interest to us is ISprite. At BREW 2002, one of the presenters demonstrated an impressive example application of this new sprite interface. It was a mock-up of a Mario-style platform game, complete with a scrolling background and animated sprites. The performance was impressive. I was skeptical. After all, this was running in an emulator. We all know how completely inaccurate that is.

Later on, they showed a similar demo running on a prototype chipset designed to take advantage of BREW 2.0's features in hardware. The performance was excellent. It remains to be seen how fast the sprite engine will be on existing BREW chipsets. However, one can imagine that BREW devices will soon begin to rival Nintendo when it comes to portable gaming hardware in pure technical prowess.

So how does this ISprite interface work? What can you do with it? The ISprite interface access is what BREW terms the sprite engine. In some cases this may just be linked into more firmware code, while in other cases this may represent specific sprite hardware support. The sprite engine requires bitmap graphics to be in specific formats. This includes having all frames of animation arranged in a single strip. The ISprite interface then uses the dimensions of a single frame specified by the user to manage animation. In order to initialize a sprite after creating the interface, you must first load the bitmap and then tell ISprite where the buffer is that contains the data. That is about it. ISprite is very simple to use.

The ISprite interface is also used for rendering background tile maps. Load and construct a tile much the same way you do with a sprite. Then create a tile map in an array similar to how we did in our manual tile example. Each tile map has a small data structure associated with it that governs the size of tiles and the position of the scroll, as well as various flags for transparency and other attributes. To scroll the map, simply alter the x and y coordinates in this data structure before you draw it. You can have multiple tile maps and draw them at different times to create the illusion of depth with parallax scrolling.

The transforms are another big boon to graphics development with BREW. The ITransform interface can alter several characteristics of a bitmap's display properties for both sprites and tiles, as well as plain images. This interface uses a simple data structure that specifies a 2x2 transformation matrix using 16-bit fixed-point values. Those of us familiar with discrete math or 3D programming are well aware of the usage of transformation matrices. By using different kinds of matrices, you can perform scaling and rotation operations to enlarge or spin your image around. A number of transformation matrices can be combined to create a stack of transform operations. For instance, you might want to both scale and rotate an image. Of course, you must be careful of the order in which you stack your transforms. To draw these transformed images, you must use ITransform's own blt functions.

Another huge enhancement is the ability to retrieve a bitmap representing the contents of the screen from the IDisplay interface. Previously, it was impossible to alter the contents of the screen's frame buffer on a pixel level. Now you can call IDISPLAY_GetDeviceBitmap to retrieve the screen contents as an IBitmap interface. Via IBitmap, it is possible to get and set individual pixel values. Of course, the IBitmap interface can be used to get pixel access to any bitmap, not just the screen.

It is also worth mentioning BREW 2.0's support for full-motion video. Through the new IMedia interface, it is possible to play back MPEG4 video streams. Sure, this requires powerful hardware and a fat connection to the Internet for streaming. Regardless, any device that supports full-motion video will most likely have a decent fill rate. This should also be reflected in the speed of bitmap display functions.

The BREW 2.0 API is a major improvement for graphics programmers. Although a lot of these functions are still not fully supported by hardware and have some limitations imposed on them, it is obvious that Qualcomm has a strong commitment to the needs of game and graphics developers. There must be some monster handsets on the way if they are going to support most of these new multimedia options. The great thing is that there is more to come. At BREW 2002, Qualcomm showed an impressive new 3D graphics API that included the display of transformed and lit polygons, as well as effects like specular highlights. It will be interesting to see what other goodies Qualcomm has in store for us in the future.

Tool Enhancements

There have been many additions to the existing BREW tools. The new version of the emulator allows the emulation of network transmission speeds. This includes the ability to set the emulator's latencies for sending and receiving packets. Also, the emulator has support for GPS location services, either from a file or a GPS device connected to the host PC. The Resource Editor has better command-line support, so it can be included in batch processes. As for new tools, Qualcomm has finally included one for creating 2-bit BCI images. Also, there is extensive new documentation including a help file that can be integrated with Visual Studio.

When Can I Use It?

Most devices still use BREW 1.0 or 1.1. Therefore, any code you write in BREW 2.0 will be incompatible with the vast majority of current BREW handsets for the time being. Qualcomm has stated that BREW 2.0 hardware should start appearing by late 2003. However, you will be targeting a tiny minority of BREW users by restricting your products to only BREW 2.0. Many of these features are irresistible to us game programmers. Perhaps it is a good idea to make an enhanced BREW 2.0 version of your game for newer handsets.

The Future of BREW

What lies beyond BREW 2.0? Qualcomm has demonstrated a few new features and revealed upcoming developments that are slated to make it into BREW 3.0 in late 2003. Perhaps the most significant of all of these is the addition of a 3D API. BREW 3.0 handsets will supposedly be able to display around 3,000 polygons per second using Qualcomm's new 3D technology. 2D display speed will also be boosted by handsets with LCDs that can hit refresh rates approaching 15 fps. Also, in addition to a revamped toolset, BREW 3.0 will include the use of GNU's free GCC ARM compiler. This is a major win for hobbyists and small shops that want to develop on the platform but can't afford the expensive ARM tool suite. More features are slated to appear in the revision, including advanced sound mixing capabilities, multiple keypress control, and more. BREW 3.0 is going to be a major advance for mobile game developers.

Conclusion

BREW 2.0 shows Qualcomm's solid commitment to the future of mobile application development. For game programmers, the new graphics features are invaluable. Again, the same compatibility issues apply to BREW 2.0 that did to BREW 1.1. You sacrifice users for features. However, as the BREW platform gains momentum, this will become less of an issue.

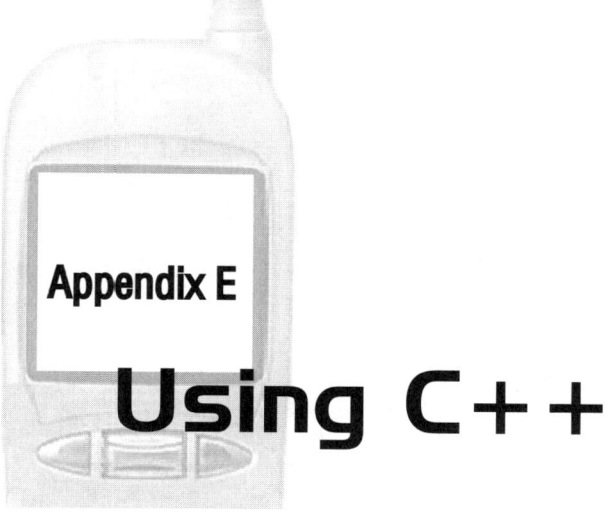

Using C++

Introduction

Ask any programmer about her feelings on C versus C++, and you are likely to start a holy war. I have seen work literally stop for hours while programmers ramble on about the pros and cons of either language. I definitely have feelings on the issue, but in the case of writing such small programs as games on wireless phones, the arguments on both sides become rather moot.

Is it overkill to use something like C++ on such a small project? Perhaps, but if programmed correctly, there is no reason why a C++ program cannot be nearly as compact as its C equivalent. A poorly written C program can be just as bloated and inefficient as a poorly written C++ program. And vice versa.

I have written most of the code in this book in C to be consistent with the examples and SDK conventions. However, in this appendix, I delve into the usage of C++ with BREW. C++ not only can be more convenient at times, but it can also make it easier to port your code to J2ME, if the need arises.

Is It C, or Is It C++?

All of the code in the book so far has been written in C. In fact, the vast majority of the sample code that comes with the SDK is written in C. The BREW API itself is heavily rooted in C, with most of its function calls actually defined as C macros that access BREW interfaces at a lower level.

However, by looking at the structure of most of these function calls, you can see that BREW is just crying out to be used with C++. As you may have noticed, most function calls have the first argument as the pointer to the interface to which the function belongs. In a C++ environment, the interface pointer would be implicit. From the lessons learned in Chapter 15, we see that, indeed, BREW objects and interfaces are somewhat analogous to C++ function tables and classes.

Hello, Again

I have converted the Hello World applet that we developed early in this book to a C++ program. In the process, I have added a few C++ features that you will need if you are planning to go the C++ route later on. This way, you can easily see the comparison between a C and C++ applet. Let's begin the examination.

The first difference between a C and C++ applet is the applet structure. If you recall in our previous examples, we defined a custom structure with the predefined **AEEApplet** structure declared first and our own data structures and variables underneath. In the case of a C++ applet, define instead an applet class derived from **AEEApplet**, like so:

Listing E-I: Applet class

```
class MyApp : public AEEApplet
{
public:
    static boolean  HandleEvent(MyApp * pMyApp, AEEEvent eCode,
            uint16 wParam, uint32 dwParam);

    static void freeAppData(MyApp * pMyApp);

protected:

        //these functions replace the standard event/message handlers from a C applet
        void OnAppfreeData();
        boolean  OnEvent(AEEEvent eCode, uint16 wParam, uint32 dwParam);
};
```

This is standard C++ stuff. We derive from the **AEEApplet** object instead of including the data structure in our own custom applet structure. Then we put all the variables and function definitions that we want in the object.

Now, let's move on to the code. We still need to implement the **AEEClsCreateInstance** function. However, the code inside is a little different:

Listing E-2: Applet creation function

```
extern "C"
int AEEClsCreateInstance(AEECLSID ClsId,IShell * pIShell,IModule * pIModule, void **
   ppApplet)
{
```

```
AEEApplet* pMe = 0;
*ppApplet = NULL;

//This is similar to applet initialization in C, but we pass pointers
//to static member functions instead
if(ClsId == AEECLSID_CPPHELLO_BID )
{
        if(!AEEApplet_New(sizeof(MyApp), ClsId, pIShell,
                pIModule, (IApplet**)ppApplet,(AEEHANDLER)MyApp::HandleEvent,
                (PFNFREEAPPDATA)MyApp::freeAppData))
            return EFAILED;

        pMe = (AEEApplet*)(*ppApplet);

        if (!pMe)
            return(EFAILED);

        return(SUCCESS);
}
else
      return(EFAILED);
}
```

This is similar to the **AEEClsCreateInstance** in all of our other C code; however, we pass pointers to member functions of our class instead of just standard function pointers. These functions must be static and are called in the same fashion as they are in our C examples.

As you can see, we pass a cleanup function, **freeAppData**, and a message handler, **HandleEvent**, to **AEEApplet_New**. Let's look at both:

Listing E-3: Applet data freeing function

```
void  MyApp::freeAppData(MyApp *pApp)
{
      pApp->OnAppfreeData();
}
```

There is not much to clean up here, as we do not allocate memory. But this function calls the **OnAppfreeData** member function of the applet class, just in case. That function is empty in our example, but it is where you would put any cleanup and resource deallocation code necessary.

The message handler also calls a custom message handler defined inside our class:

Listing E-4: Event handler

```
boolean  MyApp::HandleEvent(MyApp *pApp, AEEEvent eCode, uint16 wParam, uint32 dwParam)
{
      return pApp->OnEvent(eCode, wParam, dwParam);
}
```

Looking at **OnEvent**, you will see a standard message handler similar to all the other ones that we have written in the past:

Listing E-5: Event handler

```
boolean MyApp::OnEvent(AEEEvent eCode, uint16 wParam, uint32 dwParam)
{
    AECHAR szBuf[] = {'H','e','l','l','o',' ','W','o', 'r', 'l', 'd', '\0'};
//wide-character string

    switch (eCode)
    {
        case EVT_APP_START:

        //clear screen (default color is white)
        IDISPLAY_ClearScreen(m_pIDisplay);

        //draw text
        IDISPLAY_DrawText(m_pIDisplay, AEE_FONT_BOLD,
            szBuf, -1, 0, 0, 0, IDF_ALIGN_CENTER | IDF_ALIGN_MIDDLE);

        //update screen
        IDISPLAY_Update(m_pIDisplay);

        return(TRUE);
        break;
    }

    return(FALSE);
}
```

This is essentially the same as the C one; it just happens to reside inside our applet class. If you look at the function calls, such as **IDISPLAY_ClearScreen**, we no longer have to access the **AEEApplet** interfaces through a pointer to our applet. Instead, this pointer is implicit, since the message handler resides inside our own applet class. Therefore there is no annoying **pApp** pointer to pass around and member-select all over the place.

Now, there is one last difference that you must know about. In the header, above the applet class, we also overload the new and delete operators:

Listing E-6: New and delete operator override

```
void *operator new ( size_t size)
{
    return MALLOC (size) ;
}

void operator delete(void * ptr)
{
    FREE(ptr) ;
}
```

We do this because the Standard C Library version of new and delete calls malloc and free. Since BREW does not allow Standard C Library calls, we need to overload new and delete to use BREW's own MALLOC and FREE macros. One good thing about overloading new and delete is that you can also create your own memory allocation system, which may track memory blocks and provide debugging information useful for tracking memory leaks. Right now, we are just doing the basics.

C++ Makefiles

To compile your C++ code for the handset, you need to use the ARM BREW Builder's C++ compiler. Therefore, instead of armcc, you must use armcpp to compile C++ source files. The same goes for the thumb compiler if you are using 16-bit instructions. I have created a modified version of the C++ makefile that Qualcomm includes with the 1.1 version of the BREW SDK to compile the C++ version of Hello World in the Source\Appendix E\cpphello folder of the companion files. You can modify this makefile just as you would the C one to add more source files to your project. If you examine it closely, you will notice that it is largely similar to the C version, so modifying it or your project should be relatively trivial.

Conclusion

In this chapter you have learned how to use C++ with BREW. In many ways it can be more convenient than using C. What is best for you? That is for you to decide. I tend to enjoy using C++ on rather large projects and C on smaller things. However, C++ provides many goodies, such as polymorphism, operator overloading, and such—not to mention C++'s object-oriented style makes it easier to port your code over to J2ME if necessary.

Appendix F

Fixed-point Math

Introduction

Because BREW is designed to run on miniscule hardware such as the common mobile phone, expensive features like floating-point math operations have been omitted. Today, floating-point math is commonplace. There was a time not too long ago when using floating-point values for math calculations on PCs was a slow and expensive operation. In fact, CPUs used to require the purchase of additional floating-point units, or FPUs, in order to support what passed for fast floating-point operations in those days. Today, blindingly fast FPUs are part of most any modern PC processor. However, for many in the mobile realm, FPUs are still just a dream. This appendix focuses on how to deal with the lack of floating-point support in BREW when you need to use fractional values.

Fixed-point Math

So, how is it possible to represent a fractional value when all we can use in BREW are integers? The key is in the usage of fixed-point math. Floating point is called so because the decimal (or radix) point can move around. This means it is possible to represent values such as 4.2 and 40.251 with the same system. We have a value with precision in the tenths with 4.2 but in the thousandths with 40.251. With fixed-point math, the decimal never moves. So if we were to represent both of these values in a fixed-point system, it would look something like 4.2 and 4.251.

Fractioning the Integer

How do we represent a fixed-point number in code? We can break an integer into two parts. A portion is for the whole number and another is for the fractional (or mantissa) value. The first thing that you must find out is how many bits you want to reserve for the mantissa. The more bits you set aside for the mantissa, the greater precision you have. For now, we will split the integer down the middle, reserving 16 bits for the mantissa. Our integer will be partitioned like this:

| 0000000000000000 | 0000000000000000 |

Figure F-1: A 16-bit integer split in half. The left side is the whole number, while the right side is the mantissa.

As you can see, we actually have 15 bits for the whole portion and 16 for the mantissa. This is because the integer is signed. When a signed integer becomes negative, the first bit is flicked on in the process of using the two's complement system to represent negative numbers. If you make this an unsigned integer, you do not have to worry about this extra bit.

So, now that we know how we want to split up our integer, how do we actually set values that adhere to this partitioning? All we need to do is shift. Assuming we are using 16-bit precision, shift the integer values by 16, that is, if we want to set our fixed-point value to 400, do it like this:

```
int value = 400 << 16;
```

Now we have 400 represented in fixed point with 16-bit precision. If we want to convert this value back to a standard integer, simply shift the other way:

```
int converted = value >> 16;
```

What if we have some bits in the fractional part? How do we read this? Well, if we had access to floating-point variables, we can simply mask off everything but the mantissa and divide it by 2 raised to the number of bits of precision we have. Looking at a fixed-point integer with the following bits:

| 0000000000011101 | 0000100010010011 |

Figure F-2: A 32-bit fixed point integer with both a mantissa and whole part

Mask off the whole number by performing an AND with 0xFFFF0000. Now we are left with this:

| 0000000000000000 | 0000100010010011 |

Figure F-3: After masking it, we are left with only the mantissa.

Reading this as a standard integer, we have a value of 2195. Now take this value and divide by 2^{16}. The result is 0.0334930419921875. The integer part of the initial fixed-point value was 11101 in binary and 29 in decimal. Therefore, the full value of the fixed-point number is 29.0334930419921875. Because BREW lacks native floating-point support, you cannot do this in code. However, it is a useful debugging technique that can be used when running code in an emulator. Otherwise, you can do it the hard way and use a calculator instead.

As you can see, the more bits that you reserve for the mantissa, the more accurate the fractional part. Obviously, the more bits that you add to your mantissa, the fewer bits you have available for the whole number. Therefore, you may want to use different bit precision, depending on the kind of data that you anticipate. If you are going to have many small fractional numbers, you may want to have even more than 16 bits reserved for the mantissa. If you are planning on mostly dealing with large integers where the fractional component is relatively unimportant, you may want to reserve fewer bits.

Now that we know how to convert integers back and forth and make sense of the fractional bits, how do you perform mathematical operations on fixed-point values? Because these are stored as regular integers, you can use regular math operations on fixed-point values with a few cautions.

Adding and Subtracting

First, addition and subtraction work normally. To add or subtract fixed-point values, simply add or subtract them as normal integers.

```
int nValue1 = 400 << 16;
int nValue2 = 10 << 16;

int nSum = nValue1 + nValue2;
```

In this case, **nSum** shifted to the right by 16 will be equal to 410. Subtraction works in the same way:

```
int nValue1 = 10 << 16;
int nValue2 = 6 << 16;

int nDifference = nValue1 – nValue2;
```

Now, the variable **nDifference** shifted to the right by 16 is equal to 4. Addition and subtraction work without a hitch if both values are fixed point. If you are trying to add a fixed-point value to a standard integer, you must convert that integer to fixed point. The same goes for subtraction.

Multiplication and Division

Multiplication and division are a little trickier. When you multiply fixed-point numbers, the mantissa moves to the left. For instance, multiplying 10 by 100 gives you 1000. But you are not multiplying normal integers here. Therefore, the decimal moves to the left when you multiply as you tack on zeroes to the end. You simply need to shift the result of a multiplication or division back to the right by the number of bits of precision that you are using. Here is a simple example:

```
int nValue1 = 10;
int nValue2 =  2;

int nResult = nValue1 * nValue2;
nResult = nResult >> 16;
```

As you can see, after we multiply the two values, we shift it down by 16. The end result is a legitimate fixed-point value. Well, most of the time. What I mean by that is sometimes with multiplication and division, you get what is called overflow.

Overflow

Overflow happens when you multiply two numbers that are so large that they cannot be represented by a standard integer. This can happen quite easily with fixed-point numbers because you are really multiplying and dividing the integer representation of them. For instance, the fixed-point value of 5000.0 looks like this:

```
0001001110001000 0000000000000000
```

Figure F-4: 5000 represented as a 32-bit
fixed-point value

However, the CPU does not know anything about fixed-point values. It is simply operating on two integers. Let's say that you are multiplying 5000.0 by 20.0. Using standard integers or floating-point values, you would arrive at a figure of 1000000.0. However, read as integers, the fixed-point values of 5000 and 200 are much larger. This is because they are shifted to the left by 16, which is in essence multiplying them by 2^{16}. Therefore, you are actually multiplying 327680000 by 1310720. This equals 429496729600000. The result of our multiplication is well out of the range of values that an integer can hold. This situation is called overflow.

There are a number of different solutions to the overflow problem. The two that I frequently use are either reducing the precision of the mantissa or using shorts instead of full, 32-bit integers.

If your mantissa is not as precise, your values are not shifted over as much after a multiplication or division. Hence, the integer values are much smaller and overflow is less likely to occur. As described previously, the fewer bits that you have for your mantissa, the less precise the fractional values. It is up to you whether or not you need a high degree of accuracy in your fractional values.

If you use 16-bit shorts instead of full, 32-bit integers, you can store the result of your multiplication of two shorts in an integer and then shift that 32-bit integer back down into a valid 16-bit short. Hence, overflow occurs in 32-bit integer space, but you convert it back to a short for the end result. Also, by using 16-bit shorts, you reduce the memory footprint of your fixed-point values by half. Because you have only 16 bits to use, the precision of your mantissa and range of whole numbers can be severely limited.

Conclusion

You have seen in this appendix how to deal with BREW's lack of floating-point operations. Fixed-point values offer a few inconveniences but can pretty much perform most any floating-point operation suitable for a simple game. In addition to the lack of floating-point values, BREW is also missing many other basic mathematical operations that we are used to from the Standard C Library. This includes the square root and various trigonometry functions, such as sine and cosine. By using fixed-point values, integer approximations, and lookup tables, it is possible to create reasonable substitutes for these missing blocks of math functionality.

Appendix G

The BMP File Format

Introduction

BREW 1.0 relies entirely on the BMP format for bitmap graphics. BREW 1.1 and later revisions utilize a variety of different file formats. However, BMP remains a very simple and convenient bitmap file format supported by just about any current art package. This appendix describes the internals of the BMP file format for those who want more intimate knowledge of BREW's internals.

Why Bother?

In general, there is no real need to know the internals of the BMP file format. BREW provides all the functions that you need to load and draw BMPs. One thing BREW lacks, however, is pixel-level access to bitmap data. You can only blt blocks of images to the frame buffer; you cannot access the frame buffer's or a bitmap's individual pixels. Or can you?

Although quite complicated, it is possible to load a BMP into memory and modify the contents of the image before converting it to a device-dependent bitmap suitable for blting. With this method, you can modify any information in the BMP, including the individual pixels, before converting it to a device-dependent bitmap. Unfortunately, if you are trying to do complicated graphics operations, such as drawing circles or lines, you will have to create your own drawing functions that act upon the internals of a BMP. This is because BREW's geometry drawing functions only operate on the device's frame buffer—not an external memory chunk such as the space occupied by a loaded BMP.

As you may recall from Chapter 8, the conversion of a bitmap from a BMP to a device-dependent bitmap is a slow process. When modifying the pixels of a BMP before displaying, you must reconvert the BMP to a native bitmap every time you change the contents of the image. This makes pixel-level access to BMPs very slow. So, not only is this process clumsy, but it is slow, too. Still, it has its purposes (that is, until future versions of BREW give you pixel-level access to off-screen bitmap buffers).

The Anatomy of a BMP

The BMP was created by Microsoft and IBM as a PC-centric file format for bitmap graphics. There have been several revisions to the BMP over the years. However, this appendix focuses on the most common version of the BMP used by most applications and BREW itself.

A BMP file contains the following structures, in this order:

```
BITMAPFILEHEADER   bmfh;
BITMAPINFOHEADER   bmih;
RGBQUAD            aColors[];
BYTE               aBitmapBits[];
```

Both **BITMAPFILEHEADER** and **BITMAPINFOHEADER** are fixed-size structures that are actually defined in the Win32 SDK headers. Because you cannot use any Windows SDK or Standard C Library functions, if you need to access these structures, you will have to redefine them in your own headers. As defined by Microsoft, the **BITMAPFILEHEADER** structure looks like this:

```
typedef struct tagBITMAPFILEHEADER { // bmfh
    WORD    bfType;
    DWORD   bfSize;
    WORD    bfReserved1;
    WORD    bfReserved2;
    DWORD   bfOffBits;
} BITMAPFILEHEADER;
```

The first member, **bfType**, is 16 bits of data that, when read as a pair of characters, form the string "BM." If the file does not have "BM" as the first two bytes of the file, it is not a valid bitmap. This is a handy way to identify the file as a bitmap if, for some reason, you cannot depend on the file extension. Next, **bfSize** specifies the size, in bytes, of the entire BMP file, including this header. Both **bfReserved1** and **bfReserved2** are unused variables earmarked for future use. They both should be set equal to 0. Finally, **bfOffBits** is the offset, in bytes, from the **BITMAP-FILEHEADER** to the start of the actual pixel data. Directly following the **BITMAPFILEHEADER** structure is the **BITMAPFINFOHEADER**. This structure is also defined by Microsoft like so:

```
typedef struct tagBITMAPINFOHEADER{ // bmih
    DWORD  biSize;
    LONG   biWidth;
    LONG   biHeight;
    WORD   biPlanes;
    WORD   biBitCount
    DWORD  biCompression;
    DWORD  biSizeImage;
    LONG   biXPelsPerMeter;
    LONG   biYPelsPerMeter;
    DWORD  biClrUsed;
    DWORD  biClrImportant;
} BITMAPINFOHEADER;
```

This structure describes the actual bitmap stored in the file. The first member, biSize, specifies the size, in bytes, of this structure. The biWidth and biHeight members specify the width and height in pixels of the bitmap image. The biPlanes member specifies the number of planes in the target device. This is a legacy member from the era of planar graphics hardware. Therefore, it is always set to 1. The biBitCount member specifies the bit depth of the BMP. This is how many bits are used for each pixel. The valid values are 0, 1, 4, 8, 16, 24, and 32. For current BREW applications, you are really only concerned with 1-, 4-, and 8-bit BMPs. These are 2-, 16-, and 256-color images, respectively. The 0 bit depth is used for JPEG-compressed BMPs, which we are unconcerned with at the moment.

The biCompression member specifies which compression method, if any, is used in this file. If the value is set to 0 (defined as BI_RGB in the Windows headers), the file is not compressed. Compression saves disk space; however, the bitmap must be decompressed when drawn on the screen, not to mention the fact that writing pixel data to a compressed RGB is a real pain. You should stick with uncompressed BMPs when creating images for BREW applets.

Next, biSizeImage contains the size of the image in bytes. This is often set to 0 when using uncompressed bitmaps, since you can calculate the size of the bitmap by multiplying the width and height by the bit depth of the image. The following biXPelsPerMeter and biYPelsPerMeter members represent the number of pixels per meter in each dimension. This information is largely useless for our purposes. The member biClrUsed specifies how many individual colors are used in this image. If it is set to 0, it uses the maximum number of colors the bit depth allows. For instance, if we have an 8-bit image and biClrUsed is equal to 0, this image uses 256 different colors. Finally, biClrImportant represents the number of colors that are necessary for displaying the image. If this is set to 0, all colors must be used.

Directly after this structure is the actual bitmap color palette previously referred to as aColors. The palette is stored as an array of RGBQUAD structures. An RGBQUAD is a simple series of bytes:

```
typedef struct tagRGBQUAD { // rgbq
    BYTE    rgbBlue;
    BYTE    rgbGreen;
    BYTE    rgbRed;
    BYTE    rgbReserved;
} RGBQUAD;
```

Aside from the final member, rgbReserved, each byte represents a value from 0 to 255 for its respective color component. The array of RGBQUADs, representing the color palette, is as large as the member biClrUsed in the BITMAPINFOHEADER structure.

After the color array is an array of bytes previously described as aBitmapBits. This array is as large as the member biSizeImage in the BITMAPINFOHEADER structure, or equal to the width times the height times the number of bits per pixel. It is this array that you must modify to get pixel-level access to the BMP.

Changing Pixels

Now that we know the location of the pixel data, we just need to get a pointer to it and start modifying values. Assuming we have a buffer with a BMP file loaded in, simply cast that pointer to a BITMAPINFO-HEADER and advance it by the value stored in the biOffBits member.

```
unsigned char * pPixels;
pPixels = pBuffer + ((BITMAPFINFOHEADER*)pBuffer)->biOffBits;
```

Assuming that pBuffer points to a buffer containing raw BMP data, simply set the pPixels pointer to the pBuffer pointer advanced by the number of bytes stored in the biOffBits member. We now have a pointer to the individual pixels.

Modifying the pixels of an 8-bit image is the easiest method. With 8-bit color, each pixel is represented as a byte. You can simply access the buffer as an array of bytes and set values accordingly. When dealing with 1- or 4-bit BMPs, you will need to handle multiple pixels per byte. This involves masking off values and carefully setting nybbles or individual bits for per-pixel modifications. After these modifications are made, simply call CONVERTBMP, as we did in our slide show example to create a device-dependent bitmap from this newly modified BMP.

Conclusion

This short appendix has focused on the internals of the BMP format. By knowing the specifics of the file format, it is possible to modify individual pixels, essentially giving you per-pixel access to bitmaps for advanced graphic effects. Given fast BREW hardware, it is possible to use BMP manipulation for impressive results.

Index

www.GameInstitute.com
A Superior Way to Learn Computer Game Development

The Game Institute provides a convenient, high-quality game development curriculum at a very affordable tuition. Our expert faculty has developed a series of courses designed to teach you fundamental and advanced game programming techniques so that you can design and develop your own computer games. Best of all, in our unique virtual classrooms you can interact with instructors and fellow students in ways that will ensure you get a firm grasp of the material. Whether you are a beginner or a game development professional, the Game Institute is the superior choice for your game development education.

Quality Courses at a Great Price

- **Weekly Online Voice Lectures** delivered by your instructor with accompanying slides and other visuals.

- **Downloadable Electronic Textbook** provides in-depth coverage of the entire curriculum with additional voice-overs from instructors.

- **Student-Teacher Interaction** both live in weekly chat sessions and via message boards where you can post your questions and solutions to exercises.

- **Downloadable Certificates** suitable for printing and framing indicate successful completion of your coursework.

- **Source Code** and sample applications for study and integration into your own gaming projects.

"The leap in required knowledge from competent general-purpose coder to games coder has grown significantly. The Game Institute provides an enormous advantage with a focused curriculum and attention to detail."

—Tom Forsyth
Lead Developer
Muckyfoot Productions, Ltd.

3D Graphics Programming With Direct3D

Examines the premier 3D graphics programming API on the Microsoft Windows platform. Create a complete 3D game engine with animated characters, light maps, special effects, and more.

3D Graphics Programming With OpenGL

An excellent course for newcomers to 3D graphics programming. Also includes advanced topics like shadows, curved surfaces, environment mapping, particle systems, and more.

Advanced BSP/PVS/CSG Techniques

A strong understanding of spatial partitioning algorithms is important for 3D graphics programmers. Learn how to leverage the BSP tree data structure for fast visibility processing and collision detection as well as powerful CSG algorithms.

Real-Time 3D Terrain Rendering

Take your 3D engine into the great outdoors. This course takes a serious look at popular terrain generation and rendering algorithms including ROAM, Rottger, and Lindstrom.

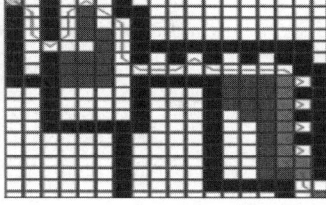

Path Finding Algorithms

Study the fundamental art of maneuver in 2D and 3D environments. Course covers the most popular academic algorithms in use today. Also includes an in-depth look at the venerable A*.

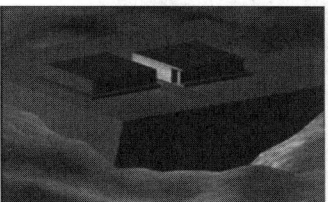

Network Game Programming With DirectPlay

Microsoft DirectPlay takes your games online quickly. Course includes coverage of basic networking, lobbies, matchmaking and session management.

MORE COURSES AVAILABLE AT

www.GameInstitute.com

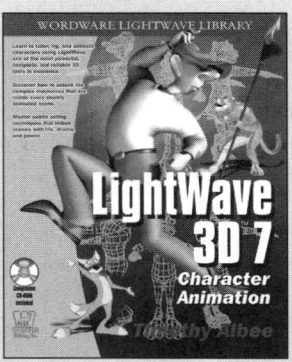

About the Companion Files

The companion files can be downloaded from www.wordware.com/brew. These files include all of the source code discussed in the book, as well as Robin Burrow's useful Mappy tile map editor and a demo of Cosmigo Pro Motion. There are also a few other useful files, including bitmaps with various palettes that may be used by some BREW handsets.

The code is organized in folders according to chapters. Simply copy the folders to your hard drive and you should be able to compile them with Visual C++ or, in some cases, the ARM BREW Builder. This book assumes you have installed the BREW SDK to the default location on your C drive. BREW Developer Studio projects require the inclusion of a few source files from the folder in which the BREW SDK is installed. If you have put the BREW SDK anywhere other than the default installation folder, you will have to edit the workspaces to reflect this.